Molecular Pathogenesis of Gastrointestinal Infections

FEDERATION OF EUROPEAN MICROBIOLOGICAL SOCIETIES SYMPOSIUM SERIES

Recent FEMS Symposium volumes published by Plenum Press

1990 • MOLECULAR BIOLOGY OF MEMBRANE-BOUND COMPLEXES IN PHOTOTROPHIC BACTERIA
Edited by Gerhart Drews and Edwin A. Dawes
(FEMS Symposium No. 53)

1990 • MICROBIOLOGY AND BIOCHEMISTRY OF STRICT ANAEROBES INVOLVED IN INTERSPECIES HYDROGEN TRANSFER
Edited by Jean-Pierre Bélaich, Mireille Bruschi, and Jean-Louis Garcia
(FEMS Symposium No. 54)

1990 • DENITRIFICATION IN SOIL AND SEDIMENT
Edited by Niels Peter Revsbech and Jan Sørensen
(FEMS Symposium No. 56)

1991 • *CANDIDA* AND CANDIDAMYCOSIS
Edited by Emel Tümbay, Heinz P. R. Seeliger, and Özdem Anğ
(FEMS Symposium No. 50)

1991 • MICROBIAL SURFACE COMPONENTS AND TOXINS IN RELATION TO PATHOGENESIS
Edited by Eliora Z. Ron and Shlomo Rottem
(FEMS Symposium No. 51)

1991 • GENETICS AND PRODUCT FORMATION IN *STREPTOMYCES*
Edited by Simon Baumberg, Hans Krügel, and Dieter Noack
(FEMS Symposium No. 55)

1991 • THE BIOLOGY OF *ACINETOBACTER*: Taxonomy, Clinical Importance, Molecular Biology, Physiology, Industrial Relevance
Edited by K. J. Towner, E. Bergogne-Bérézin, and C. A. Fewson
(FEMS Symposium No. 57)

1991 • MOLECULAR PATHOGENESIS OF GASTROINTESTINAL INFECTIONS
Edited by T. Wadström, P. H. Mäkelä, A.-M. Svennerholm, and H. Wolf-Watz
(FEMS Symposium No. 58)

A Continuation Order Plan is available for this series. A continuation order will bring delivery of each new volume immediately upon publication. Volumes are billed only upon actual shipment. For further information please contact the publisher.

Molecular Pathogenesis of Gastrointestinal Infections

Edited by

T. Wadström

University of Lund
Lund, Sweden

P. H. Mäkelä

National Public Health Institute
Helsinki, Finland

A.-M. Svennerholm

University of Gothenburg
Gothenburg, Sweden

and

H. Wolf-Watz

University of Umeå
Umeå, Sweden

Springer Science+Business Media, LLC

Library of Congress Cataloging-in-Publication Data

Molecular pathogenesis of gastrointestinal infections / edited by T.
  Wadström ... [et al.].
      p.   cm. -- (FEMS symposium ; no. 58)
    Based on the proceedings of the 58th Symposium of the Federation
  of European Microbiological Societies, held in Helsingør, Denmark,
  Sept. 2-4, 1990.
    Includes bibliographical references and index.

    1. Gastrointestinal system--Infections--Molecular aspects-
  -Congresses. 2. Gastrointestinal system--Infections--Pathogenesis-
  -Congresses.   I. Wadström, Torkel. II. Federation of European
  Microbiological Societies. Symposium (1990 : Helsingør, Denmark)
  III. Series.
    [DNLM. 1. Bacteria--pathogenicity--congresses. 2. Enterotoxins-
  -congresses. 3. Gastrointestinal Diseases--congresses.
  4. Gastrointestinal Diseases--etiology--congresses.
  5. Gastrointestinal Diseases--microbiology--congresses.   W3 F21 no.
  58 / WI 100 M718 1990]
  RC840.I53M65 1991
  616.3'3--dc20
  DGPO/DLC
  for Library of Congress   91-32037
                                                                     CIP

Proceedings of a symposium held under the auspices of the
Federation of European Microbiological Societies,
on September 2-4, 1990, in Helsingør, Denmark

ISBN 978-1-4684-5984-5   ISBN 978-1-4684-5982-1 (eBook)
DOI 10.1007/978-1-4684-5982-1

©1991 Springer Science+Business Media New York
Originally published by Plenum Press, New York 1991
Softcover reprint of the hardcover 1st edition 1991

PREFACE

     The meeting that provided the material for this book was the 58th
Symposium of the Federation of European Microbiological Societies (FEMS)
entitled MOLECULAR PATHOGENESIS OF GASTROINTESTINAL INFECTIONS which was
held in Helsingor, Denmark from 2nd to 4th September, 1990. The aim of this
meeting was to bring together scientists from a range of disciplines -
microbiology, cell biology, molecular biology and immunology - to consider
how microbes, including parasites, colonize and infect the gastrointestinal
tract. The programme was designed to focus particular attention on the
range of strategies whereby enterovirulent bacteria and parasites colonize
the gastrointestinal mucin layer, how they adhere to and penetrate the
epithelial layer by entering the cells or passing between them, and how
various protein toxins may facilitate these processes. Speakers were
especially encouraged to highlight the recent expansion in our knowledge of
the molecular mechanisms by which enterotoxigenic and enteropathogenic
Escherichia coli, shigellae, salmonellae and Yersinia enterocolitica cause
intestinal disease. There were also discussions of recently-discovered
gastrointestinal pathogens such as Clostridium difficile and Helicobacter
pylori as well as accounts of how virulent determinants can be used to
develop new diagnostic methods based on DNA gene probes and the polymerase
chain reaction (PCR). These presentations provided the basis for the
chapters in this book.

     The organizers of the meeting owe special thanks to the following
sponsors who contributed to the realization of this Symposium: FEMS, the
Swedish Board for Technical Development, ASTRA Läkemedel AB, Beckman,
Beecham SmithKline & French, DAKO, Gist-Brocade AB, Glaxo Läkemedel AB,
Orion Diagnostika, Pharmacia AB, Rocheproduckter, and Wellcome Diagnostics.

                                                        T Wadström
                                                        H Wolf-Watz
                                                        P H Mäkelä
                                                        A-M Svennerholm

CONTENTS

THE MUCOSAL SURFACE

The Enterocyte and Its Brush Border   1
    R.W. Lobley

Characteristics of the Recognition of Host Cell
        Carbohydrates by Viruses and Bacteria   9
    K.-A. Karlsson, J. Ångström, and S. Teneberg

Structure and Properties of Rat Gastrointestinal Mucins   23
    I. Carlstedt, S. Elmquist, I. Ljusegren, and G.C. Hansson

The Role of Large Intestine Mucus in Colonization of the
        Mouse Large Intestine by Escherichia coli F-18 and
        Salmonella typhimurium.   29
    P.S. Cohen, B.A. McCormick, D.P. Franklin, R.L. Burghoff,
        and D.C. Laux

Germfree Animals Intestinal Glycoconjugates and Colonization. . . .   33
    T. Midvedt, G. Larson, and C. Svanborg

Bacterial Glycosidases and Degradation of Glycoconjugates in
        the Human Gut   37
    L.C. Hoskins

THE PARASITE

The Clone Concept and Enteropathogenic Escherichia coli (EPEC). . . .   49
    I. Ørskov and F. Ørskov

Genetics of Histone-Like Protein H-NS/H1 and Regulation of
        Virulence Determinants in Enterobacteria.   55
    B.E. Uhlin, B. Dagberg, K. Forsman, M. Göransson, B. Knepper,
        P. Nilsson, and B. Sondén

Regulation of Expression of Fimbriae of Human
        Enterotoxigenic Escherichia coli   61
    W. Gaastra, A.M. Hamers, B.J.A.M. Jordi, P.H.M. Savelkoul,
        G.A. Willshaw, M.M. McConnell, J.G. Kusters,
        A.H.M. van Vliet, and B.A.M. van der Zeijst

Function and Molecular Architecture of Escherichia coli
        Adhesins.   71
    H. Hoschützky, T. Bühler, R. Ahrens, and K. Jann

Newly Characterized Putative Colonization Factors of
    Human Enterotoxigenic Escherichia coli. 79
    M.M. McConnell

Molecular Biology of Escherichia coli Type 1 Fimbriae 87
    P. Klemm and K.A. Krogfelt

Intestinal Colonization by Enteropathogenic Escherichia coli 93
    S. Knutton

Diffuse Adherence of Enteropathogenic Escherichia coli Strains. . . . 103
    I. Benz and M.A. Schmidt

ENTEROTOXIGENIC ORGANISMS

Intracellular Mechanisms Regulating Intestinal Secretion. 107
    H.R. de Jonge

Cholera Toxin: Assembly, Secretion and In Vivo Expression. 115
    J. Holmgren, S.J.S. Hardy, T.R. Hirst, S. Johansson,
    G. Jonson, J. Sanchez, and A.-M. Svennerholm

Heat-Stable Enterotoxins Produced by Enteric Bacteria 125
    Y. Takeda, S. Yamasaki, T. Hirayama, and Y. Shimonishi

Molecular Analysis of Potential Adhesions of Vibrio cholerae O1 . . . 139
    P.A. Manning

Shigella Toxin and Related Proteins – Translocation to the
    Cytosol and Mechanism of Action 147
    S. Olsnes, K. Sandvig, and B. van Deurs

Vero Cytotoxins (Shiga-Like Toxins) of Escherichia coli 155
    S.M. Scotland

Structure and Function of Clostridium difficile Toxins. 161
    S.P. Borriello

On the Cytotoxic Modes of Action of Clostridium difficile Toxins. . . 169
    M. Thelestam, M.C. Shoshan, and C. Fiorentini

INVASIVE ORGANISMS

Salmonella as An Invasive Enteric Pathogen. 175
    P.H. Mäkelä, M. Hovi, H. Saxén, A. Muotiala, P. Riikonen,
    M. Nurminen, S. Taira, S. Sukopolvi, and M. Rhen

Experimental Salmonellosis in Retrospect and Prospect: 1990 185
    J. Stephen, G.R. Douce, and I.I. Amin

Colonization of the Murine Gastrointestinal Tract by
    Salmonella typhimurium. 197
    R. Curtiss III and J. Galán

Plasmid Encoded Virulence of Yersinia 207
    H. Wolf-Watz, Å. Forsberg, R. Rosqvist, I. Bölin, K. Erickson,
    L. Norlander, M. Rimpiläinen, T. Bergman, and S. Håkansson

Interactions of Yersinia with Collagen. 213
    Z. Kienle, L. Emödy, T. Wadström, and P. O'Toole

Thermoregulation of Invasion Genes in Shigella flexneri
    Through the Transcriptional Activation of the virB
    Gene on the Large Plasmid 217
    C. Sasakawa, T. Tobe, S. Nagai, N. Okada, B. Adler,
    K. Komatsu, and M. Yoshikawa

Association of Invasive Shigella Strains with Epithelial Cells. . . . 223
    T. Pál and A.A. Lindberg

Haemagglutinating Shigellae 231
    I. Ciznar, F. Qadri, S. Haq, and S.A. Hossain

Molecular Pathogenesis of Giardia lamblia: Adherence
    and Encystation 237
    G.T. Keusch, H.D. Ward, E. Ortega-Barria, N. Galindo,
    and M.E.A. Pereira

The Pathogenic Mechanisms of Helicobacter pylori - A Short
    Overview. 249
    T. Wadström, J.L. Guruge, S. Wei, P. Aleljung, and A. Ljungh

Superficial Components of Helicobacter pylori, in Relation
    to Adherence to Epithelial Cells. 257
    J.L. Fauchère and M. Boulot-Tolle

POLYMERASE CHAIN REACTION AND OTHER DNA BASED
DIAGNOSTICS OF ENTERIC INFECTIONS

Detection of Virulence Determinants in Enteric Escherichia coli
    Using Nucleic Acid Probes and Polymerase Chain Reaction 267
    Ø. Olsvik, E. Hornes, Y. Wasteson, and A. Lund

DNA Probes and PCR Analysis in the Detection of Clostridium
    difficile and Helicobacter pylori 273
    B.W. Wren, C.L. Clayton, and S. Tabaqchali

VACCINE DEVELOPMENT

The Development of Genetically Defined Live Bacterial Vaccines. . . . 279
    S. Chatfield, N. Fairweather, J. Tite, I. Charles,
    M. Roberts, M. Posada, R. Strugnell, and G. Dougan

Development of an Oral Vaccine Against Enterotoxigenic
    Escherichia coli Diarrhea 287
    A.-M. Svennerholm, C. Åhrén, C. Wennerås, and J. Holmgren

POSTER SESSION

Comparison of Virulence Factors in Different Freshly Isolated
    Strains of Entmaoeba histolytica. 295
    G.D. Burchard and D. Mirelman

A System for Production and Rapid Purification of Large
    Amounts of the Shiga Toxin/Shiga-Like Toxin B Subunit. 299
  A. Donohue-Rolfe, D.W.K. Acheson, G.T. Keusch,
  M.B. Goldberg, S.A. Boyko, and S.B. Calderwood

Expression and Possible Biological Functions of Curli on
    Infantile Diarrhoea Escherichi coli Isolates. 303
  L. Emödy, Å. Ljungh, T. Pal, G. Sarlós, and T. Wadstrom

Studies on Heat-Stable Enterotoxin Type II (STII or STb)
    from Escherichia coli 307
  C. Handl and J.-I. Flock

Immune Response to Vibrio cholerae Infection in Rabbits
    with Special Reference to Antibodies Against In Vivo
    Specific Antigens 313
  G. Jonson, A.-M. Svennerholm, and J. Holmgren

Expression of Type 1 Fimbriae by Escherichia coli F18 in
    the Streptomycin-Treated Mouse Large Intestine. 317
  K.A. Krogfelt, B.A. McCormick, R.L. Burgoff,
  D.C. Laux, and P.S. Cohen

Flagellar Components of Helicobacter pylori 321
  C.J. Luke, T.S.J. Elliott, and C.W. Penn

The Influence of Intestinal Mucus on Plasmid Encoded Adhesion
    and Surface Hydrophobicity of Yersinia enterocolitica 323
  A. Paerregaard, O.M. Jensen, and F. Espersen

The Binding of Bacteria Carrying CFAs and Putative CFAs to
    Rabbit Intestinal Brush Border Membranes 327
  C. Wennerås, J. Holmgren, M.M. McConnell, and A.-M. Svennerholm

Factors Contributing to the Persistence of Escherichia coli
    in the Human Large Intestinal Microflora. 331
  A.E. Wold, D. Caugant, G. Lidin-Jansson, P. de Man,
  and C. Svanborg

The Role of Piglet Intestinal Mucus in the Pathogenicity of
    Escherichia coli K88. 335
  P.L. Conway, L. Blomberg, A. Welin, and P.S. Cohen

INDEX 339

THE ENTEROCYTE AND ITS BRUSH BORDER

Robert W. Lobley

Department of Gastroenterology
Royal Infirmary
Manchester, UK

The close relationship between physical structure and physiological function is nowhere more apparent than in the small intestine. The mucosal surface is lined with a virtually continuous epithelial monolayer of columnar enterocytes that constitute the primary cellular barrier between the intestinal lumen and the internal milieu. These cells are therefore a prime target for microbial attack during infections of the gut, and many aspects of the pathogenesis of gastrointestinal infections can be properly understood only in the context of molecular interactions between the micro-organism and the enterocyte.

The enterocytes themselves are adapted in both structure and function to meet the conflicting requirements of facilitating the absorption of nutrients from the intestinal lumen and at the same time protecting against the unwanted entry of noxious luminal agents such as pathogenic micro-organisms and their products. First, they are highly polarized cells, the specialized apical brush border having numerous microvilli that considerably increase the luminal surface area. The microvillus membrane forms a digestive-absorptive surface whose functions are complementary to and integrated with those of the basolateral membrane, ensuring effective transfer of food products across the enterocyte from the intestinal lumen to the interstitial fluid and blood. Second, the external surface of the microvilli is covered with a carbohydrate-rich coat, the glycocalyx, which appears to protect the membrane proteins against luminal proteolysis. Third, adjacent cells are closely apposed to each other, their lateral membranes being connected by desmosomes and their apical poles meeting in an electron-dense tight junction that constitutes a permeability barrier to the diffusion of solute and water molecules between the cells.

The main functions of the enterocyte fall into three categories: the efficient digestion and absorption of nutrients, the provision of a physical defensive barrier between the intestinal lumen and the internal environment, and the regulation of these processes. The digestive functions of the enterocyte include the breakdown of oligosaccharides, peptides and organic phosphates, which are all hydrolysed at the brush border[1], and lipids which are metabolised within the cell. The enterocyte is able to absorb the end products of digestion, namely monosaccharides, amino acids, di- and some tri-peptides, as well as bile salts, short chain fatty acids, and water soluble vitamins. They are absorbed across the brush border by specific, carrier-mediated transport processes which are

Molecular Pathogenesis of Gastrointestinal Infections
Edited by T. Wädstrom *et al.*, Plenum Press, New York, 1991

1

Table 1. Digestive enzymes of the brush-border membrane

Function	Protein	Molecular mass (kDa)	Carbohydrate content	Membrane anchor[a]
Glycosidase	Maltase-glucoamylase	330/125 + 135[e]	+	P
	Sucrase-isomaltase	145 + 151	+	P
	Lactase-phlorizin hydrolase	160 (x2)	+	P
	Trehalase	80 (x2)	+	L
Peptidase	Aminopeptidase A	170 (x2)	+	P
	Aminopeptidase N	162 (x2)	+	P
	Aminopeptidase W	130	+	P
	Carboxypeptidase P	130	+	?
	Dipeptidyl aminopeptidase IV	136 (x2)	+	P
	Peptidyl dipeptidase[b]	180	+	P
	Pteroyl polyglutamate hydrolase[c]	91	?	?
	Enteropeptidase	300[d]	+	P?
	Endopeptidase-24.11	96 (x2)	+	P
	Endopeptidase-2[d]	100 (x2)	+	P?
	Glutamyl transferase	2 + 21	+	P
Phosphatase	Alkaline phosphatase	86 (x2)	+	L
	Phosphodiesterase-I		?	L

[a]P denotes a hydrophobic peptide anchor, L a phospholipid anchor; [b]angio-tensin converting enzyme; [c]'folate conjugase'; [d]'PABA peptidase'; [e]species variable. Where possible, data is for human enzymes; for original literature, see references 1,2.

Na^+-dependent and require metabolic energy[1]. The cell can also transport ions across the microvillus membrane either by means of specific carrier proteins or via ion channels, while vitamins such as B_{12} and, in the neonate, maternal immunoglobulins, are transported by binding to membrane receptors, followed by receptor-mediated endocytosis[1]. Lipophilic molecules such as triglycerides and fat soluble vitamins are able simply to diffuse across the lipid bilayer.

The physical barrier functions of the brush border membrane, its glycocalyx and tight junction have already been mentioned but the enterocyte has some additional defensive functions, for example the secretion of immunoglobulins, and it may play a part in antigen presentation to the gut immune system. The enterocyte must also be able to regulate its absorptive and secretory processes, particularly in the context of ion transport[3], where signals such as peptide hormones, neurotransmitters, and intracellular calcium and pH may be involved, and it must also be able to regulate the bulk flow of solutes and water across the epithelium by controlling paracellular permeability, although the regulating factors here are largely unknown. Since the brush border plays a central role in the majority of these enterocyte functions, it follows that interference by micro-organisms with brush border structure or function will invariably lead to impaired enterocyte function.

As indicated above, much of the digestive function of the enterocyte takes place at the brush border and the microvillus membrane contains a large number of glycosidases, peptidases and phosphatases, as shown in the Table. All are relatively large proteins, most comprising two or more subunits, and so far as is known all are glycosylated. They appear to be essentially globular structures, attached to the external face of the membrane by a small anchoring segment embedded in the lipid bilayer[2]. Many of the glycosidases and peptidases can be released from the membrane largely intact in a soluble form by limited proteolysis. Biosynthetic studies, amino acid sequencing and, more recently, gene sequencing studies indicate that for these microvillar enzymes, the membrane anchor consists of a short (2-5kDa), hydrophobic, N-terminal peptide sequence which traverses the lipid bilayer, as depicted in the left-hand portion of Fig. 1 for the neutral aminopeptidase[2]. Between the outer face of the membrane and the globular part of the enzyme molecule there appears to be an exposed region of the peptide stalk that is vulnerable to attack by proteases such as papain or elastase[2], or by bacterial proteases of similar specificity[4].

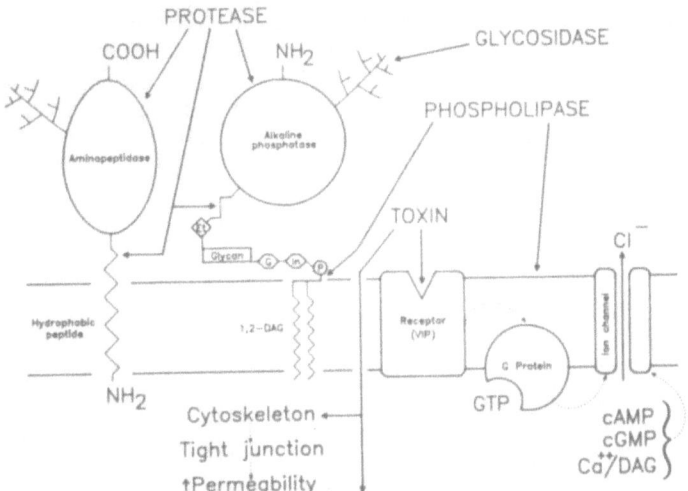

Fig. 1. Schematic diagram illustrating the organization of the microvillus membrane and some pathogenic mechanisms exploited by enteric micro-organisms. DAG, diacyl-glycerol; Et, ethanolamine; G, glucosamine; In, ino-sitol; P, phosphate ester. Branched structures represent surface oligosaccharides, large zigzags indicate peptide chains, and small zigzags denote the hydrocarbon skeleton of long-chain fatty acids. Primary microbial effects are shown with solid lines, secondary effects with broken lines.

Recently a second mode of anchorage of brush-border hydrolases to the membrane was described, as shown in Fig. 1 for alkaline phosphatase. Here the enzyme is anchored to the membrane through the diacylglycerol moiety of a phosphatidylinositol glycan attached covalently to the C-terminal part of the molecule[5,6]. Enzymes anchored in this way are generally resistant to release from the membrane by proteases, but are susceptible to release by phospholipases. In addition to alkaline phosphatase, phos-phodiesterase-I and trehalase have so far been identified with this mode of attachment to the brush-border membrane[1]. Since many microorganisms are known to produce both proteases and phospholipases having appropriate specificities, it is to be expected that bacterial colonization of the intestine will frequently lead to enzyme loss and consequent malabsorption. These effects are liable to be augmented through the action of microbial glycosidases which could remove the protective carbohydrate from the exposed, globular portion of the microvillar hydrolases, rendering them more susceptible to digestion by luminal proteases[7].

The structure, and probably also the permeability barrier function, of the apical pole of the enterocyte is maintained by the brush-border cytoskeleton, comprising the microvillus cores and the terminal web[8,9]. As shown schematically in Fig. 2, each microvillus contains an ordered bundle of actin microfilaments running longitudinally from the microvillus tip into the underlying terminal web. These microfilaments are linked into a regular hexagonal array by cross-bridges containing two actin-binding proteins, fimbrin and villin. Fimbrin is an ubiquitous 68kDa globular protein which cross-links actin filaments into highly ordered bundles. Villin is a 95kDa globular protein of more restricted tissue distribution and interacts with actin in a Ca^{2+}-regulated manner, promoting filament bundling at submicromolar Ca^{2+} concentrations but causing fragmentation at higher Ca^{2+} levels. The microvillus cores are attached laterally to the membrane through a complex of the 17kDa Ca^{2+}-binding protein calmodulin and two much larger proteins, one an unnamed 110kDa protein and the other a 140kDa glycoprotein that binds to the inner face of the membrane[8,9].

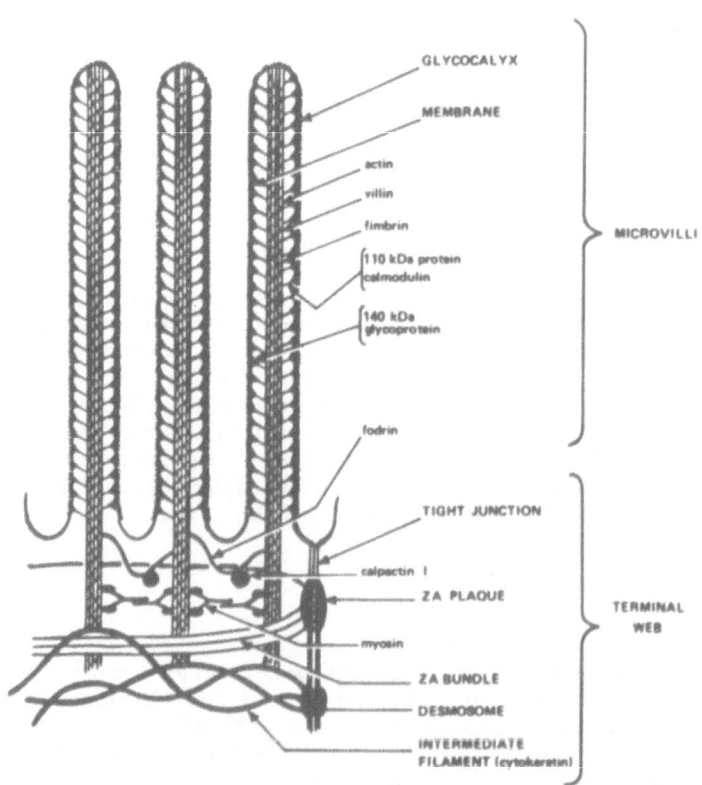

Fig. 2. Organization of the main structural components of
the brush border and its cytoskeleton. Reproduced
from reference 1 with permission.

Within the terminal web the inter-rootlet zone consists of a dense meshwork of fine, non-actin filamentswhich appear to interconnect the core rootlets of adjacent microvilli (Fig. 2). These microfilaments contain non-muscle myosin and fodrin, a calmodulin-binding, spectrin-related protein[8,9]. At the level of the zonula adherens (ZA), the junctional complex at the lateral margin of the cell consists of a circumferential bundle of actin filaments, beneath which lies a network of thicker, inter-mediate cytokeratin filaments. The ZA bundle filaments, which resemble the stress fibres of cultured fibroblasts, contain myosin, tropomyosin and &-actinin and appear to be closely associated with the ZA tight junctions through adhesion plaque-like structures[8,9]. Like stress fibres, the ZA bundle is contractile through an ATP-dependent process, and this provides an attractive candidate mechanism to control tight-junctional resistance between adjacent enterocytes and thus regulate paracellular permeability.

There is now considerable evidence that the attachment between elements of the cytoskeleton and the tight junction may be functionally important. For example, exposure of guinea pig ileum to a variety of agents which disrupt actin microfilaments within the terminal web causes changes in tight-junctional resistance and in paracellular permeability[10]. Furthermore, exposure of cultured monolayers of enterocytes to Clostridial enterotoxin A has recently been shown to produce a progressive fall in transepithelial resistance and a corresponding increase in paracellular permeability, effects that were accompanied by changes within the brush-border cytoskeleton[11]. This suggests that that micro-organisms are able to alter paracellular permeability and render the epithelium leaky at least in part through interactions with the brush-border cytoskeleton.

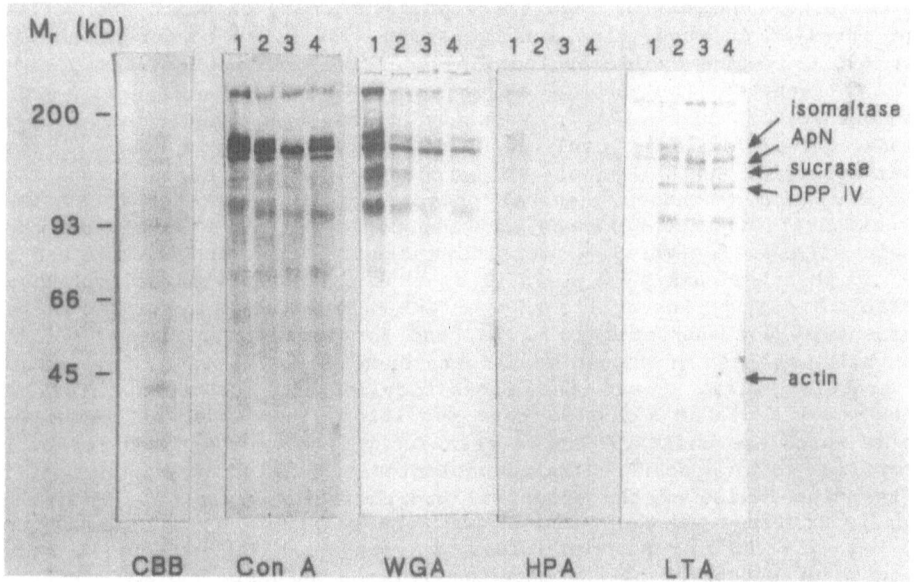

Fig. 3. Glycosylation of microvillar glycoproteins in four individual dogs. The glycoproteins were separated from isolated microvillus membranes by SDS gel electro-phoresis and detected after western blotting by lectin staining using an immunogold technique. M_r, relative molecular mass; CBB, Coomassie brilliant blue; Con A, conconavalin A; WGA, wheat germ agglutinin; HPA, Helix pomatia agglutinin; LTA, Lotus tetragonolobus agglutinin; DPP IV, dipeptidyl peptidase IV.

An important pathogenetic factor in gastrointestinal infection is microbial adherence to the mucosal surface. Considerable evidence indicates that an early event in the colonization of the intestine by pathogenic micro-organisms is attachment to the carbohydrate-rich glycocalyx of the exposed brush-border surface through binding of microbial lectins to specific microvillar oligosaccharides[12,13]. The microvillus membrane contains both glycolipids and glycoproteins that have been 'implicated in microbial adherence, and details of the precise carbohydrate structures necessary for recognition are beginning to emerge. We and others have found that there is considerable individual variation in the glycosylation of microvillar glycoproteins, as illustrated in Fig. 3. Here concanavalin A, which detects glucose and mannose residues and is relatively unspecific, stains most of the large microvillar proteins with little distinction between the individual animals. Conversely Helix lectin, which detects N-acetylgalactosamine and is blood group A specific, binds appreciably only to dog 1, while Lotus lectin, which binds to fucose and is blood group H specific, stains only proteins of dogs 2, 3 and 4. However individual variability is not restricted to differences in blood group, as shown by differences in the staining of individual glycoprotein bands between animals of the same blood group, for example between dogs 2, 3 and 4 stained with Lotus lectin. Similar results have also been obtained for human microvillar glycoproteins, and it would be interesting to know whether this variation between individuals could be important in determining their susceptibility to intestinal colonization. Host variability is certainly a factor in the host-pathogen interaction, and differences in microvillar glycosylation may have to be considered when investigating microbial pathogenesis.

In addition to its complement of digestive enzymes the brush border also contains a variety of regulatory proteins, many of which are believed to be involved in regulating ion transport. The brush border contains at least two Ca^{2+}- and cyclic nucleotide-sensitive transport systems, namely Na-Cl cotransport (by linked Na^+/H^+ and Cl^-/HCO_3^- exchange) and an electrogenic Cl^- channel[3,14]. Putative microvillar regulatory proteins include guanyl cyclase; cyclic AMP-dependent protein kinase; a brush border-specific, 86kDa G-kinase; 25 and 21kDa phosphoprotein cosubstrates for the A- and G-kinases; several phosphoprotein substrates for a Ca^{2+}- and calmodulin-dependent kinase and the Ca^{2+}- and phospholipid-dependent protein kinase C; several G-proteins including the oncogene-related p21 ras; and phospholipase A_2[3,14]. There is thus the potential for both direct regulation, e.g. by the cyclic AMP- or GMP-stimulated phosphorylation of a component of a transporter molecule, and for indirect processes mediated through the release of second messengers such as Ca^{2+} acting, for example, via protein kinase C and the phoshatidylinositol cascade system. Two cytoskeletal proteins may also have regulatory functions, although their precise roles are unclear. One, a relatively minor 80kDa component of the microvillus core, appears to be immunologically related to a substrate for the tyrosine kinase of the epidermal growth factor receptor[8]. The other, which is confined to the terminal web, is calpactin I, a 36kDa protein that has calmodulin-, phospholipid-, and actin-binding properties and is also a major substrate of the oncogene-related pp60 src tyrosine kinase[8]. The presence of these proteins suggests the possibility of functional regulation through hormone or growth factor receptor-induced protein phosphorylation of the brush-border cytoskeleton.

Although the inter-relationships between these regulatory pathways and their putative target processes are only beginning to be unravelled, it is already clear that these pathways may be severely disrupted by pathogenic micro-organisms and their toxins[3,14,15]. For example E. coli and Yersinia heat-stable enterotoxins bind to specific microvillar glycoproteins and activate brush-border guanyl cylase, although Ca^{2+}, calmodulin and the phosphatidylinositol cascade appear also to be involved in the secretory response. Binding of Clostridial enterotoxin A to a microvillar glycoprotein has also been demonstrated and evidence that its secretory effects may be mediated partly through the cytoskeleton has already been mentioned[11]. In cholera, after binding of the B subunits of the enterotoxin to specific microvillar glycolipids (G_{M1}-ganglioside), the A subunit penetrates the membrane and activates adenyl cyclase at the basolateral membrane, thereby initiating the secretory diarrhoea[3,14,15].

Fig. 1 attempts to illustrate some of these pathogenetic mechanisms and to summarize the major kinds of microbial interaction with the apical surface of the enterocyte. After binding of a micro-organism to the carbohydrate of the brush border surface or the mucus layer, the enterocyte may be damaged in a number of ways. Microbial proteases or phospholipases may act on the membrane to degrade or release microvillar ectoenzymes, leading to maldigestion and osmotic diarrhoea. Phospholipases may also theoretically attack the membrane lipids directly, with the release of such intracellular messengers as calcium and diacylglycerol. These intracellular messengers may also be released by microbial toxins which may act indirectly through cell surface receptors and G-proteins, or by directly activating microvillar regulatory proteins such as guanyl cyclase or phospholipase A_2, or by penetrating the membrane and activating adenyl cyclase or protein kinase C within the cell, for example. Ion channels may then open and a secretory diarrhoea ensues. Toxins may also act intracellularly on the cytoskeleton, leading to a fall in tight-junctional resistance and a rise in paracellular permeability, thereby allowing toxins to penetrate into the mucosa and a bulk flow of electrolytes and water out into the lumen. The overall effect is thus an impairment of digestive function, a reduction, disruption or reversal of many absorptive processes, and a secretory diarrhoea.

As outlined above, a description of many aspects of the structure and functions of the enterocyte, and especially of its brush border, at the molecular level is now available. Despite the inevitable omissions and simplifications, it is hoped that this brief overview has demonstrated that such information is essential to a full understanding of the pathogenesis of gastrointestinal infections and may thereby contribute to the development of new therapeutic strategies.

REFERENCES

1. R. Holmes and R. W. Lobley, Intestinal brush border revisited, Gut 30:1667-1678 (1974).
2. G. Semenza, Anchoring and biosynthesis of stalked brush border membrane proteins: glycosidases and peptidases of enterocytes and renal tubuli, Ann. Rev. Cell Biol. 2:255-313 (1986).
3. M. Donowitz and M. J. Walsh, Regulation of mammalian small intestinal electrolyte secretion, in: "Physiology of the gastrointestinal tract, vol. 2," L. R. Johnson, ed., Raven Press, New York (1987).
4. A. Jonas, C. Krishnan, and G. Forstner, Pathogenesis of mucosal injury in the blind loop syndrome. Release of disaccharidases from brush border membranes by extracts of bacteria obtained from intestinal blind loops in rats, Gastroenterology 75:791-795 (1978).
5. M. G. Low, Biochemistry of the glycosyl-phosphatidylinositol membrane anchors, Biochem. J. 44:1-13 (1987).
6. M. G. Low, The glycosyl-phosphatidyinositol anchor of membrane proteins, Biochim. Biophys. Acta 988:427-454 (1989).
7. S. P. Riepe, J. Goldstein, and D. H. Alpers, Effect of secreted Bacteroides proteases on human intestinal brush border hydrolases, J. Clin. Invest. 66:314-22 (1980).

8. M. S. Mooseker, Organisation, chemistry and assembly of the cytoskeletal apparatus of the intestinal brush border, Ann. Rev. Cell Biol. 1:209-241 (1985).

9. S. Maroux, E. Coudrier, H. Feracci, J. P. Gorvel, and D. Louvard, Molecular organisation of the intestinal brush border, Biochimie 70:1297-1306 (1988).

10. J. L. Madara, D. Barenberg, and S. Carlson, Effects of cytochalasin D on occluding junctions of intesinal absorptive cells: further evidence that the cytoskeleton may influence paracellular permeability and junctional charge selectivity, J. Cell Biol. 102:2125-2136 (1986).

11. G. Hecht, C. Pothoulakis, J. T. La Mont, and J. L. Madara, Clostridium difficile toxin A perturbs cytoskeletal structure and tight junction permeability of cultured human intestinal epithelial monolayers, J. Clin. Invest. 82:1516-1524 (1988).

12. C. P. Cheney, and E. C. Boedeker, Adherence of an enterotoxigenic Escherichia coli strain, serotype 078:H11, to purified human intestinal brush borders, Infect. Immun. 39:1280-1284 (1983).

13. G. W. Jones, and R. Freter, Adhesive properties of Vibrio cholerae: nature of the interaction with isolated rabbit brush border membranes and human erythrocytes, Infect. Immun. 14:240-245 (1976).

14. H. R. De Jonge, and S. M. Lohmann, Mechanisms by which cyclic nucleotides and other intracellular mediators regulate secretion, in: "Microbial toxins and disease," (Ciba Foundation Symp. 112), D. Evered and J. Whelan J, eds., Pitman Books, London (1985).

15. K. J. Moriarty and L. A. Turnberg, Bacterial toxins and diarrhoea, Clinics Gastroenterol. 15:529-543(1986).

CHARACTERISTICS OF THE RECOGNITION OF HOST CELL

CARBOHYDRATES BY VIRUSES AND BACTERIA

Karl-Anders Karlsson, Jonas Ångström and
Susann Teneberg

Department of Medical Biochemistry, University
of Göteborg, P.O. Box 33031, S-400 33 Göteborg,
Sweden

INTRODUCTION

Microbial interactions with host tissues is at present a field in a fascinating development both basically and concerning potential applications. Our knowledge of virus receptors is more advanced than that of bacterial systems, in part based on the crystal conformation of the influenza virus hemagglutinin in complex with the receptor, sialic acid (1), and the crystal structures of picornaviruses (see 2). The "Canyon hypothesis" (2) suggests that one strategy for viruses to escape immune surveillance is to protect the receptor attachment site in a surface depression which is too narrow to be reached by antibodies. This site is conserved on the otherwise hypervariable surface of the viruses. Development of protective vaccines may therefore get very difficult or impossible, and efforts are growing to prepare soluble receptor analogues that may inhibit viral attachment, based on carbohydrate (1,3,4) or protein (2,5,6) receptors, and there appears to be good opportunities for rational drug design.

No receptor-binding protein for bacterial systems has yet been obtained in three dimensions. However, several cloned toxins and adhesins are at present being expressed for crystallization and X-ray analysis. Meanwhile a number of protein and carbohydrate receptors are being chemically characterized. The purpose of the present short overview is to summarize the main characteristics of carbohydrate receptors. Due to limit in space we will not gather all known carbohydrate receptors but rather select examples to illustrate the points. An impressive list of specificities for bacteria has been presented (7) and the status for virus receptors was recently summarized (8,9). The essence of the present discussion was elaborated in more detail elsewhere (10).

COMPLEXITY AND DIVERSITY OF CELL-ASSOCIATED GLYCOCONJUGATES

The main port of entry of infections are the mucous membranes, which form very large total surfaces in the respiratory and gastro-intestinal tracts. Fig. 1 defines the types of glycoconjugates associated with epithelial cells. Due to complexity only one epithelial cell, of human small intestine, has so far been defined for both lipid-linked (11) and peptide-linked (12) membrane-bound glycans. However, the complex secreted mucins await detailed analysis. Corresponding studies on isolated colon epithelial cells are going on (13; Holgersson, Jovall, Breimer, in preparation).

Although bacteria may colonize in secretions, there may be an advantage to adhere to membrane-bound structures, and viruses and bacterial toxins have to attach membrane-close to be able to penetrate into cells. In this respect the glycolipids are of special interest since they are usually strictly membrane-bound and do not appear in secretions, except for shedded epithelial cells in the intestine. As will be repeatedly illustrated below a number of glycolipids are known to bind microbes.

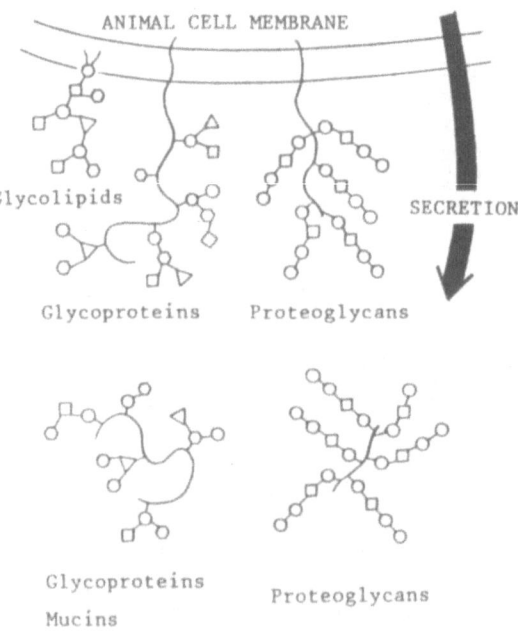

Fig. 1. The microbial recognition and binding of glycoconjugates at the host epithelial cell surface may be a stepwise process through multiple binding proteins (e.g. bacteria carrying several adhesin specificities). Primary association may be to secreted mucus, followed by penetration to the cell membrane, and invasion through the cell layer to the basal membrane (often protein receptors). The balance between secreted and strictly membrane-bound receptors may determine the outcome of an infection (possible elution). This may explain why lipid-linked oligosaccharides (glycolipids) have been repeatedly selected as receptors (do not usually appear in secretions).

It has been estimated that about 6.000 different saccharide structures are known in nature (calculation for a recently delivered PC-based retrieval system, ref. 14). Of these, about 500 are glycolipids, but many more remain to be characterized. There are distinct differences in glycoconjugate expression, especially for glycolipids, between animal species, between individuals, and tissues and cells, as a possible basis for the tropism of various infections (10).

PROXIMITY TO THE CELL MEMBRANE: A PROPERTY OF LIPID-LINKED OLIGOSACCHARIDES

Many bacteria carry several adhesion specificities, which may be used for successive passage over different host cells (e.g. ascending urinary tract infection) or for step-wise penetration through the secreted mucus layer to attach to epithelial cells for invasion and binding to the basal membrane. Need for proximity to membrane to be functional has been nicely demonstrated recently for cholera toxin (15). Early attempts to demonstrate peptide-linked oligosaccharide receptors were unsuccessful and the only proven receptor substance is the pentaglycosylceramide, the ganglioside GM1 (see also Fig. 3 below). The two paraffin chains anchor the binding epitope close to the membrane bilayer and a mechanism has been proposed where the five binding subunits of the toxin create the proximity for the catalytic subunit to penetrate into the membrane. Neoglycoprotein was prepared (15) containing the GM1 saccharide and this protein was covalently linked to receptor-deficient cells. The cells were shown to bind the toxin well but did not respond in contrast to cells coated with GM1 glycolipid. This is a convincing demonstration of the proximity rule.

A second example of some interest, although still preliminary, is our finding of a glycolipid receptor in common for several viruses (16,17). Using the thin-layer chromatography overlay technique (16) and a number of reference mixtures of glycolipids and isolated substances we found an identical binding for such diverse viruses as influenza virus, adeno virus, reo virus, Sendai virus, Rota virus, rabies virus, EB virus and mumps virus, with a common denominator of monoglycosylceramide with a certain ceramide composition (17). The conclusion from epitope dissection (see below) was a membrane-close epitope as illustrated in Fig. 2. The hypothesis was that the virus particles recognize their respective host cells through a first-step receptor and thereafter reaches the second-step (and imbedded, cryptic and bilayer-close) receptor for penetration. The proposal was that the relatively conserved fusion peptide is the ligand for this receptor. Thus several viruses with separate first-step receptors and host cells may use a common fusion receptor, which is unaccessible by direct contact from the outside due to the membrane-close location. Theoretically it would be possible to design a common soluble receptor analogue for the treatment or prophylaxis of several virus infections. However, the biological relevance of these binding data has first to be shown.

One may also add the lactose specificity as an example of membrane proximity. As summarized elsewhere a number of bacteria show a specificity for lactosylceramide (10). This is noteworthy since lactose (long known as a free oligosaccharide secreted in milk) in bound form is known only in glycolipids (may also be extended with additional saccharides). This means an ideal cell adhesion site with no competing receptor analogues appearing in secreted glycoproteins at the cell surface (compare Fig. 1). However, this specificity cannot explain the varying tissue tropism for the bacteria carrying this specificity. Therefore lactosylceramide, which is not directly accessible from the outside (deeply imbedded on the membrane surface) may be a second-step receptor (compare viruses above) to establish and ascertain the attachment after binding to a first-step receptor which may vary (carbohydrate or protein). Recently, information on the lactosylceramide binding has been supplemented for *Propionibacterium* (18), *Actinomyces* (19), *Bordetella pertussis* (20), *Pseudomonas aeruginosa* (21), and clinically important yeast cells like *Cryptococcus* and *Candida* (22).

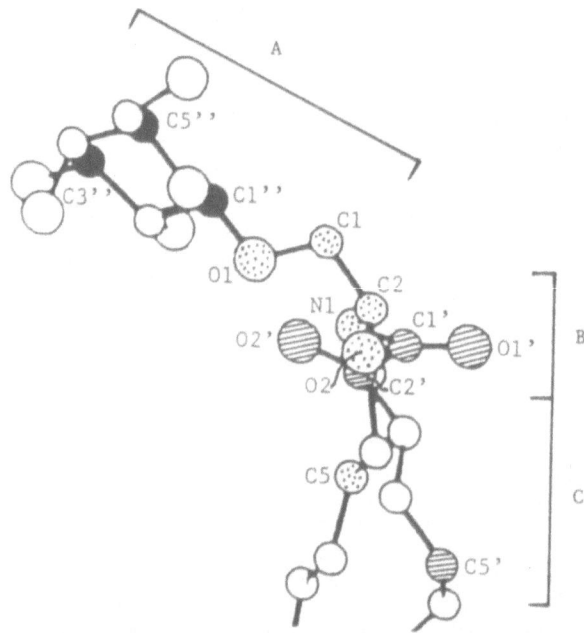

Fig. 2. Dissected binding epitope (compare Fig. 3) on a proposed (16,17) second-step receptor in common for several separate viruses (see text).

The two paraffin chains of the one-sugar glycolipid (GalβCer) have been cut at C7 and C6, respectively (pointing downwards). The three-part epitope is just at the interface of the cell surface monolayer involving the non-polar side of the hexose (A), a hydrogen-bond-forming region (B) and a non-polar part (C). The hexose may be substituted with up to four sugars. The epitope is inaccessible by direct contact from the outside due to the complex cell surface (see Fig. 1), and the hypothesis predicts a two-step penetration process. The virus first selects host cell through association to a more peripheral first-step receptor (carbohydrate or protein). This may induce a lateral reorganization of surface components to allow binding to the second-step receptor, mediated by the fusion peptide. Binding data for epitope dissection were obtained by overlay of viruses on thin-layer plates (16).

The membrane-proximity of the lactosylceramide epitope may mediate invasion for some of these bacteria, although the only evidence for this is that several bacteria carrying this specificity are invasive. One may note that the adhesin of *Neisseria gonorrhoea* with lactosylceramide specificity recently was genetically cloned (23).

INTERNAL BINDING AND RECEPTOR-BINDING VARIANTS

One early finding was that bacteria (24) and bacterial toxins (25) were able to recognize internally placed epitopes of sacharide receptors. This differs from most mammalian anti-carbohydrate antibodies which bind to terminal parts, e.g. blood group determinants. The internal binding seems to be a general property of microbes and the biological meaning may be an easier mutational shift of receptor specificity, including the change of host cell (10). A recent example are the cloned adhesin variants of uropathogenic *E. coli* of human and dog (26) showing different binding patterns to Galα4Gal-containing isoreceptors (see also below and Table 1). Although the minimum requirement for these adhesins seems to be this disaccharide, they show different preferences for saccharide extensions. Human isolates prefer GalNAcβ3 (dominating receptor in the human urinary tract), while dog isolates prefer GalNAcα3GalNAcβ3 (dominating receptor in the dog urinary tract) as extensions. Very likely, the two adhesin proteins differ only slightly in their receptor-binding sites, although the rest of the molecules may be rather dissimilar due to antigenic drift. This may be analogous to the sialic acid-binding variants of influenza virus, where only one substitution of amino acid of the hemagglutinin binding site is enough to shift between NeuAcα3 and NeuAcα6 receptors (1,9).

It is likely that such variant receptor-binding proteins have slightly separate binding epitopes on the saccharide receptor surface, resulting in different dependencies for neighboring groups of the minimally required sequence. Another example of variant specificity is the lactosylceramide binding discussed above. Most members require lactosylceramide with more hydroxylated ceramide and tolerate some saccharide additions to lactose (10). However, others like *Propionibacterium freudenreichii* (18,24,27) and yeast cells (22), prefer lactosylceramide with less hydroxylated ceramide and tolerate no extensions to lactose. The first type of receptor is present in epithelial cells and the second in non-epithelial cells (10).

The most extensively known variants of one basic specificity are uropathogenic *E. coli* (see also above) and Shiga-like toxins, which play on the Galα4Gal specificity. In Table 1 we have gathered selected data for these systems. The interested reader has to look up the original papers for details. From the Table it is clear that the ligands must differ at their binding sites. Although sequences of several of the binding proteins are known, analysis of homologies are still premature (32), without knowledge of the binding sites. Very likely, all ligands shown require Galα4Gal (or Galα), but

Table 1. Variety of binding patterns of *E. coli* and Shiga-like toxins that recognize Galα4Gal-containing isoreceptors. Binding data were obtained by overlay of ligand on thin-layer plates with separated glycolipids (except Pap-2, see below columns). A limited number of isoreceptors have been included only to illustrate the differences in binding.

Glycolipid Structure	PapG	PrsG	Pap-2	Shiga Toxin	Pig Edema Toxin	SLT	SLT vp	SLT vh
Galα4GalβCer	+							
Galα4Galβ4GlcβCer	+	-	+	+		+	-	+
GalNAcβ3Galα4Galα4Galβ4GlcβCer	+	(+)		(+)	+	+	+	+
GalNAcα3GalNAcβ3Galα4Galβ4GlcβCer	+	+		-	+	(+)	+	-
Galβ3GalNAcβ3Galα4Galβ4GlcβCer	-	+	+	-	(+)	-	(+)	-
GalNAcβ3Galα3Galβ4GlcβCer		-				-	+	-
Galα3Galα4Galβ4GlcβCer				+	+			
Reference	26	26	28	25,29	30	31	31	31

PapG, PrsG and Pap-2 were cloned adhesins from *E. coli*. The other ligands are Shiga toxin or Shiga-like toxins. Pap-2 was not analyzed by overlay and the two positive signs were added by us based on the reported inhibition with GalNAcβ3Gal (28).

A positive sign, +, means a clear positive binding.
A positive sign within parenthesis, (+), means a weak binding.
A negative sign, -, means no binding.

may recognize separate parts of the convex surface of the disaccharide (10,33,34). Also differences in amino acid sequence at the border of the binding site may influence the interaction with neighboring groups to this disaccharide. The disaccharide was underlined in Table 1 to indicate the internal binding. Of interest is Pap-2 which binds selectively to Bowman´s capsule of the kidney and which is possible to inhibit with GalNAcβ3Gal, possibly having lost the need for Galβ of Galα4Galβ. It may therefore also recognize the glycolipid added to the Table: GalNAcβ3Galα3Galβ4GlcβCer, isogloboside. This sequence was recently shown to be bound by pig edema disease toxin (Table 1). The optimal binding epitope in this case may therefore be GalNAcβ3Galα, since Galα4Galβ4GlcβCer was a weaker binder (30). In case of *E. coli* PrsG Galα4GalβGlcβCer does not bind, but addition of GalNAcβ3 produces a clear binding. However, GalNAcβ3Galα3Galβ4GlcβCer does not bind, indicating that the minimum sequence in this case is GalNAcβ3Galα4Gal, thus differing from the edema toxin and possibly Pap-2. Altogether these examples document receptor-binding variants which will be very interesting to elucidate concerning detailed protein-carbohydrate interactions, preferably through X-ray analysis, to learn about their evolution. There is no evidence in case of these toxins for peptide-linked receptor sites, and these toxins are the first known ligands which use glycolipid receptors for endocytosis from coated pits (see 32).

LOW-AFFINITY INTERACTIONS

In case of Shiga toxin (Table 1) it was not possible to inhibit the binding of the toxin to Vero cells by preincubation with soluble oligosaccharides containing Galα4Gal (29). However, a good selective inhibition was obtained using albumin with covalently bound saccharide, optimal with 25 saccharides/mole albumin. Thus a multivalency was needed, and this multivalency also appears with glycolipids presented in the thin-layer overlay assay. This indicates a relatively low-affinity interaction of individual sites, in spite of a rather strong binding of the intact system. In case of *E. coli* however (Table 1), soluble saccharides do inhibit, thus indicating higher affinity. Precise figures on association constants are not yet available and have to await pure components. The lactosylceramide specificities are also of the low-affinity type since soluble lactose is unable to inhibit at high concentrations (18,19,22,27). For the classical studies of influenza virus free receptor sialic acid was a very poor inhibitor. However, this assay problem was overcome by using modified red cells with a very low density of sialic acid sites (9), allowing tests of sialyl derivatives. A recent elegant development is the use of NMR for precise measurement of oligosaccharide-protein interaction (see 4). Univalent receptors are therefore often very poor inhibitors of the natural multivalent interaction systems.

EPITOPE DISSECTION AND DRUG DESIGN

Rational drug design is based on the three-dimensional
structure of the active or binding site of the macromolecule,
and modeling soluble low-molecular weight compounds for
synthesis by advanced computer graphics (35). Due to the
multivalency of the natural micbrobial attachment systems (36)
such a drug must be superior in affinity compared to the
univalent natural receptor. At the level of the binding site
this should mean an optimized atomic fit between receptor
analogue and binding protein, which excludes a classical
empirical screening for efficient inhibitors. However, at the
present stage of development the best known inhibitor for
microbial adhesion is that for mannose-binding *E. coli*, where a
classical enzyme substrate, umbelliferyl-αMan, was 1000 times
more efficient than methyl-αMan (37).

As referred to in the INTRODUCTION, receptor-binding sites
have been localized in crystal conformations of binding
proteins of influenza virus (carbohydrate receptor) and picorna
viruses (protein receptors) but no corresponding information is
yet available for bacterial systems, although work has been
initiated (34,38). In case of influenza virus hemagglutinin a

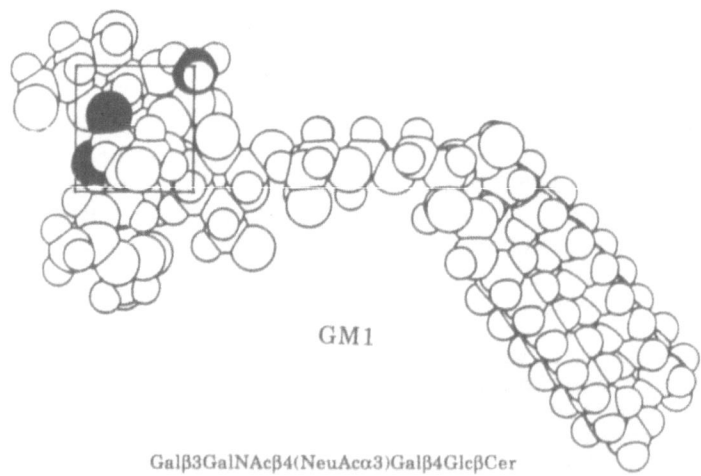

GM1

Galβ3GalNAcβ4(NeuAcα3)Galβ4GlcβCer

Fig. 3. Computer-calculated model (HSEA, GESA) of the receptor glycolipid
for cholera toxin, the ganglioside GM1. The rectangle on the
oligosaccharide defines the minimum epitope proposed to be recognized by
the protein. The two carboxyl oxygens of NeuAc (left) and the methyl
carbon of GalNAc (upper right) have been shaded. The conclusion was based
on a comparison of toxin binding preferences for natural and modified GM1
isoreceptors with calculated conformations (epitope dissection, ref. 10).
The actual epitope is limited to one side of the sequence, at the branched
interface of the three minimally required sugars, GalNAcβ4(NeuAcα3)Galβ.

The improved binding caused by the terminal Galβ3 of GM1 (the best natural
binder) may in part be due to toxin interaction with the non-polar ring
hydrogens of this extra sugar. The dissection approach rationalizes the
chemical synthesis of soluble receptor analogues for potential therapy,
without knowledge of the binding site of the protein, which otherwise is
the basis for rational drug design (35).

precise binding assay has been established based on NMR for testing various modifications of sialic acids rationalized from atomic coordinates (4). In the end the analogue may have little resemblance on the paper with the natural receptor.

We will present an example of rationalized information on binding epitope at the level of carbohydrate receptor, without knowledge of the binding site of the protein. The approach has been described and exemplified elsewhere (9,39), and has been named epitope dissection. It is based on the internal binding as defined above. Comparison of binding preferences to isoreceptors (see Table 1) with computer-calculated three-dimensional views of these isoreceptors may inform about which part of a minimum receptor sequence is the binding surface or epitope. Neighboring groups and sequences have steric extensions which may or may not hinder access for the binding. Thus negative isoreceptors (although carrying the receptor sequence) may be negative because neighboring groups may protrude on the epitope side. Positive binders should have no such hindrances. In the example of Fig. 3 for cholera toxin and its receptor it has been shown that the calculated conformation of the receptor is in perfect agreement with NMR data from solution of the released GM1 pentasaccharide (40). Therefore, this approach is valid for the approximation of a binding epitope, provided an enough number of isoreceptor preferences is available for comparison. In the case of cholera toxin and its isoreceptors we have demonstrated (Ångström, Karlsson and Teneberg, in preparation) that also this ligand recognizes an internal epitope (Fig. 3). Based on binding data both to natural isoreceptors and modified isoreceptors (data also taken from the literature) and calculations, a surprisingly limited minimum binding surface was located, just on one side of the branched interface of the three minimally required sugars, GalNAcβ4(NeuAcα3)Galβ. Thus, only a small part of the GM1 total surface is being specifically recognized by the toxin. Interestingly, this internal binding suggests an overlapping similarity in binding epitope between cholera toxin and tetanus toxin, on GM1 and GQ1b gangliosides, respectively, although their sequences, viewed from the terminals, look quite different on the paper (Ångström, Karlsson, Teneberg, in preparation).

This information from epitope dissection is a rational basis for synthesis of soluble receptor analogues, which is a laborious task in case of carbohydrates (see 34). For the example of Fig. 3 one may thus limit the modeling to the atoms and properties at the rectangle, wihout having to take care of the remaining parts of the receptor molecule.

There are several alternatives for therapeutic use of the information from microbial-host cell interactions (8). Very few productive experiments have been performed on infection models *in vivo* using receptor substances or receptor analogues. One may note recent positive effects on *E. coli* K99 diarrhea of calves using glycoprotein glycans from bovine plasma (41). The carbohydrate receptor for this bacterium is being analyzed in detail (42). The minimum part is the N-glycolyl type of sialic acid, NeuGc.

CONCLUSIONS

The rapid development at present concerning our knowledge of receptors for microbes, and the need for supplementing the traditional therapy against infections, will most certainly soon result in more solid data on the practical use of receptor analogues. One obvious problem is the need of synthetic substances which are distinctly superior to the natural receptor, if the substance in question has to be resorbed from the intestine (univalency) and be able to overcome the natural multivalency. Much efforts therefore have to be spent on drug design, optimally requiring crystal conformations of receptor-binding proteins and improved methods of synthesis.

ACKNOWLEDGEMENT

Our own work reported here was supported by grants from The Swedish Medical Research Council (no. 3967) and from Symbicom AB.

REFERENCES

1. W. Weis, J.H. Brown, S. Cusack, J.C. Paulson, J.J. Skehel and D.C. Wiley, Structure of the influenza virus hemagglutinin complexed with its receptor, sialic acid, Nature 333:426 (1988).
2. M.G. Rossman, The Canyon hypothesis, J. Biol. Chem. 264:14587 (1989).
3. M.G. Rossman, Viral receptors and drug design, Nature 333:392 (1988).
4. N.K. Sauter, M.D. Bednarski, B.A. Wurzburg, J.E. Hanson, G.M. Whitesides, J.J. Skehel and D.C. Wiley, Hemagglutinins of two influenza virus variants bind to sialic acid derivatives with millimolar dissociation constants: A 500-MHz proton nuclear magnetic resonance study, Biochemistry 28:8388 (1989).
5. S.D. Marlin, D.E. Staunton, T.A. Springer, C. Stratowa, W. Sommergruber and V.J. Merluzzi, A soluble form of intercellular adhesion molecule-1 inhibits rhinovirus infection, Nature 344:70 (1990).
6. C.R.M. Bangham and A.J. McMichael, Nosing ahead in the cold war, Nature 344:16 (1990).
7. D. Mirelman (ed), "Microbial lectins and agglutinins", John Wiley and Sons, New York (1986).
8. T.L. Lentz, The recognition event between virus and host cell receptor: a target for antiviral agents, J. Gen. Virol. 71:751 (1990).
9. J.C. Paulson, Interactions of animal viruses with cell surface receptors, in: "The receptors", Vol. 2, P.C. Conn, ed., Academic Press, Orlando (1985).
10. K.-A. Karlsson, Animal glycosphingolipids as membrane attachment sites for bacteria, Annu. Rev. Biochem. 58:309 (1989).
11. S. Björk, M.E. Breimer, G.C. Hansson, K.-A. Karlsson and H. Leffler, Structures of blood group glycosphingolipids of human small intestine. A relation between the expression of fucolipids of

epithelial cells and the ABO, Le and Se phenotype
of the donor, J. Biol. Chem. 262:6758 (1987).

12. J. Finne, M.E. Breimer, G.C. Hansson, K.-A. Karlsson,
    H. Leffler, J.F.G. Vliegenthart and H. van
    Halbeek, Novel polyfucosylated N-linked
    glycopeptides with blood group A, H, X and Y
    determinants from human small intestinal
    epithelial cells, J. Biol. Chem. 264:5720 (1989).

13. J. Holgersson, N. Strömberg and M.E. Breimer,
    Glycolipids of human large intestine: differences
    in glycolipid expression related to anatomical
    localization, epithelial/nonepithelial tissue and
    the ABO, Le and Se phenotypes of the donors,
    Biochimie 70:1565 (1988).

14. CarbBank, The Complex Carbohydrate Structure Database,
    University of Georgia, Complex Carbohydrate
    Research Center, Copyright (1990).

15. T. Pacuszka and P. Fishman, Generation of cell surface
    neoganglioproteins. GM1-Neoganglioproteins are
    non-functional receptors for cholera toxin, J.
    Biol. Chem. 265:7673 (1990).

16. G.C. Hansson, K.-A. Karlsson, G. Larson, N. Strömberg,
    J. Thurin, C. Örvell and E. Norrby, A novel
    approach to the study of glycolipid receptors for
    viruses. Binding of Sendai virus to thin-layer
    chromatograms, FEBS Lett. 170:15 (1984).

17. K.-A. Karlsson, E. Norrby and G. Wadell, Antiviral
    agents, Patent PCT no. PCT/DK/00007 (1986).

18. N. Strömberg and K.-A. Karlsson, Characterization of
    the binding of Propionibacterium granulosum to
    glycosphingolipids adsorbed on surfaces. An
    apparent recognition of lactose which is dependent
    on the ceramide structure, J. Biol. Chem.
    265:11244 (1990).

19. N. Strömberg and K.-A. Karlsson, Characterization of
    the binding of Actinomyces naeslundii (ATCC 12104)
    and Actinomyces viscosus (ATCC 19246) to
    glycosphingolipids, using a solid-phase overlay
    approach, J. Biol. Chem. 265:11251 (1990).

20. E. Tuomanen, H. Towbin, G. Rosenfelder, D. Braun, G.
    Larson, G.C. Hansson and R. Hill, Receptor
    analogues and monoclonal antibodies that inhibit
    adherence of Bordetella pertussis to human
    ciliated respiratory epithelial cells, J. Exp.
    Med. 168:267 (1988).

21. N. Baker, G.C. Hansson, H. Leffler, G. Riise and C.
    Svanborg Edén, Glycosphingolipid receptors for
    Pseudomonas aeruginosa, Infect. Immun. 58:2361
    (1990).

22. V. Jimenez-Lucho, V. Ginsburg and H.C. Krivan,
    Cryptococcus neoformans, Candida albicans, and
    other fungi bind to the glycosphingolipid
    lactosylceramide, a possible adhesion receptor for
    yeasts, Infect. Immun. 58:2085 (1990).

23. D.K. Paruchuri, H.S. Seifert, R.S Ajioka, K.-A.
    Karlsson and M. So, Identification and
    characterization of a Neisseria gonorrhoeae gene
    encoding a glycolipid-binding adhesin, Proc. Natl.
    Acad. Sci. USA 87:333 (1990).

24. G.C. Hansson, K.-A. Karlsson, G. Larson, A.A. Lindberg, N. Strömberg and J. Thurin, Lactosylceramide is the probable adhesion site for major indigenous bacteria of the gastrointestinal tract, in: "Glycoconjugates", Proc. of the 7th International Symposium on Glycoconjugates, M.A. Chester, D. Heinegård, A. Lundblad and S. Svensson, eds., Rahms, Lund, Sweden (1983).

25. J.E. Brown, K.-A. Karlsson, A.A. Lindberg, N. Strömberg and J. Thurin, Identification of the receptor glycolipid for the toxin of Shigella dysenteriae, in: "Glycoconjugates", Proc. of the 7th International Symposium on Glycoconjugates, M.A. Chester, D. Heinegård, A. Lundblad and S. Svensson, eds., Rahms, Lund, Sweden (1983).

26. N. Strömberg, B.-I. Marklund, B. Lund, D. Ilver, A. Hamers, W. Gaastra, K.-A. Karlsson and S. Normark, Host-specificity of uropathogenic E. coli depends on differences in binding specificity to Galα4Gal-containing isoreceptors, EMBO J. 9:2001 (1990).

27. N. Strömberg, M. Ryd, A.A. Lindberg and K.-A. Karlsson, Two species of Propionibacterium apparently recognize separate epitopes on lactose of lactosylceramide, FEBS Lett. 232:193 (1988).

28. J.F. Karr, B. Nowicki, L.D. Truong, R.A. Hull and S.I. Hull, Purified P-fimbriae from two cloned gene clusters of a single pyelonephritogenic strain adhere to unique structures in the human kidney, Infect. Immun. 57:3594 (1989).

29. A.A. Lindberg, J.E. Brown, N. Strömberg, M. Westling-Ryd, J.E. Schultz and K.-A. Karlsson, Identification of the carbohydrate receptor for Shiga toxin produced by Shigella dysenteriae type 1, J. Biol. Chem. 262:1779 (1987).

30. S. DeGrandis, H. Law, J. Brunton, C. Gyles and C.A. Lingwood, Globotetraosylceramide is recognized by the pig edema disease toxin, J. Biol. Chem. 264:12520 (1989).

31. J.E. Samuel, L.P. Perera, S. Ward, A.D. O'Brien, V. Ginsburg and H.C. Krivan, Comparison of the glycolipid receptor specificities of Shiga-like toxin typeII and Shiga-like toxin type II variants, Infect. Immun. 58:611 (1990).

32. M.P. Jackson, Structure-function analyses of Shiga toxin and the Shiga-like toxins, Microb. Pathogenesis 8:235 (1990).

33. K.-A. Karlsson, Animal glycolipids as attachment sites for microbes, Chem. Phys. Lipids 42:153 (1986).

34. J. Kihlberg, S.J. Hultgren, S. Normark and G. Magnusson, Probing the combining site of the PapG adhesin of uropathogenic E. coli bacteria by synthetic analogues of galabiose, J. Am. Chem. Soc. 111:6364 (1989).

35. W.G.J. Hol, Protein crystallography and computer graphics- toward rational drug design, Angew. Chem. 25:767 (1986).

36. M.N. Matrosovich, Towards the development of antimicrobial drugs acting by inhibition of pathogen attachment to host cells: a need for multivalency, FEBS Lett. 252:1 (1989).

37. N. Firon, S. Ashkenazi, D. Mirelman, I. Ofek and N. Sharon, Aromatic alpha-glycosides of mannose are powerful inhibitors of the adherence of type 1 fimbriated E. coli to yeast and intestinal epithelial cells, Infect. Immun. 55:472 (1987).
38. A. Holmgren and C.-I. Brändén, Crystal structure of chaperone protein PapD reveals an immunoglobulin fold, Nature 342:248 (1989).
39. K.-A. Karlsson, Current experience from the interaction of bacteria with glycosphingolipids, in: "Molecular mechanisms of microbial adhesion", L. Switalski, M. Höök and E. Beachey, eds., Springer-Verlag, New York (1989).
40. S. Sabesan, K. Bock and R.U. Lemieux, The conformational properties of the gangliosides GM2 and GM1 based on H1 and C13 nuclear magnetic resonance studies, Can. J. Chem.62:1034 (1984).
41. M. Mouricout, J.M. Petit, J.R. Carias and R. Julien, Glycoprotein glycans that inhibit adhesion of E. coli mediated by K99 fimbriae: Treatment of experimental colibacillosis, Infect. Immun. 58:98 (1990).
42. S. Teneberg, P. Willemsen, F.K. de Graaf and K.-A. Karlsson, Receptor-active glycolipids of epithelial cells of the small intestine of young and adult pigs in relation to susceptibility to infection with E. coli K99, FEBS Lett. 263:10 (1990).

STRUCTURE AND PROPERTIES OF RAT GASTROINTESTINAL MUCINS

Ingemar Carlstedt, Stefan Elmquist, Ingela Ljusegren and Gunnar C. Hansson[*]

University of Lund, Department of Physiological Chemistry 2, P.O.Box 94, S-221 00 Lund, Sweden and [*]University of Gothenburg, Department of Medical Biochemistry, P.O.Box 33031, S-400 33, Gothenburg, Sweden

INTRODUCTION

Mucus is a highly hydrated gel which covers the mucosal surfaces of the body and provides a protective barrier against the external environment. The major part (>95%) of mucus is water and the gel matrix is formed by large and complex glycoproteins referred to as the mucus glycoproteins or the mucins, which account for about 0.5-5% of the secretion. In addition, proteins such as secretory IgA, lysozyme, lactoferrin and proteinase inhibitors are often present in the gel.

Mucus glycoproteins have M_r of the order 5-25 million and are assembled from subunits (M_r approx. 2.5 million) joined by disulphide bonds. Subunits consist of a central protein core with a large number of oligosaccharides attached to serine or threonine residues. Most of the glycans are present in tightly spaced 'clusters', which are approximately 100-200 nm long. A subunit may contain several 'clusters' flanked by less substituted, 'naked' regions of the protein core. Mucin oligosaccharides comprise a heterogeneous family of linear and branched structures and a typical mucin is substituted with several different oligosaccharide structures. In human respiratory, human cervical and pig gastric mucins, subunits are joined end-to-end into linear and flexible chains (1). We have noticed that intestinal mucins are difficult to solubilize unless procedures that break covalent bonds are used (2), implying that these mucins are 'polymerized' to a larger extent than those from other sources and, possibly, that their macromolecular 'architecture' is different.

MATERIALS AND METHODS

Isolation of rat gastric mucins:

Mucosal scrapings were collected from rat stomachs (Sprague Dawley, inbred strain) and immediately frozen. The material was gently stirred in 6M-guanidinium chloride supplemented with proteinase inhibitors and the solubilized material was subjected to density-gradient centrifugation in CsCl/4M-guanidinium chloride followed by density-gradient centrifugation in CsCl/0.2M-guanidinium chloride. Mucins isolated with this procedure contain little protein and DNA (3) or lipid (4). Subunits were obtained after reduction of the whole mucins and high-M_r glycopeptides corresponding to the oligosaccharide 'clusters' after subsequent trypsin digestion (5).

Isolation of rat intestinal mucins:

Mucosal scrapings from rat small intestine were collected and extracted with 6M-guanidinium chloride/proteinase inhibitors as described above. The *non-solubilized* material was brought into solution by reduction/alkylation and the subunits were purified by density-gradient centrifugation as

Molecular Pathogenesis of Gastrointestinal Infections
Edited by T. Wädstrom *et al.*, Plenum Press, New York, 1991

were the gastric mucins. The oligosaccharide 'clusters' were isolated with gel chromatography on Sephacryl S-500 after trypsin digestion of the subunits.

Chromatography of mucin fragments:

Gel chromatography on Sephacryl S-500 HR was performed on a column (1.6 x 51.5 cm) eluted at 9ml/h with 0.1M-ammonium acetate. Ion-exchange h.p.l.c was performed on a Mono Q HR 5/5 column eluted at 0.5ml/min with a gradient (60mins) of $LiClO_4$ (0-0.25M in 10mM piperazine/ClO_4 buffer, pH 5.0) after an initial isocratic period (10mins) with buffer alone. Fractions from gel and ion-exchange chromatography (1.5ml and 0.5ml respectively) were analysed for hexose (2).

GC/MS of mucin oligosaccharides

The O-linked oligosaccharides from one of the major high-M_r glycopeptides (corresponding to component A, Fig. 1b) from rat small intestine were released with alkaline/borohydride treatment and separated into neutral and acidic species. The neutral oligosaccharides were then analysed by GC/MS after permethylation (6).

RESULTS AND DISCUSSION

Rat gastric mucins were readily solubilized in 6M-guanidinium chloride. The weight-average M_r was approx 10×10^6 as obtained with laser light scattering (results not shown). Gel chromatography on Sephacryl S-500 (Fig. 1a) shows that the subunits appear as a somewhat 'skewed' peak, which was significantly retarded on the column. After trypsin digestion of the subunits, the high-M_r glycopeptides were further included on the gel. The results are similar to those obtained by Dekker *et al.* (7) and show that these mucins are composed of subunits which contain a number of oligosaccharide 'clusters' in keeping with the structure of human cervical and respiratory mucins, as well as pig gastric mucins (1).

In contrast, rat intestinal mucins were not solubilized with 6M-guanidinium chloride. After repeated extractions, however, the residue was found to be a purified complex of insoluble mucins, which could be brought into solution by reduction. The resulting subunits were finally purified with density-gradient centrifugation. These structures eluted with the void volume of Sephacryl S-500 (Fig. 1b) and preliminary physical characterization suggests that their relative molecular mass is several million daltons. After subsequent trypsin digestion, two major populations of high-M_r glycopeptides were obtained, both of which are larger than the corresponding structure from the stomach. The small intestinal mucins are clearly composed of subunits which each contain a number of oligosaccharide 'clusters' but the intestinal subunits are joined into very large 'insoluble' complexes in an unknown way. Furthermore, it is not known whether the two populations of 'clusters' originate from two different subunits or from the same structure. When the high-M_r glycopeptides from the two mucins were subjected to ion-exchange chromatography (Fig. 2a,b) it was found that those from the gastric mucins were much more heterogeneous than the intestinal ones and also contained a larger proportion of 'neutral' species. The two populations (A & B, Fig. 1b) from the intestinal mucins behaved in a similar way (results not shown).

Rat intestinal and gastric mucins thus share some important structural features with mucus glycoproteins from other sources, including that the macromolecules are composed of subunits and that these contain a number of oligosaccharide 'clusters'. However, the polymerization of subunits in the intestinal mucin must be different from that in gastric mucins because the former macromolecules form an insoluble glycoprotein complex. The different behaviour of the oligosaccharide 'clusters' from the two mucins when subjected to ion-exchange chromatorgaphy is most likely due to a different oligosaccaharide substitution.

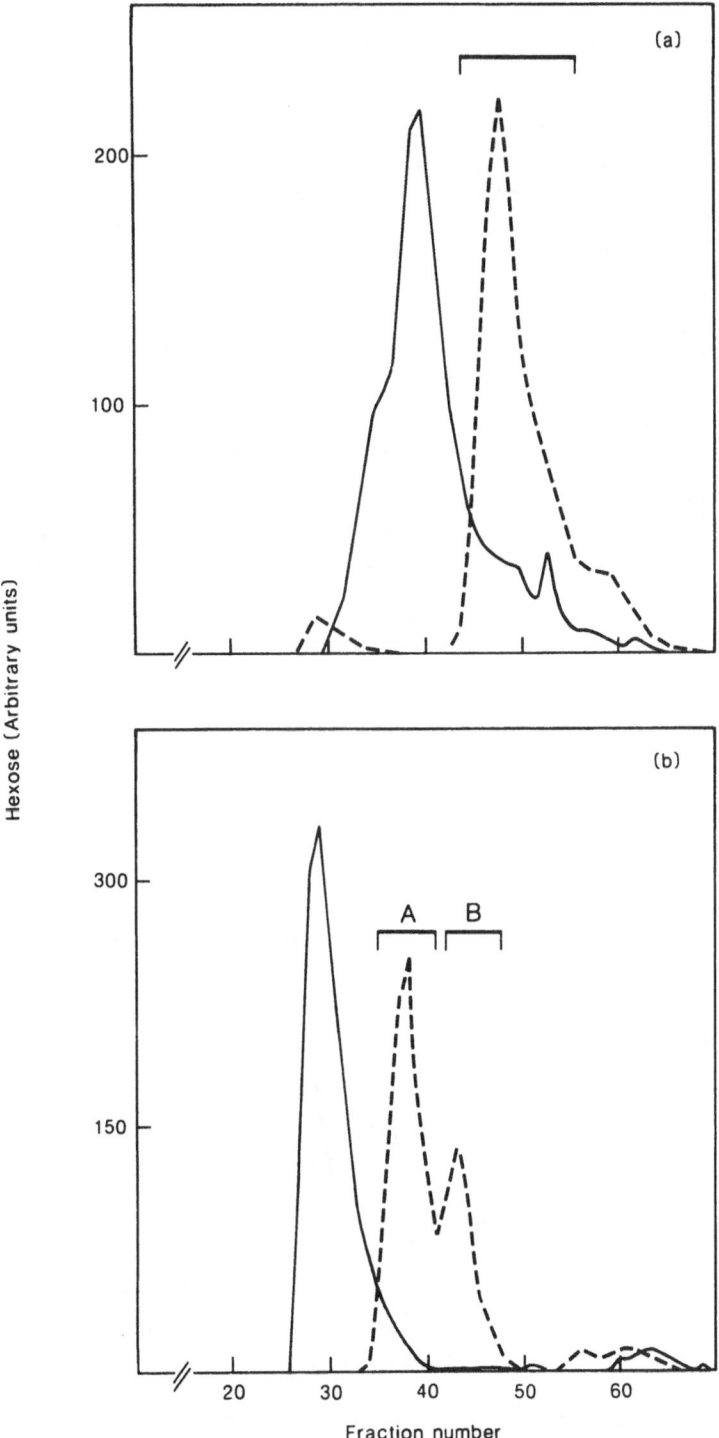

Fig. 1. Gel chromatography on Sephacryl S-500 HR of mucin fragments from (a) rat stomach and (b) rat small intestine. Subunits (——) from both sources are larger than the cognate high-M_r glycopeptides (- - - -) and are thus likely to contain more than one oligosaccharide 'cluster'. The horizontal bars indicate how material was pooled for ion-exchange h.p.l.c. and oligosaccharide analysis. The void volume is at fraction 29.

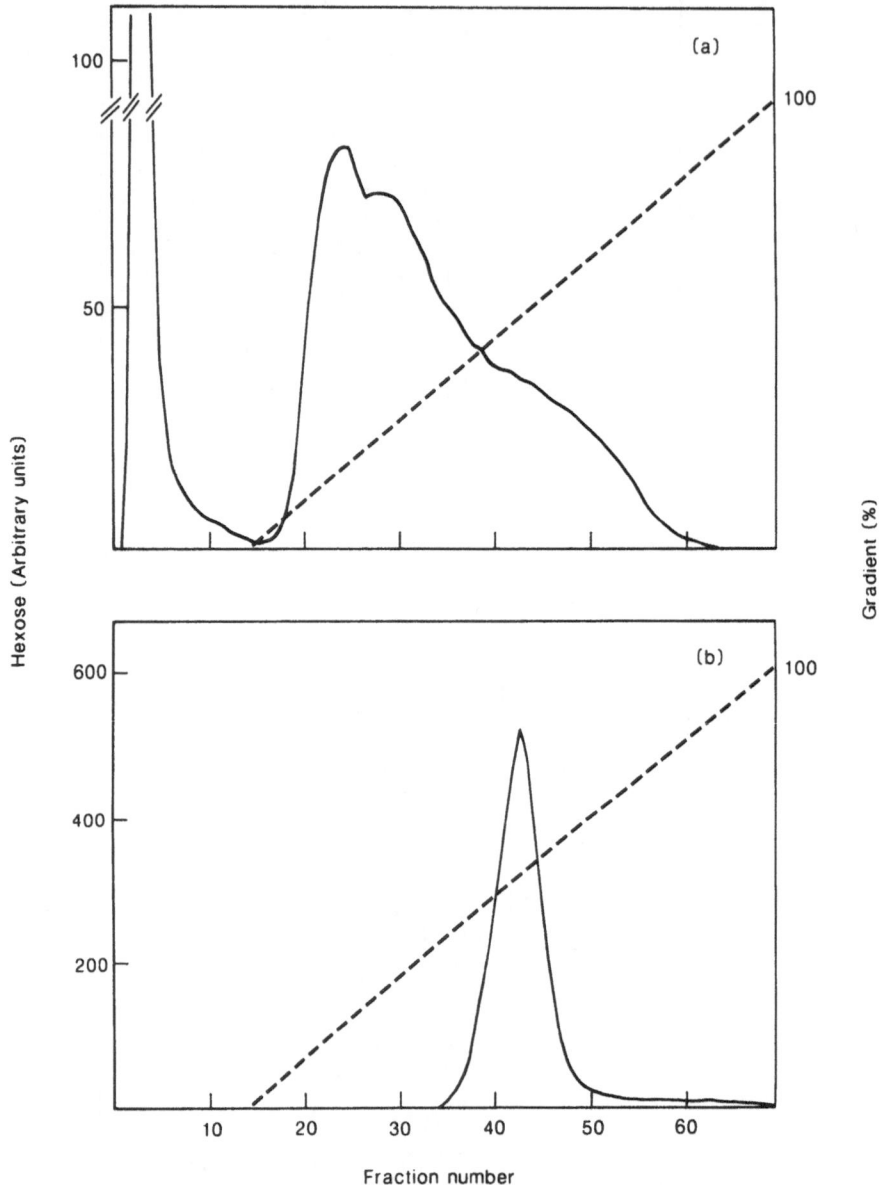

Fig. 2. Ion-exchange h.p.l.c. of high-M_r mucin glycopeptides from (a) rat stomach and (b) rat small intestine (component A). obtained as shown in Fig. 1. Note that the glycopeptides from rat stomach are less acidic than those from rat small intestine suggesting a different oligosaccharide substitution.

When the neutral oligosaccharides from the intestinal mucins were studied with GC/MS, more than 16 different structures were identified. These ranged in size from a single unsubstituted GalNAc up to complex structures with seven sugar residues (Fig 3). The GalNAc linked to the core could be substituted at C-3 with Gal or GlcNAc and at C-6 with GlcNAc. Both these substitutions could be extended into complex structures, some of which contained a blood-group H structure based on either type 1 (Gal1-3GlcNAc) or type 2 (Gal1-4GlcNAc) chains. Most of the smaller glycans could be regarded as incomplete intermediates in the biosynthesis of the larger blood group-containing structures.

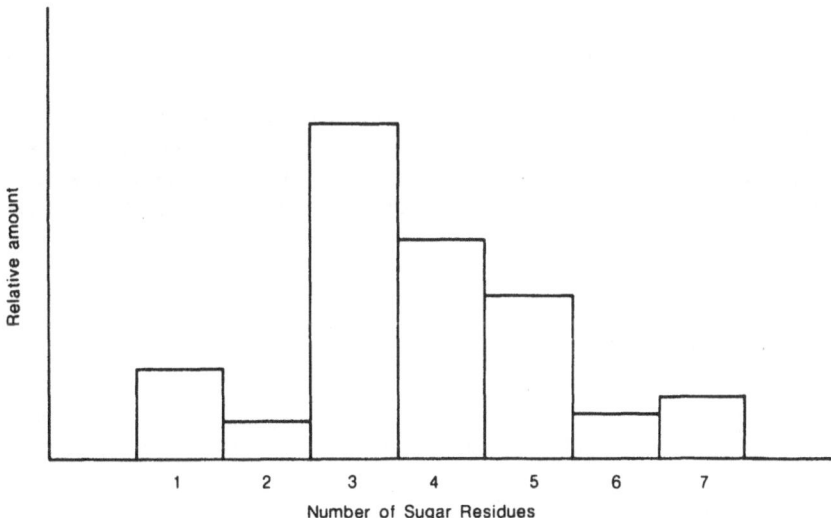

Fig. 3. Size distribution of neutral oligosaccharides from rat small intestine.

IMPLICATIONS FOR ADHESION OF MICROORGANISMS TO MUCUS

Mucins are large molecules with linear dimensions of the same order of magnitude as bacteria (Fig. 4). Most of the O-linked oligosaccharides in the molecules are located to the oligosaccharide 'clusters' and these structures project a heterogeneous population of oligosaccharides into the surroundings. The steric situation as outlined in Fig. 4 suggests that multiple weak interactions between oligosaccharide 'clusters' and fimbriae could be present, possibly via oligosaccharides that do not bind efficiently on their own. In fact, several of the known carbohydrate-lectin interactions are of relatively low affinity (8). The O-linked glycans are made by the sequential addition of single sugar residues to the linkage GalNAc and the shorter oligosaccharides are to a relatively large extent incomplete intermediates of larger structures. This ensures that mucins contain many different carbohydrate epitopes for bacterial interactions, an interesting aspect in the role of mucins in competing for bacterial interactions with the underlying epithelial surface. Because the oligosaccharide substitution could be very different between mucins from different sources, the interaction with the surroundings, including bacteria, could vary significantly from one mucin to the other.

It is not only the oligosaccharide 'clusters' that are of potential interest for bacterial adhesion. The 'naked' regions have been reported to contain N-linked oligosaccharides of the 'high-mannose' type which may provide adhesion sites for *e.g.* E.coli type 1 fimbriae (9). Furthermore, the naked regions may also contain hydrophobic sites (10) allowing for nonspecific hydrophobic interactions between bacteria and mucins.

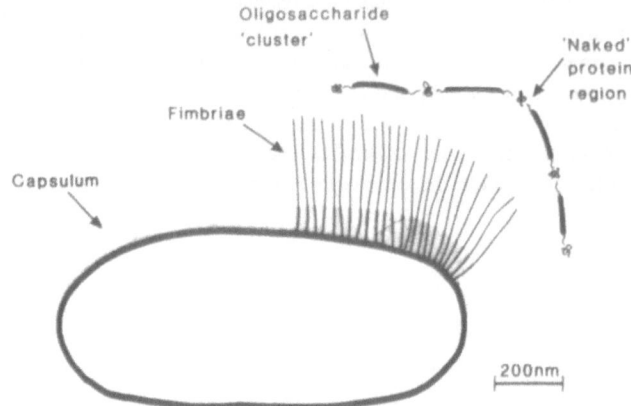

Fig. 4. Schematic drawing of a mucin subunit and a bacterium. The oligosaccharide 'clusters' are approximately 100-200nm long and are thus large enough to make contact with several fimbriae. The 'naked' stretches of protein may contain oligosaccharides of the 'high-mannose' type and/or hydrophobic sites, which, together with the O-linked oligosaccharides in the 'clusters', provide many different possibilities for interactions between mucins and bacteria. Mucin subunits may vary in size and number of oligosaccharide 'clusters'.

ACKNOWLEDGEMENTS

Grants were obtained from the Swedish Medical Research Council (7902 & 7461) and the National Swedish Board for Technical Development. B. Jönsson is acknowledged for the art-work.

REFERENCES

1. I. Carlstedt and J.K. Sheehan, Structure and macromolecular properties of cervical mucus glycoproteins, in: Mucus and Related Topics, E. Chantler and N.A. Ratcliffe, eds., The Company of Biologists Ltd, Cambridge (1989).
2. S. Mårtensson, A. Lundblad, G.C. Hansson and I. Carlstedt, Mucins obtained from patients with enterocutaneous urinary diversions, Scand J Clin Lab Invest. 48:633 (1988).
3. I. Carlstedt, H. Lindgren, J.K. Sheehan, U. Ulmsten and L. Wingerup, Isolation and characterization of human cervical-mucus glycoproteins, Biochem J. 211:13 (1983).
4. G.C. Hansson, J.K. Sheehan and I. Carlstedt, Only trace amounts of fatty acids are found in pure mucus glycoproteins, Arch Biochem Biophys. 266:197 (1988).
5. D.J. Thornton, J.R. Davies, M. Kraayenbrink, P.S. Richardson, J.K. Sheehan and I. Carlstedt, Mucus glycoproteins from 'normal' human tracheobronchial secretion, Biochem J. 265:179 (1990).
6. H. Karlsson, I. Carlstedt and G.C. Hansson, The use of gas chromatography and gas chromatography-mass spectrometry for the characterization of permethylated oligosaccharides with molecular mass up to 2300, Anal Biochem. 182:438 (1989).
7. J. Dekker, W.M.O. Van Beurden-Lamers, A. Oprins and G.J. Strous, Isolation and structural analysis of rat gastric mucus glycoprotein suggests a homogeneous protein backbone, Biochem J. 260:717 (1990).
8. K.-A. Karlsson, Animal glycosphingolipids as membrane attachment sites for bacteria, Ann Rev Biochem. 58:309 (1989).
9. S.U. Sajjan and J.F. Forstner, Role of the putative 'link' glycopeptide of intestinal mucin in binding of piliated Escherichia coli serotype O157:H7 strain CL-49, Infect Immun. 58:868 (1990).
10. B.F. Smith and J.T. LaMont, Hydrophobic binding properties of bovine gallbladder mucin, J Biol Chem. 259:12170 (1984).

THE ROLE OF LARGE INTESTINE MUCUS IN COLONIZATION OF THE MOUSE LARGE INTESTINE BY ESCHERICHIA COLI F-18 AND SALMONELLA TYPHIMURIUM

Paul S. Cohen, Beth A. McCormick, David P. Franklin, Robert L. Burghoff, and David C. Laux

Department of Microbiology, University of Rhode Island, Kingston, Rhode Island 02881 USA

INTRODUCTION

Little information is available as to the role of mucus in bacterial colonization of the intestinal tract. We are interested in defining that role.

RESULTS AND DISCUSSION

E. coli F-18 is a normal human fecal strain isolated in 1977. It contains 7 plasmids, its serotype is rough:K1:H5, it produces colicin V, and makes type 1-pili. E. coli F-18 Col⁻ has the same serotype as E. coli F-18 but lacks an 86 kilobase plasmid and does not produce the E. coli F-18 colicin. E. coli F-18(pPKL91) contains plasmid pPKL91 which is fimB⁺, bla⁺, cat⁺ and contains the parB stabilizing region to prevent plasmid segregation in vivo. The fimB gene on pPKL91 ensures that the fim operon is phase-lacked "on", i.e. that every cell makes type 1 pili. Moreover, E. coli F-18(pPRKL91) makes about twice as many pili as E. coli F-18(pPR633), a strain that contains essentially the same plasmid with the exception of the fimB gene. E. coli F-18 FimA⁻ does not make type 1 pili because it contains the tetracycline resistance gene from Tn10 inserted into the fimA gene. S. typhimurium SL5319, which contains wild-type lipopolysaccharide (LPS), is of biotype FIRN and so is unable to make type 1 pili. Because of an aroA mutation it is aromatic dependent and non-virulent. S. typhimurium SL5325 is nearly isogenic to strain SL5319, but is a spontaneous rough rfaJ mutant.

In order to colonize conventional mice with Gram negatives, the normal facultative flora must first be reduced with antibiotics. We have chosen streptomycin treatment. The mice are fed two streptomycin-resistant strains and over the next two weeks, the level of each strain in feces is monitored. Using this system, we have determined that in competition, E. coli F-18 essentially eliminates E. coli F-18 Col⁻ (1), that E. coli F-18(pPR633) essentially eliminates the type 1 overproducing strain, E. coli F-18(pPKL91), and that S. typhimurium SL5319 eliminates S. typhimurium SL5325, the LPS mutant (2). It should be noted that each of the above "bad" colonizers is poor only relative to the competing strain. All of the strains tested to date, including those listed above, can colonize the large intestine when fed alone to mice.

It appears likely that both E. coli F-18 and S. typhimurium colonize by growing in large intestinal mucus. That is, although cecal mucus scraped directly into sterile culture tubes supports growth of individually inoculated

E. coli F-18 (3), E. coli F-18 Col⁻ (3), S. typhimurium SL5319 (2), and S. typhimurium SL5325 (2), no growth is observed when cecal contents are placed in sterile culture tubes and inoculated with the same strains (2,3). Interestingly, when E. coli F-18 and E. coli F-18 Col⁻ are inoculated together in cecal mucus, E. coli F-18 grows but E. coli F-18 Col⁻ does not (3), suggesting that the inability of E. coli F-18 Col⁻ to colonize in the presence of E. coli F-18 is due either to a defect in its ability to compete with E. coli F-18 for mucus nutrients or to an E. coli F-18 metabolic product that limits E. coli F-18 Col⁻ growth. E. coli F-18 does not, however, limit E. coli F-18 Col⁻ growth in L-broth (1). In contrast, E. coli F-18(pPR633) and E. coli F-18(pPKL91) grow equally well together in cecal mucus and S. typhimurium SL5319 and S. typhimurium SL5325, the LPS mutants, grow equally well together in cecal mucus (2), suggesting that the defect in colonizing ability of E. coli F-18(pPKL91) and S. typhimurium SL5325 is not due to reduced ability to grow in mucus.

It is clear from these studies that crude mucus preparations represent a good source of nutrients for bacterial growth and we have just begun to identify the nutrients in cecal mucus which support S. typhimurium SL5319 growth. Recent work in our laboratory suggests that S. typhimurium SL5319 utilizes peptides present in cecal mucus as its source of amino acids (4). The cecal mucus fraction which contains the peptides also contains oligosaccharide consisting of glucosamine (presumably as N-acetylglucosamine), galactose, and mannose (4). It is therefore possible that it is actually glycopeptides which support S. typhimurium SL5319 growth in mouse cecal mucus.

The role of E. coli F-18 adhesion in the colonization process is difficult to discern. For example, the level of adhesion of E. coli F-18 to immobilized mouse cecal epithelial cells, brush borders, and mucus relative to E. coli F-18 FimA⁻ is about 2 to 1 in each case (5), yet both strains colonize the streptomycin-treated mouse large intestine equally well in co-feeding experiments (5). Furthermore, E. coli F-18(pPR633) is a far better colonizer than E. coli F-18(pPKL91), which overproduces type 1 pili, yet adhesion of E. coli F-18(pPKL91) to immobilized mouse cecal epithelial cell brush borders is six times that of E. coli F-18(pPR633). Clearly, in the case of E. coli F-18, in vitro adhesion experiments shed little light on the molecular basis of colonization.

When the S. typhimurium strains are tested for their abilities to bind to mouse cecal mucus and cecal epithelial cells, the results are also surprising. That is, the poor colonizing LPS mutant, S. typhimurium SL5325 binds five times better to cecal mucus and three times better to cecal epithelial cells than the far better colonizing strain, S. typhimurium SL5319 (2), again showing that an increased capacity to bind mucosal components in vitro does not necessarily mean better colonizing ability in vivo.

We have shown that S. typhimurium SL5319, the good colonizer, travels through a layer of cecal mucus rapidly in vitro, whereas S. typhimurium SL5325, the poor colonizing LPS mutant, travels through the mucus layer very slowly (2), suggesting that the LPS mutant is a poor colonizer because it penetrates mucus slowly and is therefore starved for nutrients by the faster penetrating S. typhimurium SL5319. Similarly, E. coli F-18(pPR633) and E. coli(pPKL91), which grow equally well when inoculated into cecal mucus in vitro, were tested for their abilities to penetrate a mucus layer in vitro. Again, the poor colonizer, E. coli(pPKL91), which overproduces type 1 pili, was a poor mucus penetrator relative to E. coli(pPR633), the better colonizer. These data are of particular interest in that they suggest that the ability of a strain to regulate the number of type 1 pili on its surface may determine its ability to penetrate the mucus layer. Such regulation, therefore, may be an important large intestine colonization factor.

It has been postulated that motility and chemotaxis may be necessary for colonization. Evidence is accumulating, however, which suggest that the importance of motility, chemotaxis, or both, in colonization of the mammalian intestine depends on the microorganism, whether the host has a conventional microflora, is antibiotic treated or gnotobiotic, and whether the microorganism colonizes the small or large intestine. Non-flagellated (Fla⁻) and non-chemotactic (Che⁻) coli F-18 strains were found to be just as good colonizers as E. coli F-18 (6), and the Fla⁻ and Che⁻ S. typhimurium strains were found to be just as good colonizers as S. typhimurium SL5319 (2). Interestingly, when either E. coli F-18 or S. typhimurium SL5319 were grown in cecal or colonic mucus in vitro, they became non-motile (2,6). In the case of E. coli F-18, control experiments showed that the lack of motility was not due to viscosity or to catabolite repression of flagella synthesis (6). In fact, as determined by electron microscopy, E. coli F-18 grown in cecal mucus is flagellated. It therefore appears likely that growth of E. coli F-18 and S. typhimurium SL5319 in intestinal mucus in vivo renders them non-motile, yet flagellated. Why this is so remains to be determined.

ACKNOWLEDGEMENT

This work was supported by Public Health Service grant AI16370 from the National Institute of Allergy and Infectious Diseases.

REFERENCES

1. Cohen, P.S., Rossell, R., Cabelli, V.J., Yang, S.L., and Laux, D.C. Relationship between the mouse colonizing ability of a human fecal Escherichia coli strain and its ability to bind a specific mouse colonic mucous gel protein. Infect. Immun. 40:62-69, 1983.
2. McCormick, B.A., Stocker, B.A.D., Laux, D.C., and Cohen, P.S. Roles of motility, chemotaxis, and penetration through and growth in intestinal mucus in the ability of an avirulent strain of Salmonella typhimurium to colonize the large intestine of streptomycin-treated mice Infect. Immun. 56:2209-2217, 1988.
3. Wadolkowski, E.A., Laux, D.C., and Cohen, P.S. Colonization of the streptomycin-treated mouse large intestine by a human fecal Escherichia coli strain: role of growth in mucus. Infect. Immun. 56:1030-1035, 1988.
4. Franklin, D.P., Laux, D.C., Williams, T.J., Falk, M.C., and Cohen, P.S. Growth of Salmonella typhimurium SL5319 and Escherichia coli F-18 in mouse cecal mucus: role of peptides and iron. FEMS Microbiol. Ecol. (in press).
5. McCormick, B.A., Franklin, D.P., Laux, D.C., and Cohen, P.S. Type 1 pili are not necessary for colonization of the streptomycin-treated mouse large intestine by type 1-piliated Eschericha coli F-18 and E. coli K-12. Infect. Immun. 57:3022-3029, 1989.
6. McCormick, B.A., Laux, D.C., and Cohen, P.S. Neither motility nor chemotaxis play a role in the ability of Escherichia coli F-18 to colonize the streptomycin-treated mouse large intestine. Infect. Immun. 58:2957-2961, 1990.

GERMFREE ANIMALS INTESTINAL GLYCOCONJUGATES AND COLONIZATION

Tore Midtvedt
Göran Larson
Catharina Svanborg

Department of Medical Microbial Ecology
Karolinska Institute, Stockholm, Sweden
Department of Clinical Chemistry, University
of Gothenburg, Gothenburg, Sweden
Department of Clinical Immunology, University
of Lund, Lund, Sweden

The mucus substance found in the intestinal tract of conventional mammals, including man, will be partly exogenous, originating from the diet, partly derived from the microorganisms present, and partly endogenous, mainly comprising the epithelial mucus which is secreted continuously by the epithelial lining. The latter material is usually termed as mucin, indicating a mucus consisting mainly of glycoproteins. However, some other glycoconjugates, as glycolipids, glycosphingolipids etc., might be present.

Normally, intestinal mucin is extensively degraded by the microflora of the digestive tract, leaving small amounts to be excreted in feces. The flora may of course also act upon dietary and microbial derived glycoconjugates.

By the use of animals devoid of any microflora, i.e. germfree (GF) animals, it is possible to obtain intestinal mucin in a yield far better than that obtained from conventional animals. In addition, feeding the germfree animal an entirely synthetic diet, consisting of low-molecular-weight substances only, reduces the amount of exogenous macromolecules in the intestinal contents to an insignificant level. Consequently, germfree animals have been extensively used when studying intestinal glycoconjugates. Several studies have demonstrated high amounts of mucin and hexosamines in cecal content of germfree rats and mice compared to their conventional conterparts and a similar high fecal excretion of such compounds (1,2,3,4). In a series of experiments we studied the composition of purified glycoprotein from various parts of the intestinal tracts of GF rats which were fed either an ordinary lab chow or a chemically defined, low-molecular diet (4,5,6,7). Taken together, the results demonstrate that the composition of fecal mucin is very similar to the intestinal one (Table 1).

Molecular Pathogenesis of Gastrointestinal Infections
Edited by T. Wädstrom *et al.*, Plenum Press, New York, 1991

Table 1. Composition of glycoprotein in cecal and fecal extract
from germfree rats on a chemically defined, low-molecular diet

	Cecum		Feces	
	%	umol/100 mg	%	umol/100 mg
Galactose	23.0	128	22.6	126
N-Acetylgalactosamine	18.8	85	19.0	86
N-Acetylglucosamine	18.2	82	18,2	82
Fucose	8.6	53	8.8	54
N-Glycolylneuraminic acid	7.2	22	7.2	22
N-Acetylneuraminic acid	6.3	20	6.6	21
Sulphate	1.3	13	1.4	14
Threonine	6.8	57	6.7	56
Proline	2.4	21	2.3	20
Serine	2.2	21	2.2	21
Valine	0.7	6	0.8	6
Glutamic acid	0.7	5	0.8	6
Aspartic acid	0.6	5	0.7	5
Isoleucine	0.5	4	0.5	4
Glycine	0.3	4	0.4	5
Alanine	0.3	4	0.4	4
Lysine	0.2	2	0.2	1
Leucine	0.1	1	0.1	1

For further details, see Wold, Midtvedt & Jeanloz (6).

In some experiments we have injected radiolabelled glucose
intraperitoneally into GF rats and followed its incorporation into
intestinal glycoprotein (6). Judging from the results obtained
it seems appropriate to collect intestinal contents about 24 h
after the injection of labelled sugar. However, it is possible
that the rate of mucin production is influenced by age. It is
possible, although not yet satisfactorally shown, that the
accumulation of mucin found under germfree conditions may down-
regulate the production in older animals.

It has to be underlined that there might be some strain
variations in the composition of intestinal glycoconjugates. In
some ongoing studies on another rat strain than that referred to
in the table above, the amount of N-acetylneuraminic acid is
found to be higher and the amount of N-glycolylneuraminic acid is
lower than in the data given in Table 1. (approx. the same age of
the rats). It might also be mentioned that we have started some
studies utilizing darkfield electron microscopy on intestinal
mucin fram GF rats fed a diet containg casein as the only high-
molecular weight component (Slayter, Wold & Midtvedt, to be
published).

It has been generally assumed that a complete degradation of
intestinal mucin requires a sequential action of several bacteria.
However, as recently underlined by Carlstedt-Duke from our
laboratory (8), several factors have to be taken into account when
studying microbial degradation of intestinal mucin. A proper way
of isolating microbes actually at work in vivo, might be cultivation of the microbes found by serial sectioning from the serosa

side of cecum from a conventional rat. Using this technique, a
strain of Peptostreptococcus micros, able to degrade mucin in
vivo after monocontamination of GF rats, as well as in vitro in a
special mucin-containing medium, has been isolated (9). When this
strain was established as a monocontaminant in ex-GF rats, the
number was found to be 10   both in the middle and distal part
of the small intestine and 10 - 10   in the cecum, colon and
feces (8). In these monocontamined rats, gel electrophoresis of
content from the small intestine show a mucin pattern similar to
that found in GF animals (a so-called Germfree Animal Characteristic = GAC), and a pattern similar the findings in conventional
animals in the samples taken from the large bowel and feces
(a microflora-associated characteristic = MAC). This strain of
P. micros is not influencing upon the other GAC/MAC systems so far
studied. However, when the strain is established in GF mice, it
seems to protect the animals if they are subsequently confronted
with a pathogenic strain of Clostridium difficile. In such
experiments, a reduction of the death rate from 50% to 0% has
been demonstrated (8). So far, however, the mechanism(s) behind
this protection has not been evaluated.

These (and other results) indicate that different microbes
might be responsible for physiologically occurring GAC/MAC switches
at various anatomical levels within the digestive tract and
consequently it will be differences in the microbial colonization
pattern - followed by differences in microbial interactions - at
these levels.

This variation in microbial colonization might of course be
governed by structural variations in the compound(s) produced at
various levels. Recently we have demonstrated marked qualitative
and quantitative changes in intestinal glycosphingolipids in
growing (17 - 51 days of age) GF rats (10). The amount of non-acid
glycosphingolipids and of acid glycosphingolipids change from
22.7 and 25.7 on day 17-18 to 4.6 and 3.5 on day 25-27 (data
given as mg/dry weight/feces). Our findings are in accordance
with previous findings in conventional rats (11), indicating that
these changes are regulated by non-microbial factors. To exclude
dietary influence(s), we are now raising young GF rats on rat
milk only. However, whatever the mechanism(s) behind these changes
might be, the altered composition of intestinal glycoconjugates
may in various ways influence the composition and function of the
flora in conventional animals. It is well established that
glycosphingolipids function as receptors for bacterial adhesion.
(12,13). It is also well established that microbial adhesive
proteins associated with fimbriae recognize oligosaccharide
sequences in intestinal and urinary tract glycoconjugates, leading
to attatchment to the mucosal surface. These events have been
extensively studied in so-called lipopolysaccharide responder
(C3H/HeN) and lipopolysaccharide non-responder (C3H/HeJ) mice
(14). These two mouse strains are now established under germfree
conditions and will be utilized in forth-coming studies.

It seems appropriate to conclude that germfree animals are
very suitable as sources of intestinal glycoconjugates of high
purity. They are also of value when studying in vivo breakdown
of such compounds by isolated microbial strains as well as when
studying various molecular aspects in the establishment and
maintenance of an intestinal ecosystem and of infecting microbes.

REFERENCES

1. G. Lindstedt, S. Lindstedt, and B. E. Gustafsson: Mucus in intestinal contents of germfree rats. J. Exp. Med., 121: 201-213, 1965.
2. L. C. Hoskins, and N. Zamcheck: Bacterial degradation of gastro-intestinal mucins. I: Comparison of mucus constituents in the stools of germ-free and conventional rats. Gastroenterol., 54:210-217, 1968
3. W. J. Loesche: Protein and carbohydrate composition of cecal contents of gnotobiotic rats and mice. Proc. Soc. Exp. Biol. Med., 128:195-199, 1968.
4. J. K. Wold, R. Khan, and T. Midtvedt: Intestinal glycoproteins of germfree rats. Acta path. microbiol. scand., 79:525-530,1971.
5. J.K.Wold, T. Midtvedt, and R. Winsnes: Intestinal glycoproteins of germfree rats. II. Further studies on the composition of water-soluble extracts from intestinal mucus. Acta Chem. Scand., 27:2997-3002, 1973.
6. J. K. Wold, T. Midtvedt, and R. W. Jeanloz: Intestinal glyco-proteins of germfree rats. III. Characterization of a water-soluble glycoprotein fraction. Acta Chem. Scand., 28:227-284, 1974.
7. J. K. Wold, B. Smestad, and T. Midtvedt: Intestinal glycoprotein of germfree rats. IV. Oligosaccharides obtained by chemical degradation of a water-soluble glycoprotein fraction. Acta Chem. Scand., 29:703-709, 1975.
8. B. Carlstedt-Duke: The normal microflora and mucin. In: The regulatory and protective role of the normal microflora. R. Grubb, T. Midtvedt and E. Norin, eds. The Macmillan Press Ltd. Basingstoke, Great Britain, 1989.
9. B. Carlstedt-Duke, T. Midtvedt, C. E. Nord, and B.E. Gustafsson: Isolation and characterization of a mucin-degrading strain of Peptostreptococcus from rat intestinal tract. Acta path. microbiol. scand., 94:292-300, 1986.
10. G. Larson, and T. Midtvedt: Glycosphingolipids in feces of germfree-rats as a source for studies of developmental changes of intestinal epithelial cell surface carbohydrates. Glycoconjugate, 5:285-292, 1989.
11. D. Bouhours, and J-F. Bouhours: Developmental changes of rats intestinal glycolipids. Biochem. Biophys. Res. Commun., 99:1384-1389, 1981.
12. H.Leffler and C. Svanborg-Eden: Chemical identification of a glycosphingolipid receptor for Escherichia coli attaching to human urinary tract epithelial cells and agglutinating human erythrocytes. FEMS Lett., 8:127-134, 1981.
13. K-A. Karlson: Animal glycolipids as attachment sites for microbes. Chem. Phys. Lipids, 42:153-174, 1989.
14. P. de Man, C. van Kooten, L Aarden, I. Engberg, and C. Svanborg-Eden: Interleukin-6 induced at mucosal surface by gram-negative bacterial infection. Infect. Immun., 57:3383-3388, 1989.

BACTERIAL GLYCOSIDASES AND DEGRADATION OF GLYCOCONJUGATES IN THE HUMAN GUT[1]

Lansing C. Hoskins

Veterans Affairs Medical Center 10701 East Blvd.
Cleveland, Ohio 44106; and Case Western Reserve
University School of Medicine Cleveland, Ohio

I. INTRODUCTION

When examining the molecular pathogenesis of gastrointestinal infections investigators should realize that colon contents contain a variety of hydrolases including proteases, phospholipases, and glycosidases. Their action on cell surface and surface-related structures in the intestine presumably influence both normal and pathological host-microbial associations. This is especially true of intralumenal glycosidases, since oligosaccharide moieties of cell membrane glycolipids and glycoproteins appear to have roles as molecular recognition sites for bacterial attachment (1) while those of mucin glycoproteins confer the viscoelastic gel properties to the overlying mucus layer (I. Carlstedt, this symposium) which is a habitat for some kinds of enteric bacteria (2).

This paper reviews features of intraluminal glycosidases and their action on gut mucins and mucosal glycosphingolipids in the gastrointestinal tract with emphasis on its implications for the molecular pathogenesis of intestinal infections.

II. Glycosidases in colon contents

A typical oligosaccharide side chain of gut mucin glycoproteins is shown in Fig. 1 which shows one of several types of chains in human colonic mucins. A feature common to complete mucin oligosaccharide chains and important to their degradation is that the terminal monosaccharides at the chain's non-reducing end and monosaccharides attached to the chain as branching sugars are in *alpha* anomeric linkage while the interior glycosidic linkages of the chains' backbone are in *beta* anomeric configuration. A similar structural feature is seen in glycosphingolipids, but the chain is linked to ceramide via the sphingosine moiety. Degradation of these chains requires glycosidases with the appropriate anomeric and monosaccharide specificities to catalyze hydrolytic cleavage of each glycosidic linkage sequentially down the chain beginning at the outer, non-reducing end. There are many glyco-

---

[1]*Abbreviations used in text:* Gal – galactose; Gal'ase – galactosidase; GalNAc – N-Acetylgalactosamine; GalNAc'ase – N-Acetylgalactosaminidase; Glc – glucose; GlcNAc – N-Acetylglucosamine; GlcNAc'ase – N-acetyl-glucosaminidase; GSL's – Glycosphingolipids; Fuc'ase – L-fucosidase; HGM – hog gastric mucin; NeuAc – N-Acetylneuraminic (sialic acid).

Molecular Pathogenesis of Gastrointestinal Infections
Edited by T. Wädstrom *et al.*, Plenum Press, New York, 1991

```
            αFuc
             1
             2
   GalNAcα1-3Galß1-3,4GlcNAc
                        \ ß1-3
                        Galß1-4GlcNAcß1-3GalNAcα1-Ser,Thr
                        / ß1-6              6
   GalNAcα1-3Galß1-3,4GlcNAc               2
            6                           αNeuAc
            2
         αNeuAc
```

Fig.1. One of several types of oligosaccharide chains in human colonic
mucin (Podolsky D.K. J Biol Chem 1985;260:8262-71). The terminal
glycosides are in *alpha* anomeric linkage while the core monosaccharides
are linked in *beta* configuration. Each requires a linkage-specific
glycosidase for hydrolytic cleavage.

sidases in fecal extracts; those capable of degrading oligosaccharide chains
of glycoconjugates arise from indigenous enteric bacteria and not from the
host (reviewed in (3)). Most of these are exoglycosidases that act
sequentially on the glycoside linkages. Endoglycosidases that act on the
chains may also be present, but they are fewer in number compared to
exoglycosidases.

     Many of these glycosidases have been detected using monosaccharide
substrates such as *para*-nitrophenyl glycosides. It is important to emphasize
that glycosidases acting on monosaccharide glycosides such as *para*-
nitrophenyl- or 4-methyl-umbelliferyl- compounds may be different molecules
from those acting on the same linkage in oligosaccharides of natural
glycoconjugates and that both types may be produced by the same bacterium in
culture (4-6).

<u>Sources of fecal glycosidases in man that degrade intestinal glycoconjugates</u>
     Strains of several genera of enteric bacteria that are normal human
commensals including <u>Bacteroides</u>, <u>Bifidobacterium</u> and <u>Clostridium</u> have been
shown to produce glycosidases active on simple synthetic and natural
glycoconjugates (5). Sialidase and strong p-nitrophenyl-β-galactosidase and
-β-N-acetyl-glucosaminidase activities are produced by some <u>Bacteroides</u> and
<u>Bifidobacterium</u> strains that dominate human fecal bacterial populations, but
their activity against the β-galactosyl and β-N-acetylglucosaminyl moieties
in the oligosaccharide chains of mucin glycoproteins is weaker (unpublished
studies).   In addition, with the exception of very weak blood group H-
degrading α-L-fucosidase activity produced by two <u>B.vulgatus</u> strains, we have
not found that these dominant populations produce the specific α-glycosidases
required to cleave blood group ABO and Lewis antigenic determinant glycosides
from their terminal positions at the non-reducing end of complete
oligosaccharide chains (Table I). Without this ability, stepwise cleavage
of such chains does not occur; these strains therefore have little or no
ability to degrade the oligosaccharide chains containing blood group ABH and
Lewis antigens in gut mucin glycoproteins (ref. 5 and in preparation).
However, strains exist that do cleave blood group antigenic moieties from the
chains. They are a subset of man's normal enteric bacteria, and they grow
consistently from fresh fecal inocula in anaerobic culture medium containing
mucin glycoproteins as the main carbohydrate source (7,8). We have used
purified commercial hog gastric mucin (HGM) from pooled hog stomachs for this
purpose since it is readily available, closely resembles human epithelial
mucins in general structure and has terminal glycosides conferring human

blood group A and H antigen specificity. The presence of mucin glycoprotein in the culture medium enhances glycosidase production in the mixed culture of fecal populations, apparently by facilitating selective growth of glycosidase-producing strains rather than by directly inducing enzyme synthesis in these strains (7,8).

Blood group-degrading activities have been uniformly produced in the anaerobic fecal cultures from more than 30 healthly subjects tested to date, and the estimated fecal population densities of the strains that produced them (10^7 - 10^8/g wet fecal weight) appeared to be stable over time (7). Nevertheless, we observed 4 instances where blood group-degrading enzyme activity transiently disappeared from fecal extracts or fecal cultures of healthy subjects for unknown reasons. The best documented of these occurred when fecal cultures were being obtained from one subject and monitored for blood group B-degrading activity during a 5-month period (Fig. 2). In March, production of this enzyme activity in anaerobic fecal cultures ceased abruptly for a 7-day period during which no activity was detected in 3 fecal cultures nor in extracts of the same fecal samples. At this time the subject had a mild upper respiratory tract infection and experienced looser stools than usual but took no medications or antibiotics. Blood group B-degrading activity in fecal cultures returned to its former levels soon thereafter. Since no inhibitor of B-degrading enzyme activity was demonstrable in two of the fecal extracts it is likely that there was a transient but marked depression of the bacterial populations producing B-degrading enzyme activity at this time.

ABH(O) blood group-degrading enzyme activities in fecal extracts and anaerobic fecal cultures of blood group A and B secretors are greater towards the fecal donor's genetically determined ABO blood group phenotype antigen in his gut mucus secretions, suggesting bacterial enzyme adaptation to environmental substrate (7-9). In the case of strains producing blood group B-degrading activity this adaptation was associated with an increase in fecal population density that averaged 50,000-fold greater in blood group B

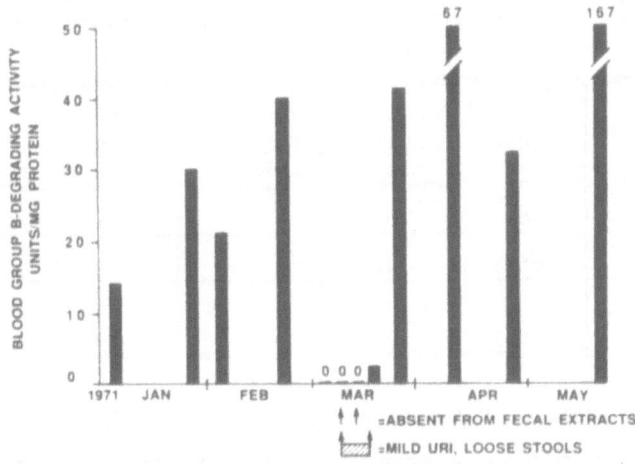

Fig 2. Transient unexplained loss of bacterial blood group B-degrading α-galactosidase activity in fecal extracts and cultures over a 10-day period during which the subject experienced mild upper respiratory tract symptoms and looser than usual stools. See text.

secretors than in blood type A or O secretors or ABH(O) non-secretors (8).

Because of their unique enzyme production bacterial strains producing ABH(O) blood group-degrading glycosidases also degrade oligosaccharide chains of mucin glycoproteins and mucosal glycolipids. That this ability was restricted to very few of the numerous kinds of human fecal bacteria was shown by Salyers and her associates (10,11): of 342 strains tested, representing 8 genera and 32 species, only 8 strains fermented hog gastric mucin with the generation of acid: these were 6 strains of Ruminococcus torques and 2 of Bifidobacterium bifidum. Work in the author's laboratory also indicated that mucin degradation was a specialized property of a subpopulation of fecal bacteria and that it was closely associated with bacterial glycosidase production. This was done by simultaneously measuring the extent of mucin degradation (loss of 60% ethanol precipitable hexoses, protein, and A and H blood group antigens) and the level of glycosidase activities in anaerobic cultures containing HGM and inoculated with serially diluted feces. Results showed that the extent of mucin degradation was directly related to the level of extracellular glycosidase activities and not to the level of cell-associated activities in any given culture, suggesting that the bacteria primarily responsible for oligosaccharide chain degradation produced the requisite glycosidases as extracellular enzymes (4). With this method it was possible to estimate the fecal population densities of these strains using a most probable number method (12). Among 11 healthy subjects, the estimated fecal population density of mucin-degrading bacteria averaged 1% of total cultivatable fecal bacteria. By use of a replicate sampling technique at limiting fecal dilutions isolates from this subset were obtained in pure culture (5). Of the 5 isolates, 2 were Ruminococcus torques, 1 was Ruminococcus gnavus (formerly termed R. AB) and 2 were Bifidobacterium species. They are gram-positive, non-sporulating, obligate anaerobes that are normal, non-pathogenic members of the human intestinal microbiota. These strains are distinguished from other enteric bacteria by their constitutive production of glycosidases which are predominantly extracellular (5). They differ from one another in the specificity of the ABH(O) blood group-degrading enzymes they produced (Table I): the 2 R.torques strains produced

Table I. Specificity towards mucin glycoprotein oligosaccharide linkages of glycosidases produced by mucin-degrading bacteria and by strains of larger populations of the human fecal microbiota, from (5) and unpublished studies. For simplicity, the relative amounts of the glycosidase activities (cell-bound plus extracellular) are listed on a scale of 0 (none) to 4+ (strong). ± = trace.

STRAINS	α1-3-GalNAc'ase	α1-3-Gal'ase	α1-2-L-fucosidase	Sial-idase	ß-Gal'ase	ß-GlcNAc'ase
R.torques	4+	0	2+	3+	4+	4+
R.gnavus	0(±)	4+	2+	2+	1+	0
Bifido-bacterium	0(±)	0	3+	3+	4+	4+
Greater Fecal Populations (10^{10}-10^{11}/gram):						
B.longum (1)	0	0	0	0	1+	1+
Bacteroides(6)	0	0	0(±)	1+	1+	1+

Table II. Degradation of hog gastric mucin (HGM) after incubation
with the partially purified enzymes in cell-free culture supernates
of mucin-degrader strains. Values are the medians and (ranges) for
each of the listed items. The enzymes in the culture supernates
were partially purified by ammonium sulfate fractionation and gel
filtration on Sephadex G-200 (13). They were incubated with 12-15
mg/ml solutions of HGM at pH 6.4 and 37°C at a ratio of 5% enzyme
protein to HGM protein. After incubation the solutions were made
60% v/v in ethanol and the precipitated residual mucin was dissolved in and dialyzed against distilled water and lyophilized. The
residual mass (lyophilized weight), hexoses and protein (12) were
compared with those of untreated HGM. Incubation with R.torques
abolished the strong blood group A and H antigen hemagglutination
inhibition titers of the untreated mucin while incubation with the
Bifidobacterium strains abolished the H antigen titer.

Strains	No. of Incuba-tions	Incuba-tion time, d	% Decrease in 60% Ethanol-Precipitable:		
			Mass	Hexoses	Protein
R.torques VIII-239 IX-70	8	5 (2-10)	75 (68-84)	95 (93-96)	28 (18-62)
Bifido-bacterium VIII-210 VIII-240	4	8 (3-11)	56 (41-66)	72 (52-87)	21 (3-11)

strong blood group A-degrading activity but no B-degrading activity, the
R.gnavus strain produced strong B-degrading but no A-degrading acitvity, the
Bififobacterium strains produced neither, but all produced H(0)-degrading
activity and sialidase. Subsequent work with glycolipid substrates (6,13)
indicated that the R.gnavus and one of the Bifidobacterium strains produced
weak A-degrading α-GalNAc'ase activity towards A(+) chains, but this activity
was minor compared to the other blood group antigen-degrading specificities
each produced. Except for the R.gnavus strain which produced very weak β-N-
acetylhexosaminidase, all also produced β-galactosidases and β-N-
acetylglucosaminidases. Thus, 4 of the 5 isolates produced enzymes capable
of extensively degrading mucin oligosaccharide chains providing that the
blood group-degrading glycosidases they produced were specific for the blood
group determinant α-glycosides at the non-reducing end of the oligosaccharide
chains. Even though it produced a blood group B-degrading α(1-
3)galactosidase, lack of strong β-N-acetylhexosaminidase activity prevented
the R.gnavus strain from degrading the backbone of blood group B chains, but
extensive degradation occurred when this strain was grown in symbiotic
association with another strain that did produce β-N-acetylhexoxaminidase
(5).

Our findings, together with the independent isolation of identical
strains fermenting HGM by Salyers et al (10), strongly suggest that the major
degradation of mucin oligosaccharide chains in human colon contents is by a
subset of Ruminococci and Bifidobacterium strains that constitutively produce
the requisite glycosidases as extracellular enzymes.

Action of glycosidases of mucin degrader strains on gut mucin glycoproteins and gut epithelial glycosphingolipids

Gut mucin glycoproteins. The extracellular glycosidases produced by the isolated strains of mucin-degrading bacteria extensively degrade the oligosaccharide chains of gut mucin glycoproteins. This is shown in Table II which lists the median percent decrease in mass, hexose and protein content of 60% ethanol-precipitable HGM after prolonged incubation with the glycosidases of the 2 R.torques and 2 Bifidobacterium strains. Loss of glycoprotein mass and hexose content was appreciable but with retention of the majority of the protein core in the ethanol-precipitable fraction. The hemagglutination inhibition titers of blood group A and H antigen of the mucin were lost during incubation with R.torques and the H antigen titer was lost during incubation with the Bifidobacterium strains. Since the R.torques strains produced strong blood group A-degrading α-GalNAc'ase activity and the Bifidobacterium strains did not, chains with terminal α(1-3)GalNAc- A determinants were degraded by the former and not by the latter, resulting in more extensive degradation of the oligosaccharide chains of HGM by the R.torques strains.

Action on Glycosphingolipids (GSL's). Drs. Goran Larson and Per Falk in collaboration with the author have performed detailed studies of substrate specificities of the enzymes produced by each strain using purified GSL's of known structure (6,13). The general results are summarized as follows:

1. Lactoseries GSL's with either type 1 (Galβ1-3GlcNAcβ1-) or type 2 (Galβ1-4GlcNAcβ1-) oligosaccharide chains were rapidly degraded, depending upon the specificity of the strain's blood group-degrading glycosidases for the ABH blood group antigen moieties at the non-reducing end of the chain (Fig. 6). Mono- and di-fucosyl groups conferring H and Lewis antigen specificities to lactoseries GSL's were also cleaved by all strains. Since all produced a sialidase acting on terminal N-acetylneuraminylα(2-3)- moieties, ganglioside G_{M3} (NeuAcα2-3-Galβ1-4Glcβ1-1Ceramide) was rapidly degraded to lactosylceramide (LacCer, Galβ1-4Glcβ1-1Cer) by all strains (Fig 4).

2. The principal end product accumulating during prolonged (48h) incubation of lactoseries GSL's with enzymes from the R.torques and Bifidobacterium strains was LacCer, with little or no further degradation of LacCer to glucosylceramide (GlcCer). By contrast, with enzymes from R.gnavus strain VI-268 the principal product from type 1 chain lactoseries

```
                     αlFuc
                     2
   A6:     GalNAcαl-3Galßl-3,4GlcNAcßl-3Galßl-4Glcßl-1Ceramide

   H                 Galßl-3,4GlcNAcßl-3Galßl-4Glcßl-1Ceramide
(Leb,Ley)            2         (4,3)
                     αlFuc     (αlFuc)




                                          │
                                          │
                                          ▼

Lactosylceramide:                 Galßl-4Glcßl-1Ceramide   (LacCer)
```

Fig. 3.   Action of enzymes from Ruminococcus torques and Bifidobacterium strains on blood group A(+), H(+) and Lewis (+) lactoseries glycosphingolipids with types 1&2 chains.

```
GM3:          NeuAcα2-3Galß1-4Glcß1-1Ceramide

                          │
                          ▼

Lactosylceramide:   Galß1-4GlcB1-1Ceramide
   (LacCer)
```

Fig. 4.   Action of enzymes from all 5 strains on GM3
           ganglioside.

GSL's was lactotetraosylceramide (Galβ1-3GlcNAcβ1-3LacCer) while from type
2 chain GSL's it was lactotriosylceramide (GlcNAcβ1-3LacCer), apparently
due to the lack of production of exo- or endo-β-N-acetylhexosaminidases
and a β(1-3)galactosidase (Fig. 5).
3.   Neither the gangliotriaosylceramide (GalNAcβ1-4LacCer), nor the
gangliotetraosylceramide (Galβ1-3GalNAcβ1-4LacCer) ganglioside core
structures were degraded by enzymes of any of the strains. However, the
sialidases from every strain rapidly cleaved the NeuAcα2-8- residues that
are linked to NeuAcα2-3- residues in gangliosides GD1b, GT1b and GQ1b and
also cleaved the terminal NeuAcα2-3- residue form GD1a, resulting in
accumulation of GM1 ganglioside (Fig.6) Of particular relevance to the
subject of this conference, the sialidases of the 2 R.torques strains were
capable of slowly cleaving the branching, interior NeuAcα2-3- moiety from
GM1 ganglioside, resulting in accumulation of gangliotetraosylceramide.
4.   Neither globoside (GalNAcβ1-4Galα1-4LacCer), globotriaosylceramide
(Galα1-4LacCer),   isoglobo-triaosylceramide   (Galα1-3LacCer),   nor
galabiosylceramide (Galα1-4Galβ1-1Cer) were degraded by the enzymes of any
strain. Since blood group active lactoseries GSL's are prominent GSL's of
intestinal epithelium while gangliosides, globosides and galabiosyl GSL's
are minor constituents, these substrate specificities are consistent with
adaptation of mucin-degrader glycosidases to the predominant substrate
GSL's derived from enterocytes comprising the epithelial mucosal lining
and shed into the gut lumen (6).

```
                        1αFuc
                        2
Type 1 B6:    Galα1-3Galß1-3GlcNAcß1-3Galß1-4Glcß1-1Ceramide

nLcOse₄Cer          Galß1-4GlcNAcß1-3Galß1-4Glcß1-1Ceramide



                                         │
                                         ▼


LcOse₄Cer:          Galß1-3GlcNAcß1-3Galß1-4Glcß1-1Ceramide
(Type 1)

LcOse₃Cer:               GlcNAcß1-3Galß1-4Glcß1-1Ceramide
(Type 2)
```

Fig. 5.   Action of enzymes from Ruminococcus gnavus VI-268 on blood
            group B(+) glycosphingolipids.

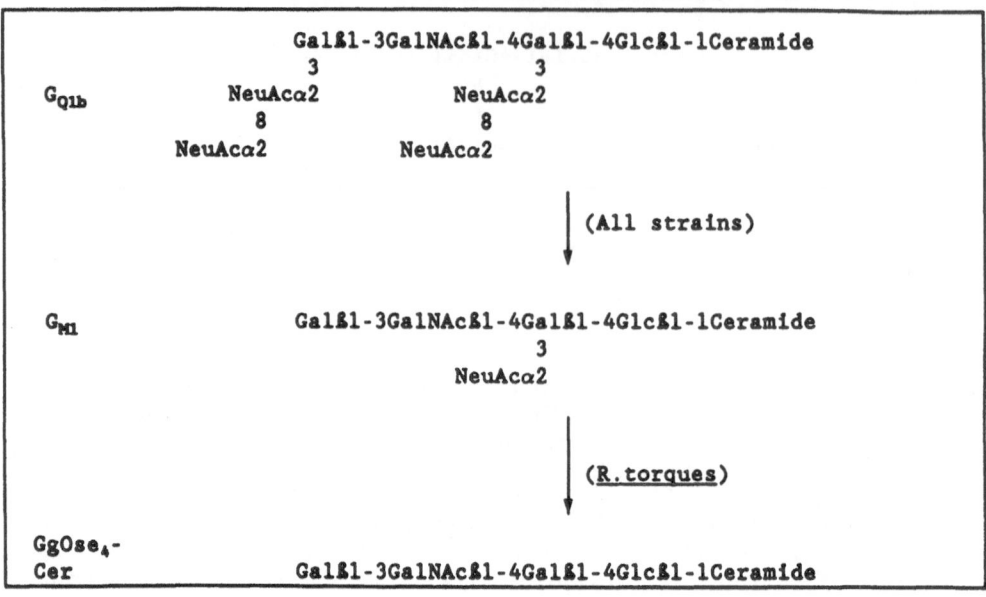

G_{Q1b}

$$Gal\beta 1\text{-}3GalNAc\beta 1\text{-}4Gal\beta 1\text{-}4Glc\beta 1\text{-}1Ceramide$$

```
                 3                            3
            NeuAcα2                      NeuAcα2
                 8                            8
       NeuAcα2                    NeuAcα2
```

↓ (All strains)

G_{M1}

$$Gal\beta 1\text{-}3GalNAc\beta 1\text{-}4Gal\beta 1\text{-}4Glc\beta 1\text{-}1Ceramide$$

```
                                        3
                                   NeuAcα2
```

↓ (R.torques)

$GgOse_4$-Cer

$$Gal\beta 1\text{-}3GalNAc\beta 1\text{-}4Gal\beta 1\text{-}4Glc\beta 1\text{-}1Ceramide$$

Fig. 6.   Action of enzmes from mucin degrader R.torques and Bifidobacterium strains on GQlb and GM1 gangliosides.

Action on other glycoconjugates in the gut lumen.   Fecal extracts and cell-free supernates of anaerobic fecal cultures cleave the human blood group B-like antigen from the cell walls of E.coli O86 with release of galactose and fucose (14).   Fecal extracts and fecal cultures of blood group B secretors cleaved the antigen more rapidly than those of blood group A or O secretors, presumably due to the greater fecal population densities of B-degrading strains in blood group B secretors.   These results indicate that glycoconjugates on the cell walls of indigenous microbiota may also be altered in their natural habitat with consequent changes in their immunological specificities.

In vivo correlates of glycoconjugate degradation in vitro by mucin-degrader glycosidases.   That lactosylceramide was the main product accumulating during degradation of lactoseries GSL's and GM3 ganglioside in vitro is consistent with observations that, along with monoglycosylceramides, LacCer becomes the dominant lactoseries GSL excreted in feces of humans beginning around the time of weaning (15).   Prior to that time LacCer is a minor component, and blood group active fucolipids and GM3 ganglioside are prominent components, of meconium and feces of breast fed infants.   Although genetically programmed changes in enterocyte GSL's occur in newborn rodents they have not yet been demonstrated in man.   Transformation from the more complex GSL's to LacCer in feces occurs in the general time period when bacteria with mucin-degrading activity become established in the fecal flora after the first few weeks of life and generally within the first year (16).   These findings suggest that the conversion to fecal excretion of LacCer occurs as the result of implantation and establishment of one or more strains of mucin-degrader bacteria in the infant's enteric ecosystem.

Potential role of bile salts on substrate specificities of mucin-degrader glycosidases in the colon lumen.   The substrate specificities of mucin-degrader glycosidases towards GSL's described above were based upon in vitro experiments using the neutral detergent, Triton X-100.   More recent work suggests that the glycosidase specificities can be significantly extended by substituting the primary and secondary bile salts or their taurine amides for Triton X-100 in the incubation mixtures (see Abstract by Falk, Hoskins & Larson, this Symposium). The more potent appeared to be deoxycholate, taurodeoxycholate, and chenodeoxycholate which at 5 mM concentrations facilitated >70% degradation of LacCer to GlcCer by the R.torques strains' enzymes after 1 h incubation as well as complete desialylation of GM1 ganglioside to gangliotetraosylceramide. 5mM bile salts also promoted cleavage of Galα1-3LacCer (isoglobotriosyl-ceramide) to LacCer by the blood group B-degrading α1-3galactosidase of R.gnavus VI-268 whereas this did not occur with 5mM Triton X-100. Enhanced activity was seen with concentrations as low as 1 mM. Since postprandial bile salt concentrations in the human small intestine range from 2 to 10 mM with lowest levels in the terminal ileum (17) this raises the interesting possibility that bile salt concentrations might modulate intralumenal bacterial glycosidase activities. However, the fragmentary evidence indicates that mucin-degrading bacteria in the gastrointestinal tract are normally confined to the distal ileum and colon (3). In the normal colon the limited evidence suggests that bile salt concentrations are less than 1 mM: in 3 healthy subjects on a regular diet fecal aqueous phase concentrations of cholate + deoxycholate + chenodeoxycholate did not exceed 0.3 mM, and in 3 subjects with excessive bile salt excretion due to ileal bypasses or resections the cecal aqueous phase concentrations of these 3 bile salts combined did not exceed 0.3 mM (18). However, one subject with ileal resection and excessive fecal bile salt excretion had aqueous phase concentrations of 1.3-1.5 mM in samples aspirated from 4 different areas of her colon (18), suggesting that concentrations that enhance glycosidase activity in vitro may sometimes occur in abnormal conditions in vivo.

Implications for the pathogenesis of bacterial infections in the intestine. The significance of intralumenal glycosidases to the pathogenesis of intestinal bacterial infections is their ability to: 1) degrade the mucus blanket overlying the gut mucosa; 2) degrade epithelial GSL's leading to uncovering of cryptic LacCer binding sites for autochthonous enteric bacteria, and 3) alter putative receptors for exotoxins on the mucosal epithelium. Transient losses of mucin-degrader strains, as shown in Fig 2, or following adminstration of broad spectrum antibiotics (19), might have pathogenetically important consequences for these activities.

1) While the mucosal mucus blanket may comprise a barrier against penetration of some pathogens to the mucosal surface it has recently become clear that this layer itself harbors a microbiota that includes bacteria not observed existing free or bound to particulates in the colon lumen (2). Glycoprotein constituents of the mucus layer have been shown to bind to pathogenic and non-pathogenic strains of E.coli, and to promote colonization of the latter in mice (see paper by Cohen, this Symposium). Therefore, degradation of mucus layer glycoproteins by the combined action of mucin degrader glycosidases and intralumenal proteases probably has important effects on the ecological balance of microbial populations in the mucus blanket as well as on binding sites for pathogenic bacteria.
2) The observations that LacCer is the principal product of intestinal epithelial GM3 and lactoseries GSL degradation by glycosidases from 4 of the 5 mucin degrader strains and that strains of several prominent genera of enteric bacteria preferentially and nonspecifically bind to LacCer in vitro (1) suggest that mucin-degrader glycosidases may have an important ecological role in the establishment of climax communities of human enteric microbiota in vivo. Glycosidase production by the newly acquired mucin degrader strains at the time of weaning could result in uncovering of LacCer binding sites by degradation of more complex GSL's on the lining colonic epithelium (20), on shed epithelial cells, on GSL-coated vesicles (21), or adhering to particulates in lumen contents. LacCer binding at these sites might act as a nidus for microcolony formation.

3) Finally, microbial glycosidases may alter mucosal receptor sites for bacterial toxins. Thus, the sialidases of the two R.torques strains cleaved sialic acid from GM1 ganglioside, the putative receptor for cholera toxin and E.coli heat labile toxin (1). That the rate of this degradation was enhanced by substituting bile salts for Triton X-100 in the incubation mixtures suggests that bile salt concentrations in intestinal contents might have a contributing role in such degradations. On the other hand, inability of glycosidases from any of the mucin degrader strains to cleave Galα1-4- from globotriaosylceramide, a receptor GSL for Shiga toxin (1), may be a factor contributing to the pathogenicity of Shigella dysentery in the human colon.

Clearly, these considerations are speculative, but they offer future directions for investigation of the molecular pathogenesis of intestinal infections and emphasize the importance of considering the total intestinal environment in such investigations.

REFERENCES

1. Karlsson K-A. Animal glycosphingolipids as membrane attachment sites for bacteria. Annu Rev Biochem. 1989;58:309-50.

2. Lee A. Neglected niches. The microbial ecology of the gastrointestinal tract. Adv Microb Ecol. 1985;8:115-162.

3. Hoskins LC. Mucin degradation by enteric bacteria: ecological aspects and implications for bacterial adherence. In Boedeker EC, ed. Attachment of Organisms to the Gut Mucosa. CRC Press Boca Raton. 1984;II:51-67.

4. Hoskins LC, Boulding ET. Mucin degradation in human colon ecosystems. Evidence for the existence and role of bacterial subpopulations that produce glycosidases as extracellular enzymes. J Clin Invest. 1981;67:163-172.

5. Hoskins LC, Agustines M, McKee WB, Boulding ET, Kriaris M, Neidermeyer G. Mucin degradation in human colon ecosystems. Isolation and properties of fecal strains that degrade ABH blood group antigens and oligosaccharides from mucin glycoproteins. J Clin Invest. 1985;75:944-953.

6. Falk P, Hoskins LC, Larson G. Bacteria of the human intestinal microbiota produce glycosidases specific for lacto-series glycosphingolipids. Biochem J (Japan). 1990;108:466-474.

7. Hoskins LC, Boulding ET. Degradation of blood group antigens in human colon ecosystems. I. In vitro production of ABH blood group-degrading enzymes by enteric bacteria. J Clin Invest. 1976;57:63-73.

8. Hoskins LC, Boulding ET. Degradation of blood group antigens in human colon ecosystems. II. A gene interaction in man that affects the fecal population density of certain enteric bacteria. J Clin Invest. 1976;57:74-82.

9. Hoskins, LC. Ecological studies of intestinal bacteria. Relation between the specificity of fecal ABO blood group antigen-degrading enzymes and the ABO blood group of the host. J Clin Invest. 1969;48:664-673.

10. Salyers AA, West SEH, Vercellotti JR, Wilkins TD. Fermentation of mucins and plant polysaccharides by anaerobic bacteria from the human colon. Appl Environ Microbiol. 1977;34:529-33.

11. Salyers AA, Vercellotti JR, West SEH, Wilkins TD. Fermentation of mucin and plant polysaccharides by strains of <u>Bacteroides</u> from the human colon. Appl Environ Microbiol. 1977; 33:319-33.

12. Miller RS, Hoskins LC. Mucin degradation in human colon ecosystems. Fecal population densities of mucin-degrading bacteria estimated by a "most probable number" method. Gastroenterology 1981;81:759-65.

13. Larson G, Falk P, Hoskins LC.   Degradation of human intestinal glycosphingolipids by extracellular glycosidases from mucin-degrading bacteria of the human fecal flora. J Biol Chem 1988;263:10790-8.

14. Cromwell CL, Hoskins LC. Antigen degradation in human colon ecosystems. Host's ABO blood type influences enteric bacterial degradation of a cell surface antigen on <u>E.coli</u> 086. Gastroenterology 1977;73:37-41.

15. Larson G, Falk P, Hynsjo L, Midtvedt AC, Midtvedt T. Fecal excretion of glycosphingolipids of breast-fed and formula-fed infants. Microb Ecol Health Dis. 1990;3:305-320.

16. Midtvedt AC, Carlstedt-Duke B, Norin KE, Saxerholt H, Midtvedt T. Development of five metabolic activities associated with the intestinal microflora of healthy infants. J Pediat Gastroent Nutr.1988;3:559-67.

17. Northfield TC, McColl I.   Postprandial concentrations of free and conjugated bile acids down the length of the normal human small intestine. Gut 1973;14:513-8.

18. McJunkin B, Fromm H, Sarva RP, Amin P.   Factors in the mechanism of diarrhea in bile acid malabsorption; fecal pH - a key determinant. Gastroenterology 1981;80:1454-64.

19. Carlstedt-Duke B,   Hoverstad T,   Lingaas E,   Norin KE,   Saxerholt H, Steinbakk M, Midtvedt T. Influence of antibiotics on intestinal mucin in healthy subjects. Eur J Clin Microbiol. 1986;5:634-8.

20. Holgersson J, Stromberg N, Breimer E.   Glycolipids of human large intestine: difference in glycolipid expression related to anatomical localization, epithelial/non-epithelial tissue and the ABO, Le and Se phenotypes of the donors. Biochimie 1988;70:1565-74.

21. Hill RH. Prevention of adhesion by indigenous bacteria to rabbit cecum epithelium. Infect Immun. 1985;47:540-3.

THE CLONE CONCEPT AND ENTEROPATHOGENIC ESCHERICHIA COLI (EPEC)

Ida Ørskov and Frits Ørskov

International Escherichia and Klebsiella Centre
(WHO),
Statens Seruminstitut,
Copenhagen, Denmark

Examination of enteropathogenic E.coli, so called EPEC strains, from outbreaks of infantile diarrhoea in both Europe and North America in the 1950's, showed that strains belonging to a limited number of O groups were associated with this disease and although strains of these O groups were found with several different flagellar antigens, i.e. H antigens, certain O:H serotypes were more often associated with outbreaks and severe cases than other O:H types belonging to the same O group. In addition these O:H serotypes had characteristic biotypes in such a way, that it was possible to predict the H type from the biotype and vice versa.

In the late 1960's EPEC strains lost interest for several reasons. The severe outbreaks disappeared in the Western World and it was difficult to diagnose a strain as an EPEC; in most cases the only test which could be carried out was the O antigen examination, while the H antigen determination was reserved for more experienced central laboratories. In the 1970's it was shown that E.coli from travellers diarrhoea produced plasmid determined heat labile (LT) or heat stable (ST) enterotoxins, and futhermore fimbriae with adhesive capacity[1,2]. These strains were of other serotypes than EPEC and were termed ETEC (enterotoxigenic E.coli). In 1976 the clone concept was formulated on the basis of geographically wide spread ETEC of a limited number of O:H serotypes representing clones, which we proposed had been selected to the special conditions in the small intestine and to carry the plasmids necessary to provoke diarrhoea[3]. The fact that classical EPEC strains apparently did not possess demonstrated virulence properties such as the ETEC's did, was another reason for the failing interest in EPEC.

However, during the last decade new abilities of EPEC were detected; they adhere in a localized manner (LA) to certain tissue culture cells. This property, which is encoded by a

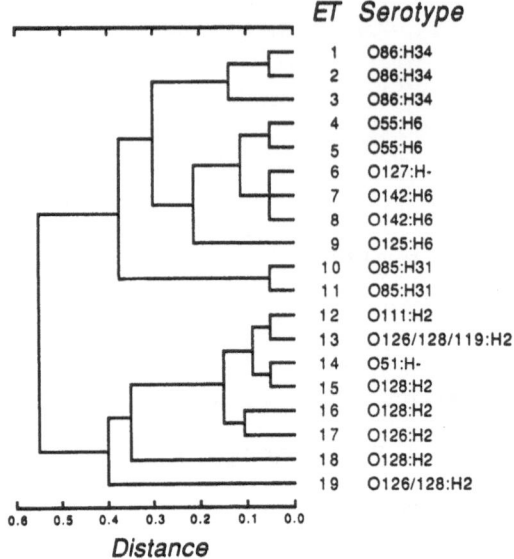

ET Serotype

1	O86:H34
2	O86:H34
3	O86:H34
4	O55:H6
5	O55:H6
6	O127:H-
7	O142:H6
8	O142:H6
9	O125:H6
10	O85:H31
11	O85:H31
12	O111:H2
13	O126/128/119:H2
14	O51:H-
15	O128:H2
16	O128:H2
17	O126:H2
18	O128:H2
19	O126/128:H2

0.6   0.5   0.4   0.3   0.2   0.1   0.0

Distance

Figure 1. Dendrogram of genetic relationships among 19 E.coli electrophoretic types (ET). (From Ørskov et al[8]).

plasmid, has been termed EPEC adherence factor (EAF) and a DNA EAF probe has been constructed[4,5]. EPEC also produce an attaching and effacing (AE) lesion in the small intestinal mucosa[6]. These findings renewed the interest in EPEC strains.

For many years we considered the widespread classical EPEC O:H-biotypes as clones, but could that idea be confirmed when the strains were analysed by multilocus enzyme electrophoresis? This technique detects genetic polymorphisms of natural selection neutral, or nearly so, enzyme coding genes by examination of the mobility of a given enzyme after electrophoresis in starch gel[7]. The genetic relationship and clonal nature of 50 strains belonging to nine classical EPEC O:H serotypes were studied by use of this method. The multilocus enzyme electrophoresis type (ET) of each isolate was defined by the combination of alleles at 20 loci[8].

The EPEC strains examined had been selected for the study because of their chatacteristic O:H-biotype. They had been isolated during the years 1950 - 1985 in many different countries and were of the following serotypes: O55:H6, O86:H34, O119:H2, O125:H6, O126:H2, O127:H⁻ or H6, O128:H2 and O142:H6. The common biotype feature was that they did not ferment dulcitol and sorbitol till after two or more days growth, a few remained negative in sorbitol even after 30 days. Three of the EPEC strains were, however, of another biotype, but included for comparative reasons. In addition to the 50 EPEC strains nine strains of the non EPEC serotypes O51:H⁻ and O85:H31 were

examined because they had been isolated from infantile diarrhoea in the 1950's in Copenhagen. Their biotypes were very
similar to the characteristic EPEC biotype.

Multilocus enzyme electrophoresis showed that 17 of the 20
enzymes assayed were polymorphic and 19 ET's were revealed.
Cluster analysis of the genetic distances between these showed
two major clusters and both contained smaller tight clusters
(fig. 1). In the upper cluster strains of serotypes O55:H6,
O125:H6, O127:H⁻ or H6 and O142:H6 are closely related and make
up a tight cluster (distance 0.12). In the lower major cluster
ET 12 - 17 are highly related (distance 0.15) and the serotypes
involved are O111:H2, O119:H2, O126:H2, O128:H2 and the non
EPEC serotype O51:H⁻. Further away from these ET types are ET 18
and 19. The O128:H2, the only strain of ET 18, was not an
isolate from diarrhoea, in contrast to the other strains, but
isolated from septicemia in a lamb in South Africa and the
strains of ET 19 were of another biotype than the remaining
strains, and were included for comparison.

Examination of the major outer membrane protein (OMP)
pattern showed that all strains of the lower major cluster had
the same pattern, while 4 other patterns were found in the
upper cluster, one in the O86:H34 (ET 1 - 3), a second in the
O55:H6, O127:H⁻ or H6 and O142:H6 (ET 4 - 8), a third in the
O125:H6 (ET 9) and a fourth in the O85:H31 (ET 10 - 11)
strains.

All strains of EPEC serotype hybridized with the EAF probe
and showed localized adherence (LA) to HEp-2 cells. The O51 and
O85 strains did not show LA, but adhered in a manner which
tentatively was called perinuclear adherence.

EPEC strains of serotype O114:H2 were examined for
multilocus enzyme genotype in another study[9], which showed that
they were members of the ET 12 - 15 cluster (fig. 1) containing
the EPEC's O111:H2, O119:H2, O126:H2 and O128:H2.

The ET examination of the 59 isolates were done blindly
and it was more or less expected that strains of the same O:H
biotype would show genetic clonality, i.e. be of the same ET or
nearly so. By and large the results came out that way; however,
we had also hypothesized that the classic EPEC O:H defined
strains had a common genetic background, i.e. would form one
big cluster. The results showed that the strains could be
grouped not in one, but in two clusters of genetically closely
related strains. Both clusters contained smaller clusters of
more closely related strains of different O groups both
serologically and chemically, but with the same H antigen, H6
in the upper tight cluster and H2 in the lower. We have other
examples of genetic clusters or lineages of strains with the
same H antigen and different O antigens. Examinations by
multilocus enzyme electrophoresis of E.coli of serotype O157:H7
causing haemorrhagic diarrhoea and haemolytic uraemic syndrome
(HUS) have shown that such strains are genetically closely
related to EPEC of serotype O55:H7[10]. The lineage formed by
O55:H7 and O157:H7 is different from the lineage formed by EPEC
O55:H6 and the other EPEC's with H6 (fig. 1). O55:H7 does not

hybridize with the EAF probe[11] and O157:H7 probably neither, but they both produce attaching and effacing lesions of the brush border microvillous membrane[11,12].

Clonality can never be absolute. All organisms are somewhat unstable genetically even without selective pressure, some genes are, however, more stable than others. Plos et al[13] showed variation in chromosomal DNA in E.coli from urinary tract infection (UTI) by hybridizing with probes specific for the P-associated-pilus (Pap) region. There was less variation in pap among UTI strains of the same specific O:K:H serotype and ET type, than among random isolates. It was, however, concluded that pap, or the DNA sequences flanking it, are evolutionary labile and it was hypothesized that the pap region had been transferred horizontally meaning that the E.coli chromosome contains segments of different ancesty, i.e. are chimeras.

The studies referred to in the present paper, which showed two genetic clusters or lineages each containing members with different O antigens, but the same H antigen, might indicate that the DNA determining the H antigen is more conserved in those strains than that determining the O antigen i.e. the O-polysaccharide side chain of the LPS molecule. It was suggested that the O antigen in these cases was transferred horizontally[8] but how and when?
In 1987 Riley et al[14] reported that the O111 O-polysaccharide side chain was expressed together with localized adherence (LA) and antibiotic resistance in transconjugants having acquired a 54-MDa plasmid from an EPEC strain of serotype O111:H⁻. To our knowledge nobody has as yet confirmed this finding.

REFERENCES

1. Sack, R.B., Gorbach, S.L., Banwell, J.G., Jacobs, B., Chattergel, B.D. and Mitra, R.C., 1971, Enterotoxigenic Escherichia coli isolated from patients with severe cholera-like disease. J. Infect. Dis., 128:378-385.
2. Evans, D.G., Silver, R.P., Evans, D.J., Chase, D.G. and Gorbach, S.L., 1975, Plasmid-controlled colinizationfactor associated with virulence in Escherichia coli enterotoxigenic for humans. Infect. Immun., 12:656-667.
3. Ørskov, F., Ørskov, I., Evans, D.J., Sack, R.B., Sack, D.A. and Wadström, T., 1976, Special Escherichia coli serotypes among enterotoxigenic strains from diarrhoea in adults and children. Med. Microbiol. Immunol., 162:73-80.
4. Cravioto, A., Gross, R.J., Scotland, S.M. and Rowe, B., 1979, An adhesive factor found in strains of Escherichia coli belonging to the traditional infantile enteropathogenic serotypes. Curr. Microbiol., 3:95-99.
5. Baldini, M.M., Kaper, J.B., Levine, M.M., Candy, D.C.A. and Moon H.W., 1983, Plasmid mediated adhesion in enteropathogenic Escherichia coli. J. Pediatr. Gastro-enterol. Nutr., 2:534-538.
6. Moon, H.W., Whipp, S.C., Argenzio, R.A., Levine, M.M. and Giannella, R.A., 1983, Attaching and effacing activities of rabbit and human enteropathogenic Escherichia coli in pig and rabbit intestines. Infect. Immun., 41:1340-1351.
7. Selander, R.K., Caugant, D.A., Ochman, H., Musser, J.M., Gilmour, M.N. and Whittam, T.S., 1986, Methods of multi-

locus enzyme electrophoresis for bacterial population genetics and systematics. Appl. Environ. Microbiol., 51:873-884.

8. Ørskov, F., Whittam, T.S., Cravioto, A. and Ørskov, I., 1990, Clonal relationships among classic enteropathogenic *Escherichia coli* (EPEC) belonging to different O groups. J. Infect. Dis., 162:76-81.

9. Beutin, L., Ørskov, I., Ørskov, F., Zimmermann, S., Prada, J., Gelderblom, H., Stephan, R. and Whittam, T.S., 1990, Clonal diversity and virulence factors in strains of *Escherihcia coli* of the classical enteropathogenic O-serogroup 114. J. Infect. Dis., in press.

10. Whittam, T.S., Wolfe, M.L., Debray, C., Wachsmuth, J.K., Ørskov, F., Ørskov, I. and Wilson, R.A. Reconstructing the evolution of a new bacterial pathogen. Science, submitted.

11. Echeverria, P., Ørskov, F., Ørskov, I., Knutton, S., Scheutz, F., Brown, J.E. and Lexomboon, U., 1990, Attaching and effacing enteropathogenic *Escherichia coli* as a cause of infantile diarrhoea. N. Eng. J. Med., submitted

12. Knutton, S., Baldwin, T., Williams, P.H. and McNeish, A.S., 1989, Actin accumulation at sites of bacterial adhesion to tissue culture cells: basis of a new diagnostic test for enteropathogenic and enterohemorrhagic *Escherichia coli*. Infect. Immun., 57:1290-1298.

13. Plos, K., Hull, S.J., Hull, R.A., Levin, B.R., Ørskov, I., Ørskov, F. and Svanborg-Eden, C., 1989, Distribution of the P-associated-Pilus (*pap*) region among *Escherichia coli* from natural sources: evidence for horizontal gene transfer. Infect. Immun., 57:1604-1611.

14. Riley, L.W., Junio, L.N,. Libaek, L.B. and Schoolnik, G.K., 1987, Plasmid-encoded expression of lipopolysaccharide O-antigenic polysaccharide in enteropathogenic *Escherichia coli*. Infect. Immun., 55:2052:2056.

GENETICS OF HISTONE-LIKE PROTEIN H-NS/H1 AND REGULATION OF

VIRULENCE DETERMINANTS IN ENTEROBACTERIA

Bernt Eric Uhlin, Björn Dagberg, Kristina Forsman, Mikael Göransson, Birgit Knepper, Peter Nilsson, and Berit Sondén

Department of Microbiology
University of Umeå
S-901 87 Umeå, Sweden

INTRODUCTION

Isolates of E. coli associated with intestinal or with extraintestinal disease are often characterized by their ability to express different properties thought to contribute to bacterial virulence. Examples of such properties are the synthesis of various types of adhesins, production of different kinds of cytotoxins (e.g. enterotoxins and hemolysins), ability to invade host tissue cells, and expression of certain capsule- and O-antigens. Changes in growth conditions may profoundly affect how different virulence-associated properties are expressed. The type of growth substrate, osmolarity, and growth temperature are examples of environmental factors found to influence the phenotypic expression of some of the virulence associated properties monitored under laboratory conditions. We have studied the expression of E. coli adhesins typically produced by many isolates from urinary tract infections in man. The P-specific adhesins are fimbrial adhesins, also referred to as pili, and they mediate binding to α-D-Gal-(1-4)-β-D-Gal-containing glycolipid structures. Genetic studies of the pap gene cluster have revealed functions of several of the gene products required for biogenesis of this type of adhesin (see Tennent et al., 1990 for a recent review). In order to elucidate mechanisms by which different growth conditions may affect expression of pili-adhesins and other virulence-associated properties we have studied transcriptional features of the pap genes. Here we will summarize some of our recent findings about environmental regulation of expression of pap genes and related virulence determinants of E. coli.

Transcriptional regulation of pap genes

Expression of Pap pili is subject to regulation depending upon growth substrate and temperature conditions. Using pap-lacZ fusions we showed that the temperature-dependence operates at the transcriptional level (Göransson and Uhlin, 1984). Two genes in the pap gene cluster, papB and papI, were found to encode products having regulatory functions (Båga et al., 1985; Göransson et al., 1988; Forsman et al., 1989). Thermoregulation of pap transcription and pili biogenesis is mediated via the regulatory genes, and the papI gene product appeared to be limiting at temperatures below 30°C (Göransson et al., 1989a). Transcription of the pap genes and expression of pili-adhesins are reduced if the bacteria are grown in the presence of glucose. Our genetic and biochemical studies established that the two divergently oriented pap promoters are dependent upon activation by the CRP-cAMP complex which binds to a defined sequence in the pap DNA (Göransson et al., 1989b). The PapB protein binds to DNA sequences directly adjacent to the CRP-cAMP binding site and the protein-DNA interactions define a UAS (upstream activating sequence) region shared by the two pap promoters (Forsman et al., 1989; Göransson et al., 1989b).

Identification of the drdX locus

In order to further define the regulatory components involved in thermoregulation we searched for mutants with altered regulation of pap transcription. Using pap-lacZ operon fusion constructs we screened for altered β-galactosidase expression under different growth conditions. A spontaneously occurring class of mutants was denoted drdX because it showed derepressed expression of pap-lacZ fusions (Göransson et al., 1990). The mutation caused an elevated level of transcription at $37^{\circ}C$, and more importantly, derepression of expression at temperatures below $30^{\circ}C$. Analysis of mRNA levels confirmed that both the monocistronic papI regulatory gene and the genes for pili-adhesin biogenesis were expressed at high levels even at the lower growth temperatures. The drdX locus was mapped to the vicinity of the trp operon in E. coli. We then made use of the phage lambda hybrid clones from the E. coli gene library described by Kohara et al. (1987) to locate the DNA region carrying the drdX locus. Using one of the lambda clones as source of DNA from the 27.5-minute region of the E. coli chromosome we cloned the wild type allele of drdX and showed that it complemented the mutation in trans (Göransson et al., 1990).

The drdX locus encodes the histone-like protein H-NS/H1

Our characterization of the cloned drdX gene revealed that it encoded a 16K protein, and nucleotide sequence analysis showed that a corresponding open reading frame was present (Göransson et al., 1990). When comparing the deduced amino acid sequence with previously published data we found it to be identical to that of a protein called H-NS or H1 in E. coli. This is one of the so called histone-like proteins found in deoxyribonucleoprotein particles of mildly lysed bacterial cells (see Drlica and Rouviere-Yaniv, 1987, for a review).

A pleiotropic locus defined by several allelic mutations

A locus denoted bglY (Defez and DeFelice, 1981) had earlier been mapped to the 27.5-minute region and as in that case we found the drdX mutation to exhibit a Bgl⁺ phenotype. Our trans-complementation studies provided clear evidence that bglY and drdX are allelic (Göransson et al., 1990). Other mutations mapped to this region are: cur (Diderichsen, 1980); pilG (Spears et al., 1986); osmZ (Higgins et al., 1988); and virR (Maurelli and Sansonetti, 1988; Hromockyj and Maurelli, 1989). Although discovered separately as loci affecting genes of different kinds (capsular polysacharide synthesis, cur; type 1 pili phase variation, pilG; osmoregulation, proU; and enteroinvasiveness, virR) the mutations are presumably allelic to each other and to drdX. Evidently this locus is involved in regulation of several unlinked genes and mutations may show pleiotropic phenotypes. Our original drdX mutation turned out to be a deletion of the structural gene and surrounding sequences (unpublished data). Using this mutant we have tested several operons for regulatory effects when protein H-NS/H1 is absent. Interestingly, a tox-cat operon fusion of the A-subunit cistron of the E. coli heat-labile enterotoxin showed effects quite similar to that of pap (unpublished data). The toxin gene transcription was derepressed both at 37°C and at lower temperatures, and as is the case of pap we may conclude that the drdX locus is involved in the thermoregulation of enterotoxin production. Similarly, we have found that expression of a plasmid-encoded cistron (virG) of an enteroinvasive E. coli strain is derepressed by a drdX mutation (unpublished data). Taken together, the results suggest that several distinct virulence factors of E. coli may be subject to similar regulatory mechanisms involving the drdX gene product as a common component.

```
E.c.    MSEALKILNNIRTLRAQARECTLETLEEMLEKLEVVV
P.v.    ---S-----------------TS--------------
S.m.    ---R---------------------------------
S.t.    -------------------------------------


E.c.    NERREEESAAAAEVEERTRKLQQYREMLIADGIDPNE
P.v.    -------Q-MQ--I---QQ---K---L--------TD
S.m.    ------D-Q-Q--I-----------------------
S.t.    -------------------------------------


E.c.    LLNSLAAVKSGTKAKRAQRPAKYSYVDENGETKTWTG
P.v.    --EAAG-S-T-*R----A----------D---------
S.m.    --QTM--N-AAG-----R-----Q-K-----L-----
S.t.    ----M--A---------A-------------------


E.c.    QGRTPAVIKKAMDEQGKSLDDFLIKQ
P.v.    ----L----R-IE-E----E----**
S.m.    -----------IE----------L**
S.t.    -----------E----Q-E-----E
```

Fig. 1   Comparison of amino acid sequence (in single letter code) of the H-NS/H1 protein from Escherichia coli (E.c.) with those deduced from gene sequences in Proteus vulgaris (P.v.), Serratia marcescens (S.m.), and Salmonella typhimurium (S.t.). See the text for references. Identical amino acids are shown by hyphens and asterisks indicate absence of residues corresponding to a particular position in the E.c. sequence.

The H-NS/H1 gene and protein appear conserved among enterobacteria

The structural gene encoding H-NS/H1 protein has been analyzed at the sequence level of a few more species in addition to E. coli: Proteus vulgaris, Serratia marcescens, and Salmonella typhimurium (La Teana et al., 1898; Marsh and Hillyard, 1990). A comparison of the amino acid sequences shows that the protein is quite conserved among these species (Fig. 1). Using the cloned gene of E. coli as a source of gene specific probes we have obtained evidence that it is also conserved among several other enterobacterial species (unpublished data).

Transcriptional silencing as a mechanism of thermoregulation

Our findings show that a member of the class of histone-like proteins in the E. coli nucleoid has a central role in thermoregulation of Pap pili-adhesin expression. Furthermore, as mentioned above, we have evidence that several virulence-associated genes are among those that are affected by mutations abolishing H-NS/H1 production. Although pleitropic, the drdX deletion mutation seemed to cause effects on a selected set of genes. Our data sofar suggest that most E. coli genes are not affected. Considering the effects on pap transcription and the normally cryptic ability to utilize β-glucosides (i.e. the change from Bgl$^-$ to Bgl$^+$ phenotype), we suggested that certain operons may be silenced in drdX$^+$ cells. It remains to be seen how the presumed interactions between the H-NS/H1 protein and the regulatory regions of pap and other operons are altered by changing environmental conditions. More or less local alterations in DNA topology (induced by e.g. a temperature change) could possibly influence protein-DNA interactions. An important question is whether or not the activity of the H-NS/H1 protein may change as a result of changing conditions. The level of drdX mRNA and the level of H-NS/H1 protein appeared to be the same in cells grown at 26°C and 37°C as judged by Northern-blot and Western-blot analyses, respectively (Göransson et al., 1990; and unpublished data). Hypothetical modifications such as phosphorylation and/or dephosphorylation of the protein itself, and thereby modification of its activity, could also be the result of some signal transduction pathway triggered by environmental changes. The genetic data indicate that the H-NS/H1 protein is a crucial component in the regulatory mechanisms of different types of environmental signals (osmolarity, temperature). Further biochemical and genetic characterization of these mechanisms should hopefully reveal details about how the bacteria regulate virulence properties such as adhesion and enteroinvasion in response to environmental conditions.

ACKNOWLEDGEMENTS

We are grateful for excellent technical and secreterial assistance by Karin Emanuelsson and Britt-Inger Strömberg, respectively. Our work was supported by grants from the Swedish Natural Science Research Council (project BU 1670), the National Swedish Board for Technical Development (project 84-5463), and the Swedish Medical Research Council (graduate fellowship 16P-07677).

REFERENCES

Båga, M., Göransson, M., Normark, S., and Uhlin, B.E., 1985, Transcriptional activation of a Pap pilus virulence operon from uropathogenic Escherichia coli, EMBO J., 4:3887.

Defez, R., and DeFelice, M., 1981, Cryptic operon for β-glucoside metabolism in Escherichia coli: genetic evidence for a regulatory protein, Genetics, 97:11.

Diderichsen, B., 1980, cur-1, a mutation affecting the phenotype of sup$^+$ strains of Escherichia coli, Mol. Gen. Genet., 180:425.

Drlica, K., and Rouviere-Yaniv, J., 1987, Histonelike proteins of bacteria. Microbiol. Rev., 51:301.

Forsman, K., Göransson, M., and Uhlin, B.E., 1989, Autoregulation and multiple DNA interactions by a transcriptional regulatory protein in E. coli pili biogenesis, EMBO J., 8:1271.

Göransson, M., and Uhlin, B.E., 1984, Environmental temperature regulates transcription of a virulence pili operon in E. coli, EMBO J., 3:2805.

Göransson, M., Forsman, K., and Uhlin, B.E., 1988, Functional and structural homology among regulatory cistrons of pili-adhesin determinants in Escherichia coli, Mol. Gen. Genet., 212:412.

Göransson, M., Forsman, K., and Uhlin, B.E., 1989a, Regulatory genes in the thermoregulation of Escherichia coli pili gene transcription, Genes & Dev., 3:123.

Göransson, M., Forsman, K., Nilsson, P., and Uhlin, B.E., 1989b, Upstream
activating sequences that are shared by two divergently transcribed
operons mediate cAMP-CRP regulation of pilus-adhesin in Escherichia
coli, Mol. Microbiol., 3:1557.

Göransson, M., Sondén, B., Nilsson, P., Dagberg, B., Forsman, K.,
Emanuelsson, K., and Uhlin, B.E., 1990, Transcriptional silencing
and thermoregulation of gene expression in Escherichia coli,
Nature, 344:682.

Higgins, C.F., Dorman, C.J. Stirling, D.A., Waddell, L., Booth, I.R.,
May, G., and Bremer, E., 1988, A physiological role for DNA
supercoiling in the osmotic regulation of gene expression in S.
typhimurium and E. coli, Cell, 52:569.

Hromockyj, A.E., and Maurelli, A.T., 1989, Identification of an
Escherichia coli gene homologous to virR, a regulator of Shigella
virulence, J. Bacteriol., 171:2879.

Kohara, Y., Akiyama, K., and Isono, K., 1987, The physical map of the
whole E. coli chromosome: application of a new strategy for rapid
analysis and sorting of a large genomic library, Cell, 50:495.

La Teana, A., Falconi, M., Scarlato, V., Lammi, M., and Pon, C.L., 1989,
Characterization of the structural genes for the DNA-binding
protein H-NS in Enterobacteriaceae, FEBS Lett., 244:34.

Marsh, M., and Hillyard, D.R., 1990, Nucleotide sequence of hns encoding
the DNA-binding protein H-NS of Salmonella typhimurium, Nucleic
Acids Res., 18:3397.

Maurelli, A.T., and Sansonetti, P.J., 1988, Identification of a
chromosomal gene controlling temperature-regulated expression of
Shigella virulence, Proc. Natl. Acad. Sci. USA, 85:2820.

Tennent, J.M., Hultgren, S., Marklund, B.-I., Forsman, K., Göransson, M.,
Uhlin, B.E., and Normark, S., 1990, Genetics of adhesin expression
in Escherichia coli, in: "The Bacteria, Volume XI, Molecular basis
of bacterial pathogenesis" B.H. Iglewski, and V.L. Clark, ed.,
Academic Press, Inc., San Diego.

REGULATION OF EXPRESSION OF FIMBRIAE OF HUMAN ENTEROTOXIGENIC ESCHERICHIA COLI

Wim Gaastra, Anja M. Hamers, Bart J.A.M. Jordi, Paul H.M. Savelkoul, Geraldine A. Willshaw[*], Moyra M. McConnell[*], Johannes G. Kusters, Arnoud H.M. van Vliet and Bernard A.M. van der Zeijst

Institute of Infectious Diseases and Immunology, Department of Bacteriology, Faculty of Veterinary Medicine, University of Utrecht, Yalelaan 1, P.O.Box 80.165, 3508 TD Utrecht, The Netherlands, and, [*]:Division of Enteric Pathogens, Central Public Health Laboratory, 61, Colindale Avenue, London NW9 5HT, England

FIMBRIAE OF HUMAN ETEC

Adhesion of bacteria to epithelial cells of the host is the first and probably the most important step in the pathogenesis of bacterial infections. In humans, adhesion of enterotoxigenic E.coli (ETEC) to the mucosa of the small intestines is mediated by serologically distinct and host-specific fimbrial antigens on the bacterial surface. The first fimbrial antigen to be identified in human ETEC strains was colonization factor antigen I (CFA/I)[1]. CFA/I is a single fimbrial antigen found on ETEC producing only heat-stable (ST) enterotoxin and on ETEC producing both heat-labile and heat-stable enterotoxin (LT,ST) of at least 15 serogroups. Later CFA/II was detected[2] and was subsequently shown to consist of three distinct coli surface associated (CS) antigens, designated CS1, CS2 and CS3[3]. Nearly all CFA/II-positive E.coli produce CS3, but those of serotype O6.H16 or O6.H- also express either CS1 or CS2 depending on their biotype: biotype A strains produce CS1 whereas those of biotypes B, C and F express CS2[3]. Production of CS1 and CS3 has been reported in only one other ETEC strain. This strain was of serotype O139.H28 and differs from other O139.H28 strains that are CS3 only producers[4]. CFA/IV, was also shown to be an antigen complex, consisting of CS6 which appears to be non-fimbrial and either CS4 or CS5 fimbriae[5]. Strains of serotype O25:H42 produce CS4 together with CS6 and both ST and LT or ST only. CS5 is produced by strains of six different serogroups together with CS6 and ST[5]. CFA/I, CS1, CS2, CS4 and CS5 fimbriae are morphologically similar and

CFA/I, CS1. CS2 and CS4 are antigenically related when tested by Western immunoblotting[6] and have homologous N-terminal amino acid sequences of their subunit proteins. The subunit proteins of CFA/I and CS1 fimbriae have recently been found to be more than 70% similar (unpublished results). CS5 in contrast is not antigenically related to the other four fimbriae, its subunit protein has a different N-terminal amino acid sequence[7] and probably belongs to a different class of fimbriae.

GENETIC ORGANIZATION OF FIMBRIAL OPERONS

In the CFA/I-ST plasmid NTP113, two regions, designated CFA/I region 1 and CFA/I region 2 (Fig.1), separated by about 40kb are required for production of CFA/I fimbriae in an *E.coli* K12 host[8]. Production of the fimbrial antigens CS1, CS2 and CS3 is also plasmid controlled. It was observed in several strains of serogroup O6 that loss of antigen production was associated with loss of a plasmid which encoded production of ST and LT[9]. Subsequently it was shown that in strains of serotype O6:H16 the structural gene for the CS3 fimbrial subunit was carried on the same plasmid[10]. The plasmid also contained a regulatory sequence termed *rns* (regulation of CS1 and CS2) that encoded a DNA binding protein required for the expression of CS1 and CS2 fimbriae[11]. The plasmid did not carry the structural genes for the CS1 and CS2 fimbrial subunits and a chromosomal location for these sequences was inferred in the O6.H16 strains studied.

Nucleotide sequence determination of CFA/I region 1 has shown that this DNA fragment codes for four proteins, CfaA, CfaB, CfaC and CfaE[12,13]. CfaB is the structural subunit protein and CfaC is the large outer membrane protein found in all *E.coli* fimbrial operons studied sofar. This protein serves as a porin for transport of the fimbrial subunit. The functions of the CfaA and CfaE proteins are not yet known. The promoter in front of the CFA/I operon has a poor -35 recognition region, but a good ribosome binding site and -10 sequence are present[12]. A poor -35 region is indicative for positively regulated promoters[14].

CFAD, A POSITIVE REGULATOR OF FIMBRIAE EXPRESSION

The nucleotide sequence of CFA/I region 2 was determined to see whether a role in the expression and/or assembly of CFA/I fimbriae could be assigned to one or more proteins encoded for by this region. Three open reading frames (ORF) larger than 100 nucleotides are present on CFA/I region 2[15], of which one is preceeded by a ribosome binding site, as well as regulatory sequences homologous to the -10 and -35 regions of *E.coli* promoters. This ORF encodes a basic protein CfaD of 265 amino acid residues, which is highly basic (pI = 9.5), contains a high amount of methionine residues and a C-terminal helix-turn-helix profile observed in DNA binding proteins[16]. Extensive homology of the *cfaD* gene with the beta-lactamase gene has been observed (data not shown). The origin of this homology is not known, but it can probably interfere with hybridization experiments in which one tries to demonstrate the presence of *cfaD* like sequences in strains resistent to ampicillin. We have however, demonstrated by Southern hybridization, that DNA sequences homologous to the *cfaD* gene are present on plasmids in *Proteus mirabilis* and *Salmonella typhimurium* strains (data not shown). Based on the homology of CfaD with a number of positive regulatory proteins (see below) it was assumed that CfaD is a positive regulator[15].

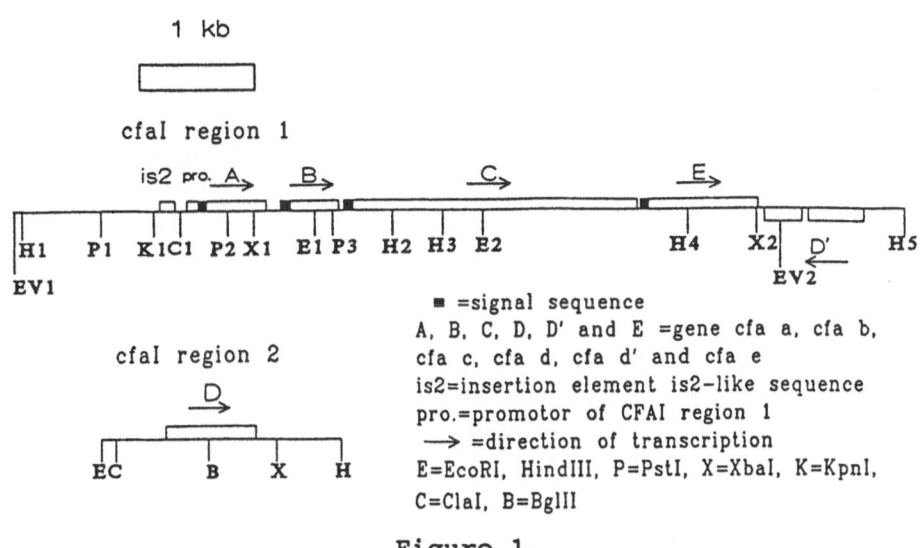

Figure 1.

HOMOLOGY OF CFAD WITH OTHER REGULATORY PROTEINS

The CfaD protein is highly similar to a number of other proteins, for example the plasmid encoded VirF protein of *Shigella flexneri* (Fig.2) and a protein encoded on plasmids in enteroinvasive *Escherichia coli*. The VirF protein is a positive regulator of the expression of four plasmid encoded proteins, involved in invasiveness of *Shigella flexneri*. The *cfaD* gene product is also homologous with the chromosomally encoded appY and envY gene products of *E.coli* (Fig. 2). The appY protein is a growth phase dependent regulatory protein for the expression of acid phosphatase. The *appY* gene product also changes the rate of synthesis of more than 30 other proteins in a growth phase dependent way and its function is influenced by anaerobiosis. The EnvY protein affects the temperature-dependent expression of a number of *E.coli* outer membrane proteins. Recently, homology of CfaD with FapR a regulatory protein of the 987P fimbrial operon of animal ETEC was demonstrated (see paper by Dr. F.K.de Graaf in this volume).

```
EC-CFAD     MDFKYTEEKEMIKINNIMIHKYTVLYTSNCIMDIYSEEEKITCFSNRLVFLERGVNISVR
EC-CFAD'    MDFKYTEEKEMIKINNIMIHITYVLYTSNCIMDIYSEEEKITCSSNRLVFLERGVNISVR
EC-PRNS     MDFKYTEEKETIKINNIMIHKYTVLYTSNCIMDIYSEEEKITCFSNRLVFLERGVNISVR
EC-FAPR     M-------KLK--N--IHLYNYVVIYTKNCEIYINKGNEQVYIPPRMVAIFEKNISFNIE
SF-VIRF     M--MDMGHKNKIDIK--VRLHNYIILYAKRCSMTVSSGNETLTIDEGQIAFIERNIQINV-
EC-APPY     MDYV----CSVV----FICQSFDLIINRRV-ISF--------IVSDKIR---RELPVCPS
EC-M5       MDYV----CSVV----FICQSFDLIINRRV-ISFKKNSL--FIVSDKIR---RELPVCPS
EC-ENVY     MQLSSSEPCVVI----LTEKEVEVSVNNHATFTLPKNYLAAFACNNNVI----ELST-LN
            *                     .     .  .   .        .
```

```
EC-CFAD     IQKKILSERPYVAFRLNGDILRHLKNALMIIYGMSKVDTNDCRGMSRKIMTTEVNKTLLD
EC-CFAD'    IQKQILSEKPYIAFRLNGDILRHLKNALMIIYGMSKIDINDCRNMSRKIMTTEVNKTLLD
EC-PRNS     MQKQILSEKPYVAFRLNGDMLRHLKDALMIIYGMSKIDTNACRSMSRKIMTTEVNKTLLD
EC-FAPR     TIRKGDDVL-YESFDMKHELLTSLRRVIEPSVKFAAESYTNKRSFKERIFKVKSCSIVID
SF-VIRF     SIKKSDSINPFEIISLDRNLLLSIIRIMEPIYSFQHSYSEEKRGLNKKIFLLSEEEVSID
EC-APPY     KLRIVDIDKK----TCLS-FFIDVNNELPGKFTLDKNGYIAEEEPPLSLVFSLFEGIKIA
EC-M5       KLRIVDIDKK----TCLS-FFIDVNNELPGKFTLDKNGYIAEEEPPLSLVFSLFEGIKIA
EC-ENVY     HVLITHINRN----RIINDYLLFLNKNLTCVKPWSRLATPVIACHSTPEVFPL--AANHS
                                           .        .   .    .
```

```
EC-CFAD     ELKNINSHDDSAF--IS-SLIYLISKIENNEKIIQSIYISSVSFPSDKVRNVIEKDLSRK
EC-CFAD'    VLKNINSHYDSVF--IS-SLIYLISKI-NNEKIIESIYISSV-FFSDKVRSVIEKDLSRK
EC-PRNS     ELKNINSHDNSAF--IS-SLIYLISKLENNEKIIESIYISSVSFPSDKVRNLIEKDLSRK
EC-FAPR     LFKRLKDNGSPEFTAIY-ELAFLVSKCENPSMFAISLFSSVAVTFSERIVTLLFSDLTRK
SF-VIRF     LFKSIKEMPFGK-RKIY-SLACLLSAVSDEEALYTSISIASSLSFSDQIRKIVEKNIEKR
EC-APPY     DSHSLW-----LKERLCISLLAMFKKRESVNSFIL----TNINTFTCKITGIISFNIERQ
EC-M5       DSHSLW-----LKERLCISLLAMFKKRESVNSFIL----TNINTFTCKITGIISFNIERQ
EC-ENVY     KQQPSRPCEAELTRALLFTVLSNFLEQSRFIALLMYILRSSVRDTVCR---IIQSDIQHY
                          .      .   .               .   .    .
```

```
EC-CFAD     WTLGIIADAFNVSEITIRKRLESE-NTNFNQILMQLRMSKAALLLLENSYQISQISNMIGI
EC-CFAD'    WTLAIIADTFNVSEITIRKRLESE-NTNFNQILMQLRMSKAALLLLENPYQISQISNMIG
EC-PRNS     WTLGIIADAFNASEITIRKRLESE-NTNFNQILMQLRMSKAALLLLENSYQISQISNMIG
EC-FAPR     WKLSDIAKEMHISEISVRKRLEQE-CLNFNQLILDVRMNQAAKFIIRSDHQIGMIASLVG
SF-VIRF     WRLSDISNNLNLSEIAVRKRLESE-KLTFQQIILLDIRMHHAAKLLLNSQSYINDVSRLIG
EC-APPY     WHLKDIAELIYTSESLIKKRLDE-GTSFTEILRDTRMRYAKKLIITSNSYSINVVAQKCG
EC-M5       WHLKDIAELIYTSESLIKKRLRDE-GTSFTEILRDTRMRYAKKLIITSNSYSINVVAQKCG
EC-ENVY     WNLRIVASSLCLSPSLLKKKLKNE-NTSYSQIVTECRMRYAVQMLLMDNKNITQVAQLCG
            *  .   ..   .* *  .* *       .      .  . . ..   .   .
```

```
YE-VIRF     WKLSKFAREFGMGLTTFKELFGTVYGISPRAWISERRILYAHQLLLNGKMSIVDIAMEAG
CF-ARAC     FDIASVAQHVCLSPSRLSHLFRQQLGISVLSWREDQRISQAKLLLSTTRMPIATVGRNVG
ST-ARAC     FDIASVAQHVCLSPSRLSHLFRQQLGISVLSWREDQRISQAKLLLSTTRMPIATVGRNVG
EC-ARAC     FDIASVAQHVCLSPSRLSHLFRQQLGISVLSWREDQRISQAKLLLSTTRMPIATVGRNVG
E.CAR-ARAC  LRIDEVARHVCLSPSRLAHLFREQVGINILRWREDQRVIRAKLLLQTTQESIANIGRVVG
EC-CELD     SALENMVALSAKSQEYLTRATQRYYGKTPMQIINEIRINFAKKQLEMTNYSVTDIAFEAG
EC-RHAR     VNWDAVADQFSLSLRTLHRQLKQQTGLTPQRYLNRLRLMKARHLLRHSEASVTDIAYRCG
YC-RHAS     WAIDKECREASCSERVIROVEROOCTGMTJNQYLRQVRVCHAOYLIQHSRILISDISTECG
                                       .         *.  .        .        *
```

```
EC-CFAD     ISSASYFIRVFNKHYGVTPKQFFTYFKGG
EC-CFAD'    ISSISYFIRVFVKHYGVAPKQFFTYFKGG
EC-PRNS     ISSASYFIRIFNKHYGVTPKQFFTYFKGG
EC-FAPR     YTSVSYFIKTFKEYYGVTPKKFEIGIKENLRCNR
SF-VIRF     ISSPSYFIRKFNEYYGITPKKFYLYHKKF
EC-APPY     YNSTSTYFICAFKDYYGVTPSHYFEKIIGVTDGINKTID
EC-M5       YNSTSTYFICAFKDYYGVTPSHYFEKIIGVTDGINKTID
EC-ENVY     YSSTSYFISVFKAFYGLTPLNYLAKQRQKVMW
            .* ****  *   **..*  ..
```

```
EC-ENVY     YSSTSYFISVFKAFYGLTPLNYLAKQRQKVMW
YE-VIRF     FSSQSYFTQSYRRRFGCTPSQARLTKIATTG
CF-ARAC     FDDQLYFSRVFKKCTGASPSEFRAGCE
ST-ARAC     FDDQLYFSRVFKKCTGASPSEFRAGCE
EC-ARAC     FDDQLYFSRVFKKCTGASPSEFRAGCEEKVNDVAVKLS
E.CAR-ARAC  YDDQLYFSRVFRKRVGVSPSDFRRRSSEINYPAAKTLPVAWGEQIPHAVSS
EC-CELD     YSSPSLFIKTFKKLTSFTPKSTRKKLTEFNQ
EC-RHAR     FSDSNHFSTLFRREFNWSPRDIRQGRDGFLQ
EC-RHAS     FEDSNYFSVVFTRETGMTPSQWRHLNSQKD
            ..   *   .    .*
```

Fig. 2. Comparison of the primary structure of CfaD with a number of
homologous regulatory proteins. Sequence data are translated from DNA
sequences taken from the EMBL-library (release 24). For Ara, Rha and Cel
proteins only the C-terminus was compared.
*=Identical residue in all sequences, .=similar residues.

Homology with DNA binding regulatory proteins of the
arabinose and rhamnose operons in *E.coli*, *Salmonella
typhimurium* and *Erwinia caratovora* was observed in the
C-terminal region containing the helix-turn-helix motive
(Fig.2). The CfaD protein may therefore well be a DNA binding
regulatory protein. The GC content of the *cfaD* gene is 28%,
which is extremely low for an *E.coli* gene. The codon usage
differs from that normally used in *E.coli*.

Evident is the frequent use of CTT (Leu), ATA (Ile), CGT, AGA, AGG (all Arg) and GGA and GGG (Gly) codons, which all correspond to minor or weakly interacting tRNAs in *E.coli*. A similar low GC content and codon usage was observed in the genes on CFA/I region 1[12,13]. This may mean that these genes do not originate in *E.coli*[11,15].

FUNCTION OF THE CFAD PROTEIN

To analyse the function of the CfaD protein, the promoter region of CFA/I region 1 (the *ClaI-PstI* fragment between nucleotides 738 and 1106 of region 1)[12] was cloned in front of the promoterless beta-galactosidase gene of the promoter probe vector pCB267 (pIVB3-107). Introduction of a compatible plasmid containing the *cfaD* gene (pIVB3-105) into the same cells as pIVB3-107 resulted in a five times higher production of beta-galactosidase[15] The production of beta-galactosidase was even higher when the *cfaD* gene was also cloned into pIVB3-107 (pIVB3-109). This extra enhancement is however due to the fact that there is no influence on the copy number of the plasmid containing the *cfaD* gene, as is the case with the combination of pIVB3-107 and pIVB3-105. These results indicate that CfaD binds to the promoter region in front of the CFA/I operon, resulting in the enhanced expression of genes under control of the CFA/I promoter. Experiments to locate the exact binding site of the CfaD protein are in progress.

FUNCTIONAL SUBSTITUTION OF *cfaD* AND *RNS* GENES

The *cfaD* gene differs at only 28 positions from the gene coding for the Rns protein[11]. Thirteen of these differences lead to a different amino acid (Fig.2). The Rns protein is a positive regulator of the expression of CS1 and CS2 fimbriae[11], produced by human ETEC strains of O-serogroup O6.H- or O6.H16[2,9]. The similarity in the nucleotide sequences is not only restricted to the genes, i.e in 260 nucleotides upstream of the gene 45 differences are found and in 160 nucleotides downstream of the gene only 9. To test whether the CfaD protein can functionally substitute the Rns protein, plasmid pIVB3-100 containing an *EcoRI-XbaI* subclone of CFA/I region 2 (Fig.1) was introduced into strains that were CS1 or CS2 negative due to loss of the plasmid encoding the Rns protein. Introduction of this plasmid restored the production of CS1 and CS2 as determined by MRHA (mannose resistant haemagglutination) and ELISA. However *rns* is not completely interchangeable with *cfaD*, since the use of *rns* in stead of *cfaD* resulted in at least 20-fold reduction of CFA/I production[17]. Thus the CfaD protein is able to substitute for the Rns protein, encoded in the parental strains on a 97kb plasmid and to regulate expression of chromosomally located CS1 or CS2 sequences. Neither *virF* nor *appY* induce production of CFA/I fimbriae from CFA/I region 1 when introduced into *E.coli* K12 cells containing CFA/I region 1 on a compatible plasmid. Expression of plasmid encoded fimbriae (CFA/I) and of chromosomally encoded fimbriae (CS1 or CS2) in strains of different serotypes, is thus regulated by the same plasmid encoded protein.

A SILENT REGULATORY GENE ON CFA/I REGION 1

Hybridization of *Hind*III digests of seven CFA/I-ST plasmids from strains of different serotypes with a probe of the *cfaD* gene revealed that *cfaD* like sequences were present on at least two different *Hind*III fragments in each wild-type plasmid[13]. A fragment also lighted up in the *Hind*III digest of the cloned region 1. This indicated that a nucleotide sequence homologous to the *cfaD* gene was present on CFA/I region 1. A nucleotide sequence (designated *cfaD'*) homologous to the *cfaD* gene was indeed found (between H_4 and H_5 in Fig.1)[13]. The homology between the *cfaD* gene and the *cfaD'* sequences extends to the flanking nucleotide sequences and the region downstream of the *cfaD* gene is homologous with the 3' end of the *cfaE* gene of CFA/I region 1. In comparison to the *cfaD* gene, the *cfaD'* sequence contains a stop codon and two deletions and therefore does not code for a protein similar to that encoded by the *cfaD* gene. It is not clear how the *cfaD'* and the *cfaD* genes have originated and whether this duplication is of evolutionary advantage. The latter is however suggested by the presence of both sequences on all seven wild- type CFA/I-ST plasmids tested. On the other hand deletion of the *cfaD'* sequence from region 1 (yielding plasmid pIVB3-205)[13] gave production of CFA/I fimbriae to the same extent as in cells that contained the intact region 1. The presence of a sequence, homologous to the *rns* gene has also been reported in CS1 and CS2 producing strains of serotype O6:H16 and O6:H- as determined by hybridization. The *cfaD'* sequence can not replace the *cfaD* and *rns* genes to induce fimbriae production, which again indicates that no functional protein is transcribed from this DNA.

REGULATORY SEQUENCES IN ETEC PRODUCING OTHER FIMBRIAE

In a single unique strain of *E.coli* O139:H28 the structural genes for CS1 biogenesis and the *rns* gene controlling expression of the structural genes are located on separate plasmids[18]. From analysis of the sequence data on plasmid pDEP23[18] containing the structural genes for CS1 production in this strain it is clear that the products of the first two genes of this operon, *csoA* and *csoB* are very homologous to the *cfaA* and *cfaB* gene products. The homology is substantially less on the nucleotide level and no homology was found in the region in front of the *cfaA* and *csoA* genes (unpublished results). This means that no consensus sequence could be identified for the promoter in front of the structural genes of the CFA/I operon and the plasmid coded CS1 operon, in spite of the fact that both can be regulated by the same positive regulatory protein. ETEC strains of serotype O25:H42 which produce CS4 fimbriae contain several plasmids. A DNA sequence, hybridizing to *cfaD*, that regulates CS4 fimbriae production is located on a plasmid that also encodes CS6 production[19]. The structural genes for CS4 production are located on another plasmid and a third plasmid encodes enterotoxin production. Enterotoxin production is therefore not linked to regulation sequences or genes encoding CS antigens in these strains.

In these strains and in the O139:H28 strain described above, the CfaD protein is also able to restore production of

66

CS1 and CS4 fimbriae but not of CS3 and CS6 fimbriae in derivatives that had lost the ability to produce these fimbriae[18,19]. Hybridization with the *cfaD* probe has also been found in ETEC strains belonging to other serogroups including O27, O159 and O166[19]. These strains do not produce CFA/I, CS1, CS2 or CS4 and the hybridization may relate to regulation of other surface structures such as PCFO166.

The fimbrial antigens CFA/I, CS1, CS2 and CS4 form a group of fimbriae that are positively regulated in human ETEC. In addition to their morphological similarity[5,9] these four fimbrial types are antigenically related when tested by Western immunoblotting6. The regulatory elements that control their expression are homologous and may function by a common mechanism. CS5 fimbriae are produced by strains of O-serogroup O115 and O167[5]. The fimbrial subunit of CS5 is not antigenically related to CFA/I, CS1, CS2 or CS4, its N-terminal amino acid sequence is also different[7] and strains producing CS5 fimbriae do not hybridize with a *cfaD* probe. This might indicate that CS5 fimbriae belong to a different class of fimbriae, whose expression might also be regulated differently. Surprisingly however, introduction of CFA/I region 1 in a wildtype strain that had lost the ability to produce CS5 fimbriae, by partial deletion of a plasmid, switched on CFA/I production in this strain. This means that other factors than cfaD can regulate CFA/I production (Hibberd, McConnell, Willshaw, Smith and Rowe, unpublished results). It is not yet known what the function of this putative regulatory element in CS5 producing strains is, nor whether it is involved in CS5 production. Studies to characterize this other factor are in progress.

DISCUSSION

The expression of fimbriae appears to be regulated in such a way that it can adapt to environmental conditions. Commercially available nutrient agars, designed for the isolation of *Enterobacteriacae* are often not suitable for the detection of fimbriae on *E.coli* isolates. Known CFA/I-positive strains are negative i.e. when grown on MacConkey agar, tergitol grown cells were either negative or gave very low production, whereas high levels of production were obtained on CFA agar. All fimbrial adhesins of ETEC studied sofar are not produced by cells grown at temperatures below 25°C except for type 1 fimbriae which appear to be produced at all temperatures. Thus ETEC produce no fimbriae at temperatures well below the body temperature of their host (i.e. when they are outside the host and no adhesion is needed). Maximal fimbriae production is observed at 37°C, the body temperature of the host (i.e when they are inside the host and adhesion is a prerequisite for survival at that niche). This thermoregulation of expression is due to the temperature-dependent transcription of regulatory genes in the case of fimbriae of *E.coli* involved in urinary tract infections in humans (see also the paper by Dr. B-E. Uhlin in this volume). Hybridization experiments of total RNA isolated from a CFA/I positive wildtype *E.coli* strain, grown at 37°C and 25°C, with a DNA probe derived from the structural subunit gene *cfaB* demonstrated that no mRNA is transcribed from region 1 at 25°C (unpublished results). Thus bacteria sense and transduce environmental information in order to respond with the optimal fimbriae production for that particular situation. The resulting bacterial population will be largely homogenous. Fimbriae production is also subject to a phenotypic control mechanism called phase variation. Phase variation is the all or none switch between two states of expression (the fimbriate and the nonfimbriate state), mediated by the inversion of a DNA fragment containing the promoter of the fimbrial operon (see also the paper by Dr. P. Klemm in this volume). The frequency of promoter switching is not influenced by environmental conditions and as a result the bacterial population will be heterogenous, containing subpopulations that are preadapted to take advantage of future situations. A third regulatory mechanism of fimbriae production is the interaction of a positive regulator protein with the promoter of a fimbrial operon, as described for the CfaD protein.

The CfaD protein belongs to a class of plasmid encoded, DNA binding regulatory proteins (of approximately 30kDa). The localisation of the structural and regulatory genes of fimbrial operons of human ETEC differs in the various systems. In strains producing CFA/I fimbriae both are located on the same plasmid[12], but on separated regions, in strains of O-serogroup O25 (producing CS4 fimbriae) and in one strain of O-serogroup O139 (producing CS1 fimbriae), they are on two different plasmids[18,19] and in strains of O-serogroup O6:H16 (producing CS1 or CS2 fimbriae) a unique requirement for the plasmid-borne regulatory sequence to control expression of chromosomally located genes is found[11]. It is not clear why it is advantageous to have a plasmid location for a regulatory gene, especially since the regulatory genes apparently are lost with a high frequency.

Regulation of gene expression in *E.coli* operates mainly at the level of initiation of transcription when RNA polymerase binds to a promoter sequence. Positively regulated promoters often have a poor -35 region. They are not transcribed unless a specific transcriptional activator binds to the DNA in the vicinity of the -35 region. The activating protein can greatly increase promoter activity. Regulatory proteins often bind DNA close to specific sequences $(CA_{4-7}T)$[20] called DNA bending sites, since the DNA bends at these regions after binding of the regulatory protein. Bending of DNA probably makes promoter regions better accessible to RNA polymerase. CFA/I region I lacks a good promoter in front of the CFA/I operon[12]. A DNA bending site CAAAAAAAAAT at the position where the -35 region of the promoter should have been was indeed observed[12]. The promoter in front of the plasmid located CS1 also lacks a good -35 sequence but here no DNA bending site is observed although this operon is also positively regulated.

REFERENCES

1. D.G. Evans, R.P. Silver, D.J. Evans Jr., D.G Chase and S.L. Gorbach, Plasmid controlled colonization factor associated with virulence in *Escherichia coli* enterotoxigenic for humans. Infect. Immun. 12:656 (1975).
2. D.G. Evans and D.J. Evans Jr., New surface-associated

heat-labile colonization factor antigen (CFA/II) produced by enterotoxigenic *Escherichia coli* of serogroups O6 and O8. Infect. Immun. 21:638 (1978).

3. C.J. Smyth, Two mannose resistant haemagglutinins on enterotoxigenic *Escherichia coli* of serotype O6:K15:H16 or,H⁻ isolated from traveller's and infantile diarrhoea. J. Gen. Microbiol. 128:2081 (1982).

4. S.M. Scotland, M.M. McConnell, G.A. Willshaw, B. Rowe and A.M. Field, Properties of wild-type strains of enterotoxigenic *Escherichia coli* which produce colonization factor antigen II, and belong to serogroups other than O6. J. Gen. Microbiol. 131:2327 (1985).

5. L.V. Thomas, M.M. McConnell, B. Rowe and A.M. Field, The possession of three novel coli surface antigens by enterotoxigenic *Escherichia coli* strains positive for the putative colonization factor PCF8775. J. Gen. Microbiol. 131:2319 (1985).

6. M.M. McConnell, H. Chart and B. Rowe, Antigenic homology within human enterotoxigenic *Escherichia coli* fimbrial colonization factor antigens: CFA/I, coli-surface associated antigens (CS)1, CS2, CS4 and CS17. FEMS Microbiol. Lett. 61:105 (1989).

7. M.W. Heuzenroeder, B.L. Neal, C.J. Thomas, R. Halter and P.A. Manning, Molecular cloning and characterization of the PCF8775 CS5 antigen from an enterotoxigenic *Escherichia coli* O115:H40 isolated in central Australia. Mol. Microbiol. 3:303 (1989).

8. G.A. Willshaw, H.R. Smith and B. Rowe, Cloning of regions encoding colonization factor antigen I and heat-stable enterotoxin in *Escherichia coli*. FEMS. Microbiol. Lett. 16:101 (1983).

9. P. Mullany, A.M. Field, M.M. McConnell, S.M. Scotland, H.R. Smith and B. Rowe, Expression of plasmids coding for colonization factor antigen II (CFA/II) and enterotoxin production in *Escherichia coli* J. Gen. Microbiol. 129:3591 (1983).

10. M. Boylan, D.C. Coleman and C.J. Smyth, Molecular cloning and characterization of the genetic determinant encoding CS3 fimbriae of enterotoxigenic *Escherichia coli*. Microb. Pathogen. 2:195 (1987).

11. J. Caron, L.M. Coffield and J.R. Scott, A plasmid-encoded regulatory gene, *rns*, required for expression of CS1 and CS2 adhesins of enterotoxigenic *Escherichia coli*. Proc. Natl. Acad. Sci. USA. 86:963 (1989).

12. A.M. Hamers, H.J. Pel, G.A. Willshaw, J.G. Kusters, B.A.M. van der Zeijst and W. Gaastra, The nucleotide sequence of the first two genes of the CFA/I fimbrial operon of human enterotoxigenic *Escherichia coli*. Microb. Pathogen. 6:297 (1989).

13. W. Gaastra, B.J.A.M. Jordi, E.M.A. Mul, A.M. Hamers, M.M. McConnell, G.A. Willshaw, H.R. Smith and B.A.M. van der Zeijst, A silent regulatory gene *cfaD'* on CFA/I region 1 in enterotoxigenic *Escherichia coli*. Microb. Pathogen. in press (1990).

14. D.K. Hawley and W.R. McClure, Compilation and analysis of *Escherichia coli* promoter DNA sequences. Nucleic Acids Res. 11:2237 (1983).

15. P.H.M. Savelkoul, G.A. Willshaw, M.M. McConnell, H.R. Smith, A.M. Hamers, B.A.M. van der Zeijst and W. Gaastra, Expression of CFA/I fimbriae is positively regulated. Microb. Pathogen. 8:91 (1990).

16. C.O. Pabo and R.T. Saurer, Protein-DNA recognition. Annu. Rev. Biochem. 53:293 (1984).

17. J. Caron and J.R. Scott, A *rns* like regulatory gene in CFA/I that controls expression of CFA/I pilin. Infect. Immun. 58:874 (1990).

18. G.A. Willshaw, H.R. Smith, M.M. McConnell, W. Gaastra, A. Thomas, M. Hibberd and B. Rowe, Plasmid-encoded production of coli surface-associated antigen 1 (CS1) in a strain of *Escherichia coli* serotype O139.H28. Microb. Pathogen. in press (1990).

19. G.A. Willshaw, M.M. McConnell, H.R. Smith and B. Rowe, Structural and regulatory genes for coli surface associated antigen 4 (CS4) are encoded by separate plasmids in enterotoxigenic *Escherichia coli* strains of serotype O25.H42. FEMS Microbiol. Lett.68:255 (1990).

20. H-M. Wu and D.M. Crothers, The locus of sequence-directed and protein-induced DNA bending. Nature 308:509 (1984).

FUNCTION AND MOLECULAR ARCHITECTURE OF E.COLI ADHESINS

Heinz Hoschützky, Thomas Bühler, Ralph Ahrens
and Klaus Jann

Max-Planck-Institute for Immunobiology
D-7800 Freiburg, FRG

INTRODUCTION

The pathogenicity of E.coli has been ascribed to different virulence factors such as toxins, cytolysins, serum resistance, O- and K-antigens or adhesive properties. The adherence of the bacteria to epithelial surfaces - e.g. cell membranes and/or mucosal surfaces - is an important early event in host parasite interactions leading to bacterial colonization and infection (1). The term adhesion as used in this context describes a specific interaction of recognition proteins attached to the bacterial surface with complex carbohydrate moieties of glycoproteins and/or glycolipids on mammalian cells. These recognition proteins (also termed adhesins, hemagglutinins or lectins) may have different appearances in the electron microscope. Structures that can be demonstrated directly by negative staining procedures have been termed fimbriae or pili (rigid, 5 - 7 nm diameter) and fibrillae (flexible, 2 - 3 nm diameter). Nonfimbrial adhesive structures can only be visualized after stabilization with specific antibodies and then have a capsule like appearance (2). It should be noted that the distinction between fimbrial and nonfimbrial adhesins may be arbitrary and may in fact be due to the limitations of electronmicroscopic resolution. In this article we describe the molecular architecture of both fimbrial and nonfimbrial adhesins expressed by pathogenic E.coli bacteria.

FIMBRIAE ASSOCIATED ADHESINS (FAC)

Genetic studies on P-, S-, and type I-fimbriae have revealed that the phenotypes of fimbriation and adhesiveness are enco-

Molecular Pathogenesis of Gastrointestinal Infections
Edited by T. Wådstrom *et al.*, Plenum Press, New York, 1991

ded by complex gene clusters and determined by different genes (for review see 3). For example, F13 (pap) fimbriae consist of a major subunit (papA), forming the fimbrial rod, and three minor proteins (papE, papF, and papG). We have purified P- and S-fimbriae and could detect the genetically predicted major and minor proteins by SDS-PAGE analysis (4,5,6). The purified fimbriae adhered to eucaryotic cells in the pH range 5 - 9 as demonstrated by immunofluorescence and a hemadhesion assay. In contrast, hemagglutination could not be observed at neutral pH. At pH values below 6 as well as in the presence of divalent cations or specific antibodies, hemagglutination could readily be demonstrated. We interpret these data, that near neutrality the fimbriae are monovalent with respect to their interaction with the corresponding carbohydrate receptors: they can adhere to eucaryotic cells but are not able to agglutinate them. At lower pH the fimbriae aggregate due to the isoelectric precipitation as can be followed by turbidity measurements. The aggregates are polyvalent and thus able to agglutinate cells.

We have developed a procedure to purify the minor and major proteins of P- and S-fimbriae (4,5). The respective adhesins of both fimbrial types could be identified as minor proteins attached to the fimbrial rod. The adhesins were identified by both receptor binding studies and with the help of neutralizing monoclonal antibodies.

Monoclonal antibodies specific for the S- or P-adhesins were used for a topographical localization of the adhesin within the fimbrial structure. For both fimbrial types it could be shown that the adhesins are predominantly located at the tips of the fimbriae. In addition in the case of F7-fimbriae, we could also show that these fimbriae carry their adhesin associated in a complex with one of the other minor subunits (fsoE) as observed in purified fimbriae (7). The immunocytochemical data are consistent with the experimental observation that the fimbrial rod serves as a supporting structure for the adhesin.

FIMBRIAE CONSISTING OF ADHESIVE SUBUNITS

The fimbriae of the uropathogenic E.coli strain 20215 (SS142 fimbriae, also termed α3000 fimbriae) and of the enterotoxigenic E.coli strain 21569 (CFAI fimbriae) have a different molecular architecture than the fimbriae described above. Both strains express rigid fimbriae with a diameter of about 5 - 7 nm consisting of a major and several minor subunits. The biochemical analysis of these fimbriae revealed that in contrast to S-, P-, and type I-fimbriae, not a minor subunit but the major subunit is the mediator of the adhesive properties.

SS142-FIMBRIAE

The SS142-fimbriae consist of a major (16 kD) and two minor subunits (17 kD and 28 kD). pH dependent hemagglutination studies with purified fimbriae indicated that these fimbriae behave as polyvalent polymers. The SS142-fimbriae could be completely depolymerized at elevated temperatures and the isolated monomeric major subunit retained receptor binding activity as could be demonstrated by an indirect agglutination assay. Since these fimbriae behave as polyvalent polymers in agglutination assays and the major subunit is the mediator of the adhesive properties we tried to localize the receptor binding sites in intact fimbriae. As a molecular probe we used ganglioside GM2 (receptor analogue) -gold complexes for immunogold-labelling of the SS142-fimbriae. In contrast to P- and S-fimbriae the receptor binding sites could not only be demonstrated at the tips but also alongside the fimbrial rods. This indicated that this type of fimbriae is built up by the adhesin as major subunit and that the accessibility of the receptor binding site is not affected by the assembly into native fimbriae (Hoschützky et al., in prep.).

CFA I-FIMBRIAE

The CFA I-fimbriae have a molecular organization quite similar to the SS142-fimbriae described above. The fimbrial polymer can be depolymerized at elevated temperatures and the isolated monomeric major subunit still exhibits receptor recognizing activity. CFA I-fimbriae behave in pH dependent hemagglutination assays as monovalent polymers (comparable to P- and S-fimbriae). These data indicate that the major subunit is the mediator of the adhesive properties, but that in contrast to SS142-fimbriae during the assembly of the major subunit into the fimbrial rod the receptor binding sites of the internal subunits are blocked, leaving only the terminal subunit accessible for the interaction with receptors. This is underlined by the observations, that both neutralizing antibodies (specific for the major subunit) and GM2 (receptor analogue) gold particles bind predominantly to the tips of CFA I-fimbriae as could be demonstrated by the immunogold labelling technique (Hoschützky et al., in prep.).

NONFIMBRIAL ADHESINS (NFA)

Adhesive E.coli bacteria which do not express fimbriae have been known for a long time. The recognition proteins surround the bacterial cell like a capsule and can only be demonstrated in the electron microscope after stabilization with specific antibodies. We have purified nonfimbrial adhesins from several E.coli strains (2,8,9). Hemagglutination induced by the purified NFAs is not dependent on pH (pH range 4 - 8), indicating polyvalency of the NFAs.
The apparent molecular weight of these adhesins is in excess of 10^7 D. Light scattering studies indicated that these adhesins are not organized into an amorphous structure but that they form highly ordered fibrillar-like polymers with a linear mass density of about 8000 D/nm. These polymers consist of a major subunit analogous to the fimbriae described above.
At elevated temperatures and/or in the presence of detergents the polymeric NFAs completely dissociate into their subunits which could be purified by chromatography. It could be shown that the isolated monomeric major subunit still has the capability to adhere to human red cells and that several monoclonal antibodies specific for the major subunit have neutralizing capacity (Hoschützky et al., in prep.). We interpret

these data that the major subunit of the analyzed nonfimbrial adhesins carries the receptor binding site. Analogous to the SS142-fimbriae, the receptor recognition site is not blocked during the assembly of the subunits into the adhesive super-structures.

DEPOLYMERIZATION AND REPOLYMERIZATION OF ADHESIVE SUBUNITS

Biochemical analysis of the purified fimbrial and nonfimbrial adhesins described above revealed that all adhesive polymers consist of a major and several minor subunits. These subunits assemble into the polymer by non-covalent interactions. Studies on the thermal stability of E.coli adhesins revealed that in the case of P-, S-, and type I-fimbriae the fimbrial rod remained intact and that the adhesin and the other minor

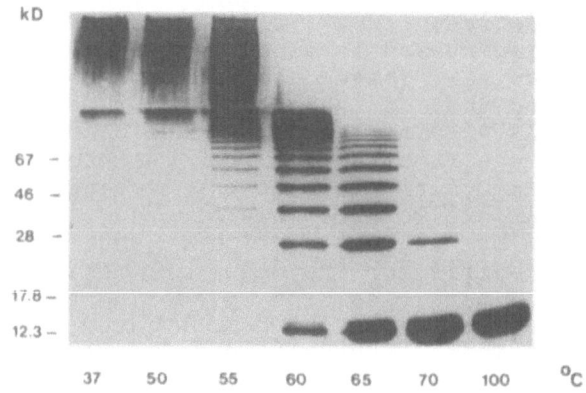

Fig. 1. Depolymerization of SS142-fimbriae

subunits dissociate from the polymer (4,5,6). All other adhesive polymers (both fimbrial and nonfimbrial) can be dissociated completely into monomers at elevated temperatures. These polymers carry the recognition protein as the major subunit, indicating that the molecular interactions between adhesive and nonadhesive major subunits are quite different. The depolymerization of polymers (consisting of adhesive subunits) into oligomers and monomers of the major subunit can be observed in the temperature range of 50 - 90 C (Fig. 1).
Isolated monomers tend to reassociate into oligomers and short polymers under defined reaction conditions. This repolymerization cannot be detected if the major subunit does not contain an intramolecular disulfide bridge or if such a SS-bridge was reduced by mercaptoethanol (Fig. 2). In addition it could be shown that one need at least tetramers to induce the agglutination of human red cells.

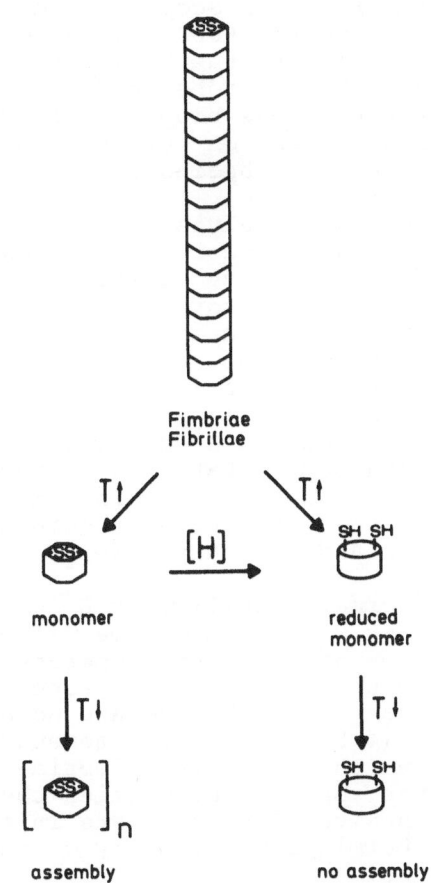

Fig. 2: Dissociation - Association
of Fimbriae/Fibrillae

DISCUSSION

The adhesive properties of E.coli bacteria are mediated by recognition proteins attached to the bacterial outer membrane. These adhesins have been characterized by serology, morphology (appearance in the electron microscope) and by their carbohydrate specificity. In this article we introduce a new view to describe E.coli adhesins: the molecular architecture of adhesive polymers.
In all cases studied so far the adhesins are organized in highly ordered polymeric structures. Light scattering studies on purified adhesins indicated that the morphological appearance in the electron microscope depends on the linear mass density of these polymers: 60000 D/nm for P-fimbriae (7 nm diameter), 20000 D/nm for the K88-fibrillae and about 8000 D/nm for the nonfimbrial adhesin NFA-1 (Ahrens et al., in prep.). We interpret these biophysical data in such a way that a distinction (based on electron microscopy) between fimbrial and nonfimbrial adhesins is tentative.

All adhesive complexes analyzed so far consist of a major subunit that is in a at least 100fold excess over several minor subunits. The adhesive polymer is built up by these subunits via non-covalent protein-protein interactions. The first hints for a differential molecular organization of these polymers came from the observation that one group of adhesins has at its isoelectric point a much higher agglutination titer than at neutral pH. Another group of adhesins exhibited no pH dependence concerning their biological activity. These data may be taken as an indicator for monovalency or polyvalency of the adhesins: monovalent (adhesive but not agglutinating) adhesins aggregate due to the isoelectric precipitation and are thus able to agglutinate eucaryotic cells.
Studies on the thermal stability of E.coli adhesins revealed that in the case of S-, P-, and type I-fimbriae the fimbrial rod remained intact at elevated temperatures while the adhesin together with the other minor proteins dissociates from the polymer leading to nonadhesive and nonagglutinating polymeric structures (4,5). Since both genetic (3) and biochemical data have shown that in these fimbriae a minor subunit is the mediator of the adhesive properties these data on thermal stability and adhesive function are in agreement with the experimentally determined monovalency of the adhesive fimbriae.

We have compared the molecular architecture of P-, S-, type I-, SS142-, and CFA I-fimbriae and of several nonfimbrial adhesins. Of these polymers, only the P-, S-, and type I-fimbriae carry distinct adhesins (minor proteins) predominantly at their tips. The fimbrial rod - built up by the major subunit - serves as supporting structure. Thus, these fimbriae are monovalent concerning their adhesive function. In contrast, the major subunit of all other polymers described is the actual mediator of the adhesive properties. The assembly of these subunits into the adhesive superstructures may arrange the receptor binding sites in such a way that only the terminal subunit (monovalent polymers: CFA I) or all subunits (polyvalent: SS142, NFAs) are accessible for receptor recognition.

76

Biochemical and biophysical analysis of E.coli adhesins has shown that there is no correlation between morphology, carbohydrate specificity and molecular architecture. Light scattering studies indicated that both fimbrial and nonfimbrial adhesins can be simply described as linear polymers of protein subunits that arrange in superstructures that may or may not be detectable in the electron microscope. The varying properties of the adhesins are due to a differential arrangement and function of major and minor subunits. Thus, some adhesive structures can be described as polymers of non-adhesive major subunits with attached terminal minor subunits as adhesins. Other can be described as polymers of adhesive subunits. In addition, these adhesins may differ in the accessibility of receptor binding sites within the polymer, leading to mono- or polyvalent adhesive superstructures.

Acknowledgement

This work was supported by the Deutsche Forschungsgemeinschaft (DFG).

LITERATURE

(1) Finlay, B.B. and Falkow, S. (1989). Common themes in microbial pathogenicity. Microbiol. Rev. 53:210-230.

(2) Jann, K. and Hoschützky, H. (1990). Nature and organization of adhesins. Curr. Top. Microbiol. Immunol. 151:55-70.

(3) Hacker, J. (1990). Genetic determinants coding for fimbriae and adhesins of extra-intestinal Escherichia coli. Curr. Top. Microb. Immunol. 151:1-28.

(4) Moch, T., Hoschützky, H., Hacker, J., Kröncke, K.D. and Jann, K. (1987). Isolation and characterization of the α-sialyl-β-2,3-galactosyl-specific adhesin from fimbriated Escherichia coli. Proc. Natl. Acad. Sci. USA 84:3462-3466.

(5) Hoschützky, H., Lottspeich, F. and Jann, K. (1989). Isolation and characterization of the α-galactosyl-1,4-β-galactosyl-specific adhesin (P adhesin) from fimbriated Escherichia coli. Infect. Immun. 57:76-81.

(6) Schmoll, T., Hoschützky, H., Morschhäuser, J., Lottspeich, F., Jann, K., and Hacker, J. (1989). Analysis of genes coding for the sialic acid-binding adhesin and two other minor fimbrial subunits of the S fimbrial adhesin determinant of Escherichia coli. Mol. Microbiol. 3:1735-1744.

(7) Riegman, N., Hoschützky, H., van Die, I., Hoekstra, W., Jann, K. and Bergmans, H. (1990). Immunochemical analysis of P-fimbrial structure: localization of the minor subunits and the influence of the minor subunit FsoE on the biogenesis of the adhesin. Mol. Microbiol. 4:1193-1198.

(8) Goldhar, J., Perry, R., Golecki, J.R., Hoschützky, H., Jann, B. and Jann, K. (1989). Non-fimbrial, mannose-resistant adhesins from uropathogenic _Escherichia coli_ O83:K1:H4 and O14:K?:H11. Infect. Immun. 55:1837-1842.

(9) Hoschützky, H., Nimmich, W., Lottspeich, F. and Jann, K. (1989). Isolation and characterization of the nonfimbrial adhesin NPA-4 from uropathogenic _Escherichia coli_. Microbiol. Pathogen. 6:351-359.

NEWLY CHARACTERIZED PUTATIVE COLONIZATION FACTORS OF HUMAN ENTEROTOXIGENIC

ESCHERICHIA COLI

Moyra M. McConnell

Division of Enteric Pathogens
Central Public Health Laboratory, Colindale
London NW9 5HT, U.K.

INTRODUCTION

Enterotoxigenic *Escherichia coli* (ETEC) are an important cause of diarrhoeal disease in infants in developing countries and in travellers to these countries. These *E.coli* belong to a large number of different serotypes but have common pathogenic mechanisms namely the ability to adhere to the epithelial surface of the small intestine by means of colonization factors and to produce a heat stable (ST) or a heat labile (LT) enterotoxin, or both. Animal studies and volunteer experiments have shown that colonization factors and some toxins induce protective immunity and could be used to make vaccines[1]. However the effectiveness of a vaccine incorporating colonization factors will depend on the identification of these antigens on the majority of ETEC strains which cause diarrhoea in man.

Colonization factors are usually fimbriae, most of which confer on *E. coli* the ability to agglutinate red blood cells from a number of animal species in the presence of mannose[1]. The best characterized of the human colonization factor antigens (CFAs) are CFA/I[2], CFA/II[3] and CFA/IV, formerly designated PCF8775[4]. CFA/I is a single antigen with a rod-like fimbrial structure of 6-7nm in diameter, while CFA/II and CFA/IV are both groups of antigens. Strains producing CFA/II possess one or other of the fimbrial antigens, coli surface-associated antigen CS1 or CS2 and a fibrillar antigen CS3, or may possess CS3 alone[5,6,7]. ETEC producing CFA/IV have either one of the fimbrial antigens CS4 or CS5, as well as the antigen CS6 which may have a fine fibrillar structure[4,8]. ETEC which produce CS6 with other fimbriae have also been identified[8]. CS1,CS2 and CS4 are rigid rod-like fimbriae similar to CFA/I[4,7]. CS5 fimbriae are semi-rigid and have a helical structure of about 5nm in diameter[8,9]. The adhesive and colonizing ability of strains carrying these fimbriae were demonstrated by the following further tests: adhesion to human enterocytes isolated from duodenal biopsy material[10], colonization in the reversible intestinal tie adult rabbit diarrhoea (RITARD) model[11,12] and attachment to membrane proteins derived from human and rabbit intestinal brush borders[13].

A number of surveys have been carried out to determine the prevalence of ETEC producing CFA/I, CFA/II and CFA/IV in particular areas. The reported prevalence of these colonization factors has varied from 29-40% of feacal isolates of ETEC in Thailand[14,15], 34-46% in Mexico[16],(Y.Lopez-Vidal,PhD thesis, University of Goteborg,1990) to 64-75% in Bangladesh[15,17]. The proportions of the ETEC producing each colonization factor varied in the different geographic regions. In Thailand and Bangladesh the largest proportion of ETEC strains produced CFA/I while in Mexico the numbers of CFA/I and CFA/IV producing ETEC were approximately equal. Very few CFA/IV producing strains were found in Thailand. The proportion of CFA/II producing strains varied from 5% in Mexico to 18% in Bangladesh. ETEC without detectable colonization factors were found in all three regions (71-25% of total ETEC). Many of these *E.coli* produced LT only and belonged to a large variety of serotypes, but ST^+ and ST^+LT^+ *E.coli* of a limited number of serotypes were also found. Therefore, if a vaccine is to be produced which would be effective against ETEC in these areas, new colonization factor antigens would need to be identified on these strains.

CFA/III AND PCFO159:H4 (PCFO159)

Two putative colonization factors (PCFs) have been reported which were not looked for in the surveys described above; CFA/III[18] and PCFO159[19]. These factors have properties in common with the previously described colonization factors, but have not been extensively tested in animals. Both were single rod-like fimbrial antigens (CFA/III, diameter 7-8nm and PCFO159, diameter 6-7nm), which did not confer on the strain the ability to cause mannose-resistant haemmagglutination (MRHA). CFA/III was first identified because of the hydrophobicity of the type strain: this property is often associated with the production of fimbriae. The type strain, of serotype O25:H-, was LT^+ and produced CS6[18,20]. The type strain producing PCFO159 was an ST^+LT^+ strain of serotype O159:H4. We have tested, by ELISA, 217 ETEC of 50 different serogroups isolated mainly in South East Asia but also in Europe, South America, Africa and the United States, for CFA/III and PCFO159 [20]. CFA/III was detected only in LT^+CS6^+ strains of serotypes O25:H16 and O25:H- and PCFO159 was found only in ST^+LT^+ strains of serotypes O159:H4 and O159:H20. The reason for the apparently limited distribution of these factors is not known. Both are encoded by plasmids; the PCFO159-encoding plasmid was transferred into *E. coli* K12 in which the fimbriae were expressed.

IDENTIFICATION AND CHARACTERIZATION OF NEW PUTATIVE COLONIZATION FACTORS

After the survey described above we still had many ETEC without known colonization factors. We have identified putative colonization factors on some of these strains by looking for those which produced fimbriae with some of the characteristics of known colonization factors. These properties included; the ability to confer MRHA on the strain, hydrophobicity, expression at 37C but not at 20C and a fimbrial subunit size of 14-30kDa. Previously identified fimbriae were encoded by plasmids which frequently encoded enterotoxin production. Putative colonization factors were further characterized for adhesive ability using human enterocytes,rabbit and human intestinal brush borders and the RITARD model.

The composition of the growth medium can be very important for the expression of colonization factors. CFA agar was developed by Evans and Evans[3] as a medium which allowed good expression of CFA/I and CFA/II. Growing strains on MacConkey agar before subculturing them onto CFA agar (L.Thomas, PhD Thesis, CNAA, 1985) or incorporating bile salts into the CFA agar[21] was also shown to improve the expression of the MRHA property of some human colonization factors. When we cultured LT^+ strains of serogroup O114 and ST^+ strains of serogroup O166 on CFA agar with bile salts, strains of both types which previously had been MRHA⁻ became MRHA⁺, though the haemagglutination patterns of the two strains were different[21,22]. The properties of the fimbriae of these strains and of two other putative colonization factors, CS7 and PCFO9 will be described.

CS17

By electron microscopy, rod-like fimbriae of diameter 6-7nm were seen on an LT⁺ strain of serotype 0114:H21[24]. The properties of these fimbriae, which we have called CS17, are listed in the table. CS17 fimbriae were antigenically distinct by enzyme-linked immunosorbent assay (ELISA) from previously described colonization factors. When they were denatured and examined on sodium dodecyl sulphate polyacrylamide gels (SDS-PAGE) a polypeptide of 17.0 kDa was seen which reacted with antiserum that had been raised against the MRHA positive 0114:H21 strain and had been absorbed by a MRHA negative derivative of the strain to make it specific for the plasmid-encoded CS. Western blotting also showed that this polypeptide like the fimbrial subunits of CFA/I, CS1 and CS2 reacted with an absorbed antiserum prepared against a strain producing CS4[25]. The fimbrial subunits of these five antigens appear to share common epitopes although the intact fimbriae are antigenically distinct. A strain producing CS17 has now been shown to colonize rabbits in the RITARD model and to attach to membrane proteins from rabbit intestinal brush borders (A.-M. Svennerholm and C. Wenneras, personal communication). CS17 has been identified on LT⁺ strains of serotypes 08:H9, 015:H-, 048:H26, and 0146:H19[23].

PCFO166

Another putative colonization factor was identified on a ST⁺ strain of serotype 0166:H27[21] . PCFO166 fimbriae were antigenically distinct from the previously described colonization factors. The fimbriae were morpho-logically similar to CFA/I fimbriae (Table). By SDS-PAGE two polypeptides of 15.5 and 17.0kDa were seen. It is more common for fimbriae to consist of one major subunit protein however two polypeptides were also seen when CS3 fimbriae were examined by SDS-PAGE[7]. By ELISA, PCFO166 was detected on ETEC of serogroups 020, 071, and 098 as well as other ST⁺ strains of serogroup 0166. Initially the adhesive properties of PCFO166 seemed similar to other colonization factors in that strains producing this factor gave MRHA of human and bovine erythrocytes and adhered to human enterocytes. However a PCFO166⁺ strain was unable to colonize rabbits in the RITARD model or to attach to membrane proteins from rabbit intestinal brush borders (A.-M. Svennerholm and C. Wenneras, personal communication) suggesting that if PCFO166 was a human colonization factor the attachment of the fimbriae to human enterocytes involved different receptors to those used by CFA/I, CFA/II and CFA/IV. Further experiments with human intestinal cells and ideally volunteer experiments are needed to confirm the status of PCFO166 as a human virulence factor.

CS7

In the late seventies attachment fimbriae of three different types were identified by Deneke et al.,[24] Tests in our laborarory (L. Thomas, PhD Thesis, CNAA, 1985) showed that two of the attachment fimbriae reacted with CFA/I or CFA/II antisera. However the attachment fimbriae of strain 334, an ST⁺LT⁺ strain of serotype 015:H11 were antigenically distinct by immunodiffusion from CFA/I and CS1 to CS6. The properties of the fimbriae of strain 334, which we have called CS7, are listed in the table[25]. CS7 fimbriae were similar to CS5 fimbriae in the MRHA pattern they conferred, their helical structure and the fimbrial subunit size[4,9,25]. By immuno-electron microscopy no cross-reaction was detected between intact CS5 and

Table 1. PROPERTIES OF NEWLY CHARACTERIZED COLONIZATION FACTORS OF ETEC

	CS17	PCFO166	CS7	PCFO9
MRHA	Bovine	Bovine Human	Bovine Human Guinea Pig	Human Chicken
Hydrophobicity	Strong	Strong	Strong	Weak
Fimbrial Morphology	Rod-like 6-7nm	Rod-like 6-7nm	Helical 3.5-6.5nm	Fibrillar
Molecular size of fimbrial subunit	17.0kDa	15.5kDa 17.0kDa	21.5kDa	27kDa
Attachment to human enterocytes	NT	+ve	+ve	NT
RITARD model	+ve	-ve	+ve	NT
Genetic location	Plasmid	Plasmid	Plasmid	Plasmid
Linkage to enterotoxin gene	LT	ST	ST-LT LT	NONE
Hybridization[a] of plasmid to *cfaD* probe	-ve	S	W	-ve

All the fimbriae were produced at 37C but not at 20C

a   S = strong hybridization, W = weak hybridization compared to the hybridization of the *cfaD* probe to a CFA/I-producing control strain NT = not tested

Data taken from references 21,22,24 and 25

CS7 fimbriae or vice-versa. However Western blotting showed that the fimbrial subunits of CS7 and CS5 reacted with both absorbed anti-CS7 and anti-CS5 serum. It appeared that the subunits of CS7 and CS5 fimbriae shared common epitopes which were not available to react on intact fimbriae. A cross-reaction was also identified in ELISAs where the reaction of the CS7-producing strain 334 with the heterologous CS5 antiserum was less than half the value of the reaction with the homologous antiserum. This may be due to some denaturing of the fimbriae by heating when the antigens were prepared which produced subunits. A CS7 producing strain was able to colonize rabbits in the RITARD model and attached to membrane proteins derived from rabbit intestinal brush borders (A.-M. Svennerholm and C. Wenneras, personal communication).By ELISA, CS7 fimbriae were detected on LT[+] producing *E.coli* of serogroups 0103 and 0114.

PCF09

A fourth newly characterized adhesin, PCFO9 has been described by Heuzenroeder et al.,[26]. This was identified on a LT$^+$ strain of serotype O9:H- isolated from a case of infant diarrhoea in Central Australia. Electron microscopy has shown the fimbriae to be of the fibrillar type similar to CS3 with a major fimbrial subunit of approximately 27kDa (table). We have identified PCFO9 on LT$^+$ strains of serotypes O143:H43 and O7:H- from Peru (unpublished).

IDENTIFICATION OF REGULATORY SEQUENCES ON PCF AND CS-ENCODING PLASMIDS

Production of CS1 or CS2 by *E. coli* of serotype O6:H16 is regulated by a sequence termed *rns* (regulation of CS1 expression) the product of which is a DNA binding protein[27]. The *rns* gene is located on a plasmid while the structural genes appear to be chromosomal. CFA/I production is also positively regulated but in this case the operons containing the fimbrial subunit gene *cfaB* and the regulatory sequence *cfaD* are located on separate regions of the same plasmid (CFA/I regions 1 and 2)[28,29]. Cloned *cfaD* was used to make a probe to test for the presence of potential regulatory sequences in ETEC producing CS17, PCFO166, CS7 and PCFO9. These strains all contain plasmids encoding fimbrial production (table). Plasmid DNA was prepared from these strains and tested for hybridization with the *cfaD* probe by Southern blotting (M.Hibberd, M.M.McConnell, G.A.Willshaw, H.R.Smith and B.Rowe, unpublished). A PCFO166-ST plasmid hybridized strongly with this probe, a CS7-LT plasmid hybridized weakly, while the CS17-LT and PCFO9 plasmids did not hybridize (table).

The ETEC with the new colonization factors were tested for the presence of functional regulatory gene sequences using a clone containing the fimbrial subunit gene *cfaB*. Transformation of this cloned *cfaB* gene into a wild-type strain producing PCFO166 resulted in expression of CFA/I as shown by MRHA, ELISA and Western blot analysis. The amount of CFA/I produced was similar to that produced by a wild-type CFA/I positive strain. There was no expression of CFA/I when a PCFO166-ST negative derivative strain was tested in a similar way. This suggested that the PCFO166-ST plasmid with a *cfaD* hybridizing sequence could functionally substitute for the *cfaD* gene sequence. Similar transformation experiments were carried out with colonization factor positive and negative colonies of ETEC producing the three other factors. Transformants of all three strains which had a colonization factor encoding plasmid expressed CFA/I while transformants of the plasmid negative derivatives did not. Transformants with the CS7-LT and PCFO9 plasmids gave as good expression of CFA/I as the control strain, while the transformant with the CS17-LT plasmid gave poorer expression, about half the value of the control strain. It appears that these plasmids contain gene sequences which can also positively regulate CFA/I production to some extent though they do not hybridize strongly with *cfaD*.

SURVEYS OF ETEC FOR NINE COLONIZATION FACTORS

We have tested, by ELISA, batches of ETEC from three different geographical areas for nine factors: CFA/I, CFA/II, CFA/III, CFA/IV, PCFO159, PCFO166, PCFO9, CS7 and CS17 (unpublished).By identifying these six extra factors other than CFA/I, CFA/II and CFA/IV the percentage of strains with a known factor was increased from 53% of ETEC to 71% in Burma, from 45 to 69% in central Africa, and 23 to 52% in Peru. Again there was variation in the type of colonization factor found in each area.

Although the proportion of ETEC with colonization factors was generally higher than in the previously described surveys, 29-48% of these ETEC carried no identifiable factor. As before the majority were LT producers of many different serotypes and it may be that some of these strains are not pathogenic. However this needs confirmation; it is likely that new factors will be identified on these strains.

REFERENCES

1. S.H.Parry and D.M.Rooke, Adhesins and colonization factors of *Escherichia coli. in*: "The virulence of *Escherichia coli*," M.Sussman, ed., Academic Press, Inc., London (1985).

2. D.G.Evans, R.P.Silver, D.J.Evans, D.G.Chase, and S.L.Gorbach, Plasmid-controlled colonization factor associated with virulence in *Escherichia coli* enterotoxigenic for humans. *Infect.Immun.*12:656 (1975).

3. D.G.Evans, and D.J.Evans,Jr., New surface-associated heat-labile colonization factor antigen (CFA/II) produced by enterotoxigenic *Escherichia coli* of serogroups 06 and 08. *Infect.Immun.*21: 638 (1978).

4. L.V.Thomas, M.M.McConnell, B.Rowe and A.M.Field, The possession of three novel coli surface antigens by enterotoxigenic *Escherichia coli* strains positive for the putative colonization factor PCF8775. *J. Gen. Microbiol.*131: 2319 (1985).

5. A.Cravioto, S.M.Scotland and B.Rowe, Hemagglutination activity and colonization factor antigens I and II in enterotoxigenic and non-enterotoxigenic strains of *Escherichia coli* isolated from humans. *Infect. Immun.* 36: 189 (1982).

6. C.J.Smyth, Two mannose-resistant haemagglutinins on enterotoxigenic *Escherichia coli* of serotype 06:K15: H16 or H- isolated from travellers' and infantile diarrhoea. *J. Gen. Microbiol.* 128: 2081 (1982).

7. M.M. Levine, P.Ristaino, G.Marley, C.Smyth, S.Knutton, E.Boedeker, R.Black, C.Young, M.L.Clements, C.Cheney and R.Patnaik, Coli surface antigens 1 (CS1) and 3 (CS3) of colonization factor antigen II positive enterotoxigenic *Escherichia coli*: morphology, purification and immune responses in man, *Infect. Immun.* 44:409 (1984).

8. S.Knutton, M.M.McConnell, B.Rowe, and A.S.McNeish, Adhesion and ultrastructural properties of human enterotoxigenic *Escherichia coli* producing CFA/III and CFA/IV. *Infect. Immun.* 57:3364 (1989).

9. P.A. Manning, G.D.Higgins, R.Lumb, and J.A.Lanser, Colonization factor antigens and a new fimbrial type, CFA/V, on 0115:H40 and H-strains of enterotoxigenic Escherichia coli in Central Australia. *J. Infect. Dis.* 156:841 (1987).

10. S.Knutton, D.R.Lloyd, D.C.A.Candy, and A.S.McNeish, Adhesion of enterotoxigenic *Escherichia coli* to human small intestinal enterocytes *Infect. Immun.* 48:824 (1985).

11. A.-M.Svennerholm, Y.Lopez-Vidal, J.Holmgren, M.M.McConnell and B.Rowe Role of PCF8775 antigen and its coli surface components for colonization, disease, and protective imunogenicity of enterotoxigenic *Escherichia coli* in rabbits, *Infect. Immun.* 56:523 (1988)

12. A.-M.Svennerholm, C.Wenneras, J.Holmgren, M.M.McConnell and B.Rowe, Roles of different coli surface antigens of colonization factor antigen II in colonization and protective immunogenicity of entero-toxigenic *Escherichia coli* in rabbits, *Infect. Immun.* 58:341 (1990)

13. C.Wenneras, J.Holmgren and A,-M.Svennerholm, The binding of colonization factor antigens of enterotoxigenic *Escherichia coli* to intestinal cell membrane proteins, *FEMS Microbiol.Lett.* 66:107 (1990).

14. S.Changchawalit, P.Echeverria, D.N.Taylor, U.Leksomboon, C.Tirapat B.Eampokalapand and B.Rowe, Colonization factors associated with enterotoxigenic *Escherichia coli* isolated in Thailand. *Infect. Immun.* 45:525 (1984).

15. M.M.McConnell,L.V.Thomas, N.P.Day and B.Rowe, Enzyme-linked immuno-sorbent assays for the detection of adhesion factor antigens of enterotoxigenic *Escherichia coli*. *J. Infect. Dis.* 152:1120 (1985).

16. A.Cravioto, R.E.Reyes, R.Ortega, G.Fernandez, R.Hernandez, and D.Lopez, Prospective study of diarrhoeal disease in a cohort of rural Mexican children: incidence and isolated pathogens during the first two years of life. *Epidem. Infect.* 101:123 (1988).

17. L.Gothefors, C.Ahren, B.Stoll, D.K.Barma. F.Orskov, M.A.Salek and A.-M.Svennerholm, Presence of colonization factor antigens on fresh isolates of fecal *Escherichia coli*: a prospective study, *J. Infect. Dis.* 152:1128 (1985).

18. T.Honda, M.Arita and T.Miwatani, Characterization of new hydrophobic pili of human enterotoxigenic *Escherichia coli*: a possible new colonization factor. *Infect. Immun.* 43:959 (1984).

19. C.O.Tacket, D.R.Maneval, and M.M.Levine, Purification, morphology, and genetics of a new fimbrial putative colonization factor of entero-toxigenic *Escherichia coli* O159:H4. *Infect. Immun.* 55:1063 (1987).

20. M.M.McConnell, and B.Rowe, Prevalence of the putative colonization factors CFA/III and PCFO159:H4 in enterotoxigenic *Escherichia coli*· *J. Infec. Dis.* 159: 582 (1989).

21. M.M.McConnell, H.Chart, A.M.Field, M.Hibberd, and B.Rowe, Characterization of a putative colonization factor (PCFO166) of enterotoxigenic *Escherichia coli* of serogroup O166. *J. Gen. Microbiol.* 135:1135 (1989)

22. M.M.McConnell, M.Hibberd, A.M.Field, H.Chart, and B.Rowe, Characterization of a new putative colonization factor (CS17) from a human enterotoxigenic *Escherichia coli* of serotype O114:H21 which produces only heat-labile enterotoxin. *J. Infect. Dis.* 161:343 (1990).

23. M.M.McConnell, H.Chart, and B.Rowe, Antigenic homology within human enterotoxigenic *Escherichia coli* fimbrial colonization factor antigens: CFA/I, coli-surface-associated antigens (CS)1, CS2, CS4 and CS17. *FEMS Microbiol .Lett.* 61:105 (1989).

24. C.F.Deneke, G.M.Thorne, and S.L.Gorbach, Serotypes of attachment pili enterotoxigenic *Escherichia coli* isolated from humans. *Infect. Immun.* 32:1254 (1981).

25. M.Hibberd,M.M.McConnell, A.M.Field and B.Rowe, The fimbriae of human enterotoxigenic *Escherichia coli* are related to CS5 fimbriae, *J. Gen. Microbiol.* In press (1991).

26. M.W.Heuzenroeder, T.R.Elliot, C.J.Thomas, R.Halter, and P.A.Manning, A new fimbrial type (PCFO9) on enterotoxigenic *Escherichia coli* O9:H-LT[+] isolated from a case of infant diarrhea in Central Australia. *FEMS Microbiol. Lett.* 66:55 (1990).

27. J.Caron, L.M.Coffield and J.R.Scott, A plasmid-encoded regulatory gene, *rns*, required for expression of the CS1 andCS2 adhesins of entero-toxigenic *Escherichia coli*, *Proc. Natl. Acad.Sci. U.S.*A 86:963 (1989).

28. G.A.Willshaw,H.R.Smith and B.Rowe, Cloning of regions encoding colonization factor antigen 1 and heat-stable enterotxin in *Escherichia coli FEMS Microbiol. Lett.* 16:101 (1983).

29. P.H.M.Savelkoul, G.A.Willshaw, M.M.McConnell, H.R.Smith, A.M.Hamers, B.A.M.van der Zeijst and W.Gaastra. Expression of CFA/I fimbriae is positively regulated, *Microbiol. Pathogen.* 8:91 (1990).

MOLECULAR BIOLOGY OF *ESCHERICHIA COLI* TYPE 1 FIMBRIAE

Per Klemm and Karen A. Krogfelt

Department of Microbiology, bld. 221, Technical University of
Denmark, DK-2800 Lyngby, Denmark

INTRODUCTION

Most strains of *Escherichia coli* are able to express
type 1 fimbriae. These are thread-like surface organelles
consisting of around 1000 subunits of a major structural
component, the FimA protein, as well as a few percent minor
components. Type 1 fimbriae mediate binding to D-mannose
containing structures and thereby enable the bacteria to
colonize various host tissues (1, 2). Inhibition of binding
of type 1 fimbriated bacteria as well as purified type 1
fimbriae to various cell types has been extensively studied.
In all cases it was found that D-mannose itself and most
derivatives of this sugar were very potent inhibitors of type
1 fimbriae mediated adhesion, whereas all saccharides not
containing D-mannose showed no inhibitory effect. It has
been proposed that the biological role of type 1 fimbriae
is to provide bacterial adhesion to mucus in the large intestine which is the natural habitat of *E. coli*. Furthermore,
several lines of evidence point to a role of type 1 fimbriae
as possible virulence factors in uropathogenic *E. coli* strains.

GENETIC BACKGROUND

All information required for the expression of type 1
fimbriae resides on a 9.5 Kb DNA fragment. The nucleotide
sequence of this fragment has been elucidated (3, 4, 5, 6,
in preparation), and consequently the total extend of and
number of genes in the *fim* gene cluster has been established.
In case of *Escherichia coli* K-12 the *fim* genes are located
at a map position on the chromosome corresponding to 98 min.
The *fim* gene cluster encompasses nine genes (Fig. 1), encoding
for regulation of expression, the export and bioassembly
machinery, as well as the actual structural components of
the fimbriae, including the actual adhesin.

PHASE VARIATION

The expression of type 1 fimbriae is phase variable i.e. a bacterial cell is either bald or fimbriated. The molecular basis of this process has only recently been determined to be transcriptionally regulated by virtue of the inversion of a 314 base pair segment of DNA. This segment is located immediately upstream of the *fimA* gene and contains a very potent promoter. The orientation of the invertible DNA-segment, or phase-switch, is determined by the products of the *fimB* and *fimE* genes. These presumably act as recombinases, conferring an "on" or "off" configuration of the switch, respectively (4). It has been shown that both the configuration as well as the rate of switching is dependent on the relative as well as total concentration of FimB and FimE (7). In addition to the recombinases, other host factors seem to be required for efficient phase switching to take place, notably IHF (8). Clinically isolated strains of *E. coli* also undergo phase variation by the same mechanism. Recently, we investigated the role of phase variation in vivo, by looking at the colonization ability of a phase-locked derivative of a human *E. coli* isolate (9). Surprisingly, the constitutive expression of type 1 fimbriae made it a poor colonizer of the large

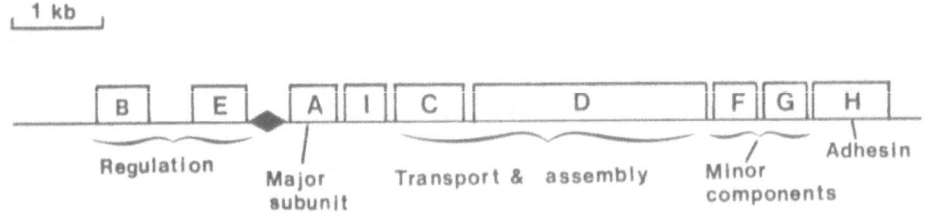

Fig. 1   The *fim* gene cluster as present on plasmid pPKL4. The legends refer to the roles of the various gene products. The diamond-shape indicates the phase-switch.

intestine compared to the parent strain. The phase-locked strain seemed to have difficulties in penetrating the mucus layer. This observation could shed some light on the biological role of phase variation of type 1 fimbriae.

THE ADHESIN AND MINOR COMPONENTS

Escherichia coli K-12 strain HB101 is non-fimbriated

and is unable to bind to guinea-pig erythrocytes or to other D-mannoside containing surfaces. When the *fim* gene cluster from *E. coli* K-12 strain PC31 was introduced into HB101 in the form of plasmid pPKL4 (Fig. 1), D-mannose binding fimbriae were produced, judged from the ability of the recombinant strain to agglutinate guinea-pig erythrocytes and to specificly adhere to Sepharose beads coated with D-mannose (5, 10). Such fimbriae contained four components viz. FimA, FimF, FimG and FimH (10). Introduction of plasmid pPKL5, a deletion of pPKL4 which misses the *fimF, fimG* and *fimH* genes, resulted in the production of pure FimA fimbriae which were devoid of mannose recognition specificity (5, 10). We have also described an HB101 host, containing two compatible plasmids, pPKL5 (*fimA*+) and pPKL53 (*fimF*+, *fimG*+), which produces fimbriae consisting of the FimA, FimF and FimG components and is unable to agglutinate guinea-pig erythrocytes, or to bind to D-mannose coated Sepharose beads (5, 10).

To establish that the FimH protein, apart from being a component of type 1 fimbriae, is uniquely responsible for the binding to D-mannose we made use of a modified Western blot, using D-mannose coupled to bovine serum albumin (BSA) as the primary reagent, and fimbriae purified from HB101 cells containing plasmid pPKL4 (10). Only the FimH protein was found to bind to D-mannose-BSA. As controls we used pure FimA fimbriae from HB101/pPKL5 and pure FimA, FimF and FimG fimbriae from HB101/pPKL5 + pPKL53 neither of which contain the FimH protein. No reaction was seen in either case. Since the D-mannose-BSA conjugate is specificly recognized by the FimH adhesin we also tried to localize the position of this protein in intact type 1 fimbriae. HB101 cells harboring plasmid pPKL4 were incubated with D-mannose-BSA, followed by anti-BSA rabbit serum and colloidal gold-labelled protein A. The cells were inspected by electron microscopy (Fig. 2). By using a direct binding assay to the natural receptor, D-mannose, for type 1 fimbriae we proved, that the FimH protein is the D-mannose specific adhesin and, that this adhesin is laterally located on the fimbriae at long intervals. In order to integrate the FimH protein in the fimbriae the two other minor components, i.e. FimF and FimG, are needed (5,10). We have previously proposed a model, based on stoichiometric data, which propose the adhesive complex to be integrated in the fimbria on the average once per every 100-200 FimA subunits (11). This would correspond to one complex on the average per 100-150 nm.
We have also studied the role of the minor fimbrial components for length regulation of the organelles. It was found that the number of minor component complexes determined the length and amount of fimbriae, presumably either acting as initiation complexes or by inducing fragile points in the fimbriae prone

to breakage (5).
Recently we showed that the major subunit protein, FimA, is dispensable for D-mannose specific adhesion. Only the genes encoding the adhesin, minor components as well as the export machinery seemed to be required (12). However, the adhesive potential was reduced 25-50 fold in the absence of the major structural gene.

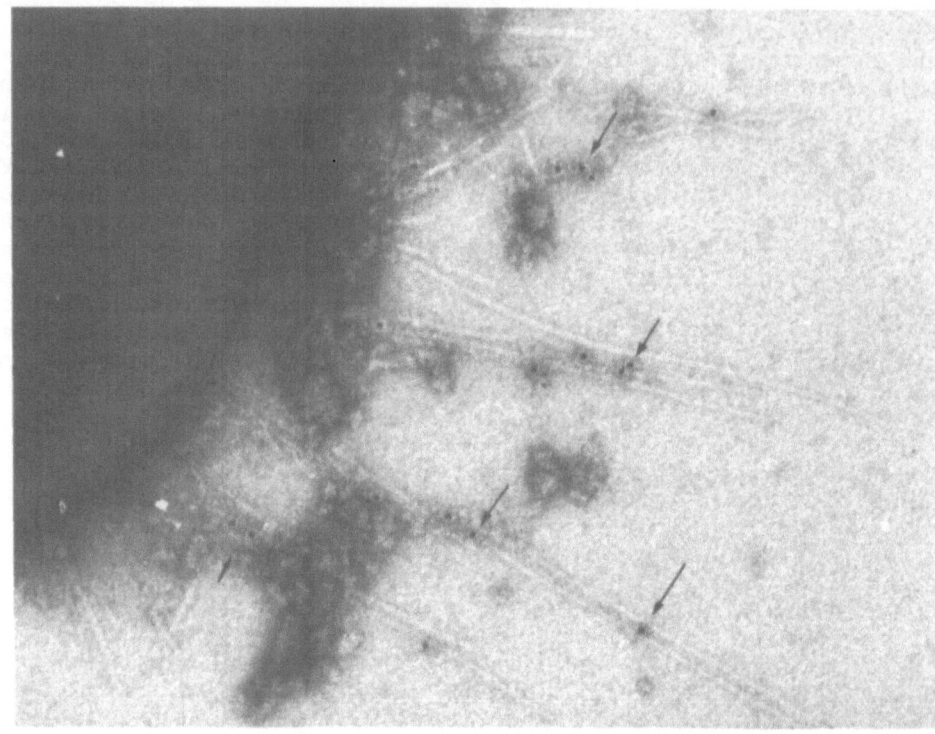

Fig. 2   Electron micrographs of *Escherichia coli* K-12 strain HB101/pPKL4 producing type 1 fimbriae after incubation with D-mannose-BSA. Binding was visualized by incubation with rabbit anti-BSA, followed by incubation with gold-labelled protein A. The localization of FimH is indicated by arrows. Bars indicate 0.25 μm.

THE EXPORT MACHINERY

The products of two genes, i.e. *fimC* and *fimD*, have been found to be required for assembly and surface localization of type 1 fimbriae. The FimC protein was found to be located in the periplasm, and is presumably acting as a shuttle for incorporation of structural elements into the fimbriae (manuscript in preparation). The FimD protein was found to be an outer membrane protein, and is probably acting as a polymerization channel. However, curiously, deletion of the *fimD* gene did not completely abolish the production of fimbriae, but reduced the number by a factor of roughly 1000 (6).

APPLICATIONS OF THE *fim* GENES

The phase switch located upstream of the *fimA* gene ensures a stochastic expression of type 1 fimbriae. We have employed this biological "flip-flop" mechanism for construction of a stochastically induced suicide cassette for biological containment of bacteria (13). The cassette contains two elements a) a small gene, *hok*, encoding a 52 amino acid polypeptide, which kills the cell when expressed; and b) the phase switch from the *fim* gene cluster and the two regulatory genes *fimB* and *fimE*. The Hok protein is lethal to the cell when expressed. The configuration of the phase switch in this constellation determines whether the cell is dead or alive instead of bald or fimbriated. One can envisage this system used, among other things, for containment of live genetically-engineered vaccines.

Another applicative use of the *fim* system is the construction of chimeric type 1 fimbriae containing foreign peptide epitopes. Fimbriae are very good immunogens, and since many fimbriae exist in several antigenic variants one might go a step further and insert totally foreign epitopes into the major structural protein. In fact we have localized three positions in the FimA protein where such inserts can be made without interfering with the normal fimbrial phenotype. Chimeric fimbriae have been made containing a range of inserts encoding various epitopes from pathogenic viral surface proteins. In several cases the chimeric proteins can be recognized by sera directed against the parent viral protein (14). One can envisage chimeric fimbriae as a new antigen-delivery system. Based on the results obtained so far we think that the system has a lot of potential.

REFERENCES

1.   Duguid, J. P., and Old, D. C. 1980. Adhesive properties of *Enterobacteriaceae, p. 184-217.in* : Bacterial adherence, receptors and recognition, series B, vol 6. E. H. Beachey, ed., Chapman and Hall, London.

2.   Klemm, P. Fimbrial adhesins of *Escherichia coli*. Rev. Infect. Dis. 7: 321-340 (1985).

3.   Klemm, P. The *fimA* gene encoding the type 1 fimbrial subunit of *Escherichia coli*, nucleotide sequence and primary structure of the protein. Eur. J. Biochem. 143:395-399 (1984).

4.   Klemm, P. Two regulatory *fim* genes, *fimB* and *fimE*, control the phase variation of type 1 fimbriae in *Escherichia coli*. EMBO J. 5:1389-1393 (1986).

5.   Klemm, P., and Christiansen, G. Three *fim* genes required for the regulation of length and mediation of adhesion of *Escherichia coli* type 1 fimbriae. Molec. Gen. Genet. 208:439-445 (1987).

6.   Klemm, P. and Christiansen, G. The *fimD* gene required for cell surface localization of *Escherichia coli* type 1 fimbriae. Molec. Gen. Genet. 220:334-338 (1990).

7.   Pallesen, L., Madsen, O. and Klemm, P. Regulation of the phase switch controlling expression of type 1 fimbriae in *Escherichia coli*. Molec. Microbiol. 3: 925-931 (1989).

8. Eisenstein, B.I., Sweet, D.F., Vaughn, V. and Friedman, D.I. Integration host factor IHF is required for the DNA inversion controlling phase variation in *Escherichia coli*. Proc. Natl. Acad. Sci. USA. <u>84</u>: 6506-6510 (1987).

9. McCormick, B.A., Krogfelt, K.A., Klemm, P., Laux, D.C. and Cohen, P. Overproduction of type 1 fimbriae by a human fecal *Escherichia coli* strain makes it a poor colonizer of the streptomycin-treated mouse large intestine. Infect. Immun. submitted for publication (1990).

10. Krogfelt, K.A., Bergmans, H. and Klemm, P. Direct evidence that the FimH protein is the mannose-specific adhesin of *Escherichia coli* type 1 fimbriae. Infect. Immun. <u>58</u>: 1995-1998 (1990).

11. Krogfelt, K. A., and Klemm, P. Investigation of minor components of *Escherichia coli* type 1 fimbriae: protein chemical and immunological aspects. Microb. Pathogen. <u>4</u>:231-238 (1988).

12. Klemm, P., Krogfelt, K. A., Hedegaard, L. and Christiansen, G. The major subunit of *Escherichia coli* type 1 fimbriae is not required for D-mannose specific adhesion. Molec. Microbiol. <u>4</u>: 553-559 (1990).

13. Molin, S., Klemm, P., Poulsen, L.K., Biehl, H., Gerdes, K. and Andersson, P. Conditional suicide system for containment of bacteria and plasmids. Biotechnology <u>5</u>: 1315-1318 (1987).

14. Hedegaard, L. and Klemm, P. Type 1 fimbriae of *Escherichia coli* as carriers of heterologous antigenic sequences. Gene <u>85</u>: 115-124 (1989).

INTESTINAL COLONIZATION BY ENTEROPATHOGENIC ESCHERICHIA COLI

S. Knutton

Institute of Child Health
University of Birmingham
Birmingham B16 8ET, U.K.

INTRODUCTION

Enteropathogenic <u>Escherichia coli</u> (EPEC) strains of
classical serotypes were the first <u>E. coli</u> strains to be
identified as intestinal pathogens in association with outbreaks of infantile diarrhoea in the 1940's and 1950's. EPEC
remain a significant cause of severe and persistant disease in
infants, particularly in developing countries. The pathogenesis of EPEC diarrhoea has remained elucive since EPEC do not
possess any of the well-defined virulence characteristics of
other diarrhoea-causing <u>E. coli</u> such as enterotoxins or
epithelial invasiveness. A major breakthrough in EPEC pathogenesis came with the observation that EPEC adhere to the
small intestine and produce an 'attaching and effacing' (AE)
lesion in the enterocyte brush border membrane (Ulshen and
Rollo, 1980). The lesion (visible by electron microscopy) is
characterised by localised destruction of brush border microvilli and intimate attachment of bacteria to the apical
enterocyte membrane, often in a cup-like pedestal structure.
In another important study Cravioto et al. (1979) showed that
most EPEC strains adhered to cultured HEp-2 cells whereas non-
EPEC strains rarely adhered. EPEC were subsequently shown to
adhere to HEp-2 cells in localised microcolonies, a pattern
termed localised adhesion (LA). To study mechanisms of EPEC
adhesion we have used the HEp-2 cell model plus an <u>in vitro</u>
human small intestinal model of adhesion.

EPEC ADHESION TO CULTURED HUMAN INTESTINAL MUCOSA

The characteristic AE mode of EPEC intestinal adhesion
can be reproduced by adding EPEC to the culture medium of
human duodenal mucosal biopsies maintained in organ culture
(Knutton et al., 1987b). Small microcolonies of adherent
bacteria were observed after 2-3 hours and these increased in
size with time such that after 12 hours incubation most of the
biopsy mucosal surface was colonized by bacteria. AE lesions
identical to those seen in natural EPEC infections were seen
when infected biopsies were examined by transmission and by

scanning electron microscopy (Fig. 1). This simple model of
EPEC infection allowed us to examine large numbers of EPEC
strains and confirm that the AE adherence property, previously
only demonstrated for EPEC serogroups O26, O119 and O125, is a
characteristic feature of E. coli belonging to most of the
classical EPEC serotypes. However, our consistent inability to
detect the AE property in some EPEC serogroups (O18, O44) and
serotypes (O126:H27) suggests that EPEC, as currently defined,
does include strains other than AE EPEC.

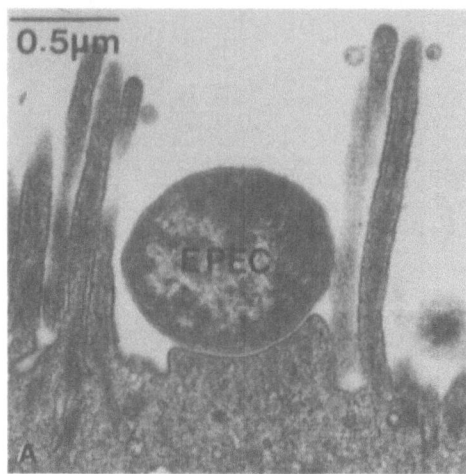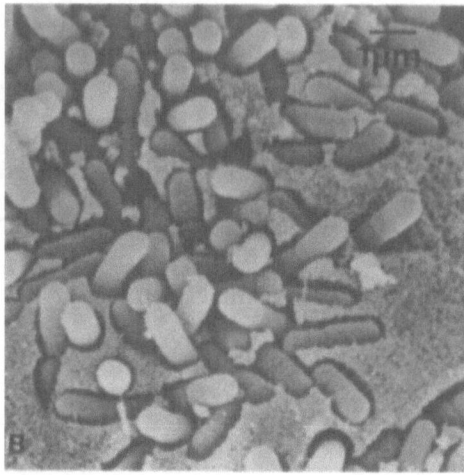

Figure 1. Transmission (A) and scanning electron micrographs
(B) showing the characteristic 'attaching and
effacing' brush border membrane lesion produced by
EPEC.

Ultrastructural observations also revealed features of AE
EPEC adhesion not previously detected. Initial adhesion of
bacteria to an intact brush border appeared to cause firstly,
an elongation of brush border microvilli (Fig. 2A) followed by
disruption of the microvillous cytoskeleton and vesiculation
of the microvillous membrane (Fig. 2B). This disruption of the
brush border cytoskeleton allows the residual apical membrane
devoid of microvilli to distort around bacteria forming an
intimate contact and producing the characteristic cup-like
pedestal structures (Fig. 1A). A dense plaque of short filaments on the cytoplasmic surface of the enterocyte membrane
beneath attached bacteria is another feature of the AE lesion
(Fig. 1A) and one that we have now used as the basis of a new
diagnostic test for EPEC (Knutton et al., 1989).

ROLE OF PLASMID AND CHROMOSOMAL GENES IN EPEC ADHESION

Virulence properties of pathogenic E. coli are frequently
encoded on plasmids. We therefore sought to investigate the
role of plasmids in EPEC adhesion. A survey of 30 EPEC strains
had shown that they all possess large plasmids in the 50-70
megadalton range (Baldini et al., 1983). EPEC E2348(O127:H6),
an outbreak strain from Taunton, England, possesed a 60 MDa
plasmid, pMAR2 and exhibited localised HEp-2 cell adhesion.
E2348 cured of pMAR2 lost the ability to adhere to HEp-2

cells; LA was aquired when pMAR2 was transformed into a non-adherent laboratory strain of E. coli. By electron microscopy, E2348 was found to produce in HEp-2 cells an AE lesion morphologically identical to that seen in intestinal cells (Fig. 3A) whereas the plasmid transformant, although adherent, did not cause an AE lesion nor did it cause any detectable change in cell surface structure (Fig. 3B) (Knutton et al., 1987a).

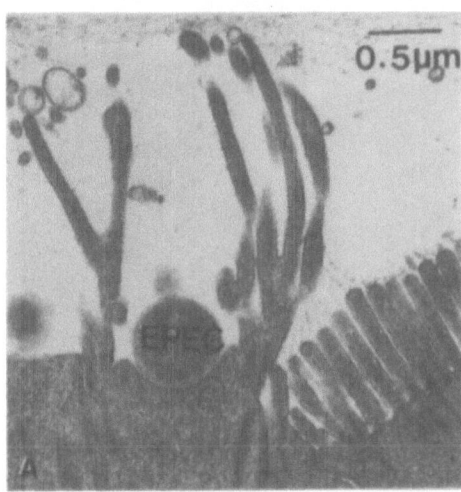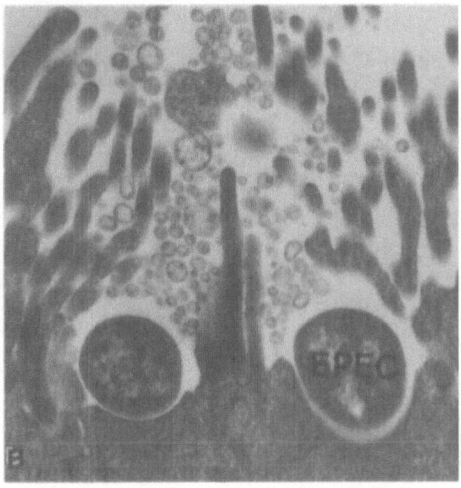

Figure 2. Electron micrographs showing EPEC-induced elongation (A) and vesiculation (B) of brush border microvilli.

Similar results were obtained when the same strains were examined in the human intestinal model except that after a 12 hour incubation plasmid-cured E2348 did show AE adhesion, albeit at very low levels compared to E2348. More recent fluorescence studies have now shown that plasmid-cured E2348 also shows low levels of AE adhesion to HEp-2 cells (Knutton et al., 1989). Thus, the process of EPEC adhesion can be dissected into two distinct stages: 1) initial non-AE adhesion promoted by plasmid-encoded adherence factors; 2) AE adhesion promoted by chromosomal-encoded factors. AE adhesion can occur at low levels in the absence of the plasmid-encoded adhesins but full expression of virulence requires both sets of factors to be present (Levine et al., 1985).

The adherence factor encoded by plasmid pMAR2 has been termed EPEC Adherence Factor (EAF) although the factor itself has yet to be identified. Nataro et al. (1985) cloned a 1 kilobase segment of plasmid pMAR2 and found it to be a highly sensitive and specific probe for detecting E. coli possessing the EAF property. In a recent study in which we screened for the AE adherence property in E. coli from infants with diarrhoea, 13 of 23 (56%) AE E. coli were EAF probe negative (Table 1). This suggests that there may be plasmid encoded adherence factors other than EAF amongst EPEC strains.

It has been reported that the AE adherence property in an O111:H- EPEC isolate is plasmid encoded (Fletcher et al., 1990). We, however, have not been able to confirm this in any of our AE adhesion assays. Furthermore, a genetic locus involved in the AE property of EPEC termed eae has recently been described and shown to be chromosomal (Jerse et al., 1990). A 1 kilobase fragment of this gene was found to be 100% sensitive and 98% specific in detecting AE EPEC. Interestingly *eae* shows significant sequence homology with the invasin gene of *Yersinia pseudotuberculosis*.

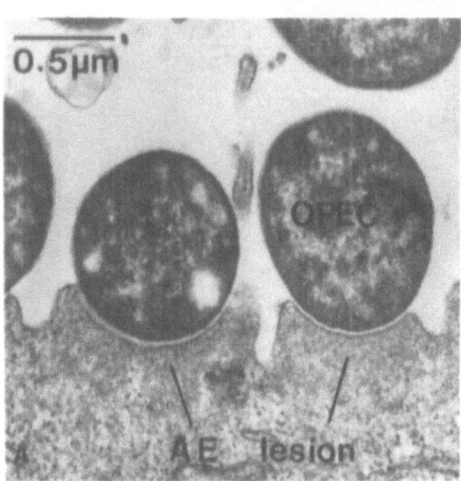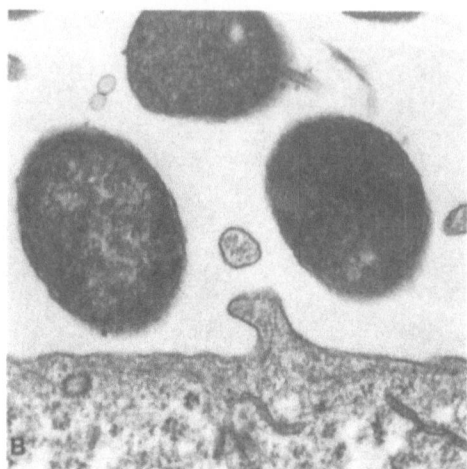

Figure 3. Electron micrographs showing the typical AE lesion produced in HEp-2 cells by EPEC strain E2348 (A) and the non-AE adherence of the plasmid transformant strain HB101pMAR15.

FLUORESCENT ACTIN STAINING (FAS) TEST FOR AE EPEC

Until recently the AE adherence property of EPEC could only be demonstrated by electron microscopy. Based on our previous observation that there is an accumulation of actin beneath attached bacteria in the AE lesion which can be visualised in a fluorescence microscope by staining with a fluorescein conjugated phallotoxin, we developed the FAS test (Knutton et al., 1989). Bacteria which we knew from electron microscopical studies to be AE EPEC all gave intense spots of actin fluorescence at sites of bacterial adhesion (Fig. 4C,D) whereas the fluorescence from cells with adherent but non-AE bacteria was indistinguishable from uninfected control cells. This and subsequent studies have shown the FAS test to be a highly sensitive and specific diagnostic test for the AE lesion, and thus for AE EPEC. The test is also diagnostic for Vero cytotoxin producing *E. coli* (VTEC) which produce an AE lesion identical to EPEC but in the large bowel. Tests for Vero cytotoxins can be used to distinguish EPEC from VTEC.

SCREENING FOR EPEC WITH THE FAS TEST

The FAS test has provided an important new tool with which to investigate EPEC (and VTEC). We have used the FAS test to screen for EPEC in infants with diarrhoea and compared the results with O serogrouping, O:H serotyping, LA to HEp-2 cells, AE adhesion to human intestinal mucosa and with the EAF gene probe. We examined two groups of strains consisting of 36 *E. coli* previously diagnosed as EPEC by O serogrouping and 297 untyped *E. coli*.

Of the 36 serotyped E. coli only 18 (50%) were FAS test positive (FAS+) and of these only 12 belonged to recognised EPEC serotypes; 8 other strains which belonged to EPEC serotypes were FAS-. Seven FAS+ strains were identified amongst the 297 untyped strains but only 2 of these were subsequently found to belong to EPEC serotypes; 5 belonged to non-EPEC serotypes or were untypable (Table 1). Fom this study we conclude that: 1) O serogrouping gives a high percentage of false positive results; 2) full O:H serotyping underestimates the incidence of EPEC because there appear to be EPEC which belong to serotypes not currently regarded as EPEC serotypes; 3) there are strains currently classed as EPEC which do not possess the AE adherence property and, therefore, probably cause diarrhoea by a mechanism different to AE EPEC.

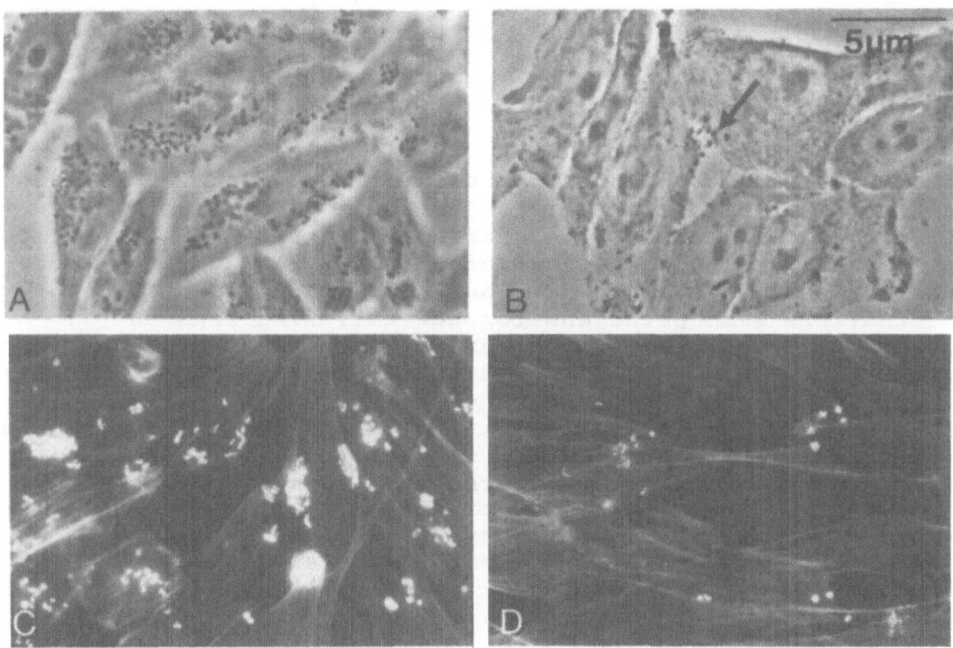

Figure 4. Phase contrast (A,B) and fluorescence micrographs (C,D) showing FAS test positive EPEC which display the good LA (A,C) and poor LA (B,D) patterns of cell adhesion.

This study also identified a previously undescribed pattern of HEp-2 cell adhesion of FAS+ E. coli which we have termed LA/P (poor LA) (Fig. 4B,D) to distinguish it from the LA pattern described by others for classical EPEC and which we have termed LA/G (good LA)(Fig. 4A,C). The 2 patterns differ only in quantitative terms but because LA/P is usually not apparent in a standard 3 hour adhesion assay and only weakly apparent in a 6 hour assay it has not previously been identified. The LA/P pattern is readily identified in the FAS test, however, because all AE bacteria are easy to detect. LA/G correlated with EAF since the 11 FAS+ LA/G+ strains were all EAF probe positive while the 14 FAS+ LA/P+ strains were all EAF probe negative.

Table 1. Detection of EPEC by O serogroup, O:H serotype, FAS test, LA, AE intestinal adhesion and EAF probe

EPEC detected	Group 1 strains (n=36)	Group 2 strains (n=297)
By O serogroup	36	10
By O:H serotype	20	2
By O:H serotype, FAS test	12	2
By non-EPEC serotype, FAS test	3	5
By FAS test	18	7
By AE intestinal assay	18	7
By LA HEp-2 adhesion		
LA/G + LA/P	18	7
LA/P	11	3
LA/G	7	4
By EAF probe	7	4

see text for definitions

Based on serotype and adherence properties this study identified 4 types of putative AE EPEC pathogen (Table 2) although an association with diarrhoea has only been confirmed for EAF+ EPEC belonging to classical serotypes (Levine, 1987). We believe, however, that all the FAS+ strains identifed in the study are EPEC pathogens for the following reasons: 1) they were all isolated from cases of severe diarrhoea where no other pathogen was identified; 2) they are not VTEC; 3) in our in vitro assay the EAF- strains colonized human intestinal mucosa as effectively as EAF+ strains known to cause diarrhoea in volunteers. Although further studies are clearly needed to confirm the role in infant diarrhoea of EAF+ E. coli belonging to non-classical serotypes and of EAF- E. coli belonging to classical or non-classical serotypes, we find it difficult to believe that strains which possess EAF or an equivalent plasmid encoded adherence factor plus the AE adherence property would not be pathogens.

Table 2. Types of putative AE EPEC

	Serotype	FAS test	EAF probe	HEp-2 cell adhesion	Association with diarrhoea
1.	Classical	+	+	LA/G	+
2.	Classical	+	-	LA/P	-
3.	Non-classical	+	+	LA/G	-
4.	Non-classical	+	-	LA/P	-

see text for definitions

The AE adherence property appears central to the ability of EPEC to cause disease. In considering possible mechanisms for the effects of EPEC on intestinal cells we noted that treatments which perturb the calcium balance of intestinal cells caused changes similar to those seen in EPEC-infected enterocytes. Increased intracellular calcium has been shown to activate villin, a calcium-dependent microvillous actin severing/bundling protein causing structural collapse of the microvillous core actin cytoskeleton with resultant vesiculation of the microvillous membrane (Matsudaira and Burgess, 1982). To determine if EPEC cause increased levels of intracellular calcium we used the HEp-2 cell model of EPEC infection and measured intracellular free calcium fluorimetrically following the loading of cells with the fluorescent calcium indicator dye, quin-2 during the last hour of the incubation with bacteria (Baldwin et al., 1989). HEp-2 cells, which possess a calcium-dependent protein gelsolin similar to villin, showed a sustained 4-6 fold increase in calcium levels compared to uninfected cells (Table 3). Chelation of extracellular calcium with EGTA had no significant effect on EPEC-induced elevation of intracellular calcium whereas dantrolene, a drug which traps calcium within intracellular stores, abolished the

Table 3. Intracellular calcium concentrations in control and HEp-2 cells infected with EPEC for 3 hours

Strain	calcium concentration (nM)
uninfected	56
HB101pMAR15[a]	50
O111 EPEC	280
O119 EPEC	250
O127 EPEC	200

[a] laboratory E. coli strain harboring an EAF plasmid

EPEC-induced increases in calcium. These results suggest that EPEC stimulate release of calcium from intracellular stores possibly by stimulating a signal transduction pathway. If calcium release was localised to sites of bacterial adhesion, local concentrations of calcium could be sufficient to activate gelsolin (or villin), cause microvillar breakdown and allow formation of the characteristic AE lesion. The presence of actin in the AE lesion at a concentration many times that normally present in HEp-2 cells, as demonstrated by the FAS test, must be due to de novo actin polymerization and may possibly occur at nucleation sites generated by dissociation of gelsolin (or villin) from the ends of actin filaments as normal calcium levels are restored.

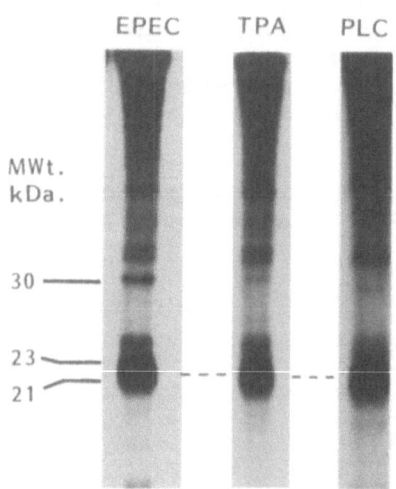

Figure 5. Effect of EPEC and known activators of proteinkinase
C on HEp-2 cell protein phosphorylation.

Since EPEC appear to stimulate increases in intracellular calcium levels and calcium is a known intracellular second messenger regulator of intestinal ion transport functioning through activation of the calcium-dependent protein kinase C (PKC) (Fondacaro, 1986), we considered the possibility that EPEC may cause diarrhoea by such a mechanism. We therefore examined alterations in the phosphorylation state of HEp-2 cell proteins by electrophoresis and autoradiography following incubation of ^{32}P treated cells with EPEC. Infection of HEp-2 cells with EPEC stimulated phosphorylation of several cell proteins, the most prominant of which had molecular weights of 21,000 and 29,000. The same proteins were phosphorylated by known activator of PKC including the phorbol ester, TPA and phospholipase C (PLC) (Fig. 5). Several features of EPEC infection of HEp-2 cells (alterations in cell morphology, elevation of intracellular calcium concentration and stimulation of PKC) resemble the effects of some hormones on various mammalian cells. Angiotensin II, for example, also promoted phosphorylation of 21,000 and 29,000 molecular weight proteins in HEp-2 cells (Baldwin et al., 1990). One attractive possibility, therefore, is that EPEC display hormone-like action promoting a calcium second message and activation of PKC, possibly through host receptor-coupled phospholipases. EPEC diarrhoea could thus be the combined result of a) loss of absorbtive surface due to the effects of increased intracellular calcium and b) intestinal secretion stimulated by PKC.

ACKNOWLEDGEMENTS

The author wishes to acknowledge the collaboration of Dr Jim Kaper, University of Maryland, Dr Peter Williams, University of Leicester, Dr Alan Phillips, Queen Elizabeth Hospital for Children, London, Dr Henry Smith, Central Public Health Laboratory, London, and Dr Alistair Aitken, National Institute for Medical Research, and their coworkers in some of these studies.

REFERENCES

Baldini, M. M., Kaper, J. B., Levine, M. M., Candy, D.C.A., and Moon, H. W., 1983, Plasmid-encoded adhesion of enteropathogenic Escherichia coli, J. Pediatr. Gastroenterol. Nutr., 2:534.

Baldwin, T. J., Knutton, S., Williams, P. H., and McNeish, A. S., 1989, Elevated calcium levels and protein kinase C activation in response to infection by enteropathogenic Escherichia coli, Paed. Res., 26:272.

Baldwin, T. J., Brookes, S. F., Knutton, S., Manjarrez Hernandez, H. A., Aitken, A., and Williams, P. H., 1990, Protein phosphorylation by protein kinase C in HEp-2 cells infected with enteropathogenic Escherichia coli, Infect. Immun., 58:761.

Cravioto, A., Gross, R. J., Scotland, S. M., and Rowe, B., 1979, An adhesive factor found in strains of Escherichia coli belonging to the traditional infantile enteropathogenic serotypes, Curr. Microbiol., 3:95.

Fletcher, J. N., Saunders, J. N., Batt, R. M., Embaye, H., Getty, B., and Hart, C. A., 1990, Attaching effacement of the rabbit enterocyte brush border is encoded on a single 96.5-kilobase-pair plasmid in an enteropathogenic <u>Escherichia coli</u> O111 strain, Infect. Immun., 58:1316.

Fondacaro, J. D., 1986, Intestinal ion transport and diarrheal disease, Am. J. Physiol., 250:G1.

Jerse, A. E., Yu, J., Tall, B. D., Gicquelais, K. G., and Kaper, J. B., 1990, Characterization of <u>eae</u>: a genetic locus involved in attaching and effacing activity of enteropathogenic <u>Escherichia coli</u>, Abstract B-158 presented at the annual meeting of the American Society for Microbiology, Anaheim, USA.

Knutton, S., Baldini, M. M., Kaper, J. B., and McNeish, A. S., 1987a, Role of plasmid-encoded adherence factors in adhesion of enteropathogenic <u>Escherichia coli</u> to HEp-2 cells, Infect. Immun., 55:78.

Knutton, S., Baldwin, T. J., Williams, P. H., and McNeish, A. S., 1989, Actin accumulation at sites of bacterial adhesion to tissue culture cells: basis of a new diagnostic test for enteropathogenic and enterohemorrhagic <u>Escherichia coli</u>, Infect. Immun., 57:1290.

Knutton, S., Lloyd, D. R., and McNeish, A. S., 1987b, Adhesion of enteropathogenic <u>Escherichia coli</u> to human intestinal enterocytes and cultured human intestinal mucosa, Infect. Immun., 55:69.

Levine, M. M., 1987, <u>Escherichia coli</u> that cause diarrhea: enterotoxigenic, enteropathogenic, enteroinvasive, enterohemorrhagic and enteroadherent, J. Infect. Dis., 155:377.

Levine, M. M., Nataro, J. P., Karch, H., Baldini, M. M., Kaper, J. B., Black, R. E., Clements, M. L., and O'Brien, A. D. 1985, Diarrhoeal response of humans to some classic serotypes of <u>Escherichia coli</u> is dependent on a plasmid encoding an enteroadhesiveness factor, J. Infect. Dis., 152:550.

Nataro, J. P., Baldini, M. M., Kaper, J. P., Black, R. E., Bravo, N., and Levine, M. M., 1985, Detection of an adherence factor of enteropathogenic <u>Escherichia coli</u> with a DNA probe, J. Infect. Dis.,152:560.

Matsudaira, P. T., and Burgess, R. A., 1982, Partial reconstruction of the microvillous core bundle: characterization of villin as a Ca^{2+}-dependent actin bundling/depolymerizing protein, J. Cell Biol., 92:648.

Ulshen, M. H., and Rollo, J. L., 1980, Pathogenesis of <u>Escherichia coli</u> gastroenteritis in man - another mechanism, N. Engl. J. Med., 302:99.

DIFFUSE ADHERENCE OF ENTEROPATHOGENIC ESCHERICHIA COLI STRAINS

Inga Benz and M. Alexander Schmidt

Zentrum für Molekulare Biologie Heidelberg (ZMBH)
6900 Heidelberg, FRG

Abstract

The adherence of enteropathogenic Escherichia coli (EPEC) to the small bowel mucosa is an important step in the pathogenesis of diarrhoeal diseases. Two distinct patterns of adherence have been reported for EPEC bacteria in tissue culture employing HeLa- or HEp-2-cells as model systems: localized (LA) and diffuse (DA) adherence. To study adhesins responsible for DA we used EPEC-strains isolated from cases of infantile diarrhoea in Germany and Brazil. In all EPEC strains investigated so far, the DA-factor has been found to be plasmid encoded. The adhesin gene of strain 2787 was cloned into pBR322. The gene is localized on a plasmid of about 100 kb. The recombinant plasmid pIB6 conferred the binding activity of the wild-type strain on E.coli K-12 strains. By deletion analysis a 6.0 kb fragment was shown to be sufficient for expression of the DA phenotype. The adhesin, designated AIDA-I, was purified by heat-shock extraction of the wildtype isolate as well as the recombinant strain. AIDA-I was shown to adhere specifically to HeLa-cells and adhesion of C600(pIB6) could be inhibited by antiserum.

Introduction

Enteropathogenic Escherichia coli (EPEC) strains are the major cause of neonatal and infantile gastroenteritis throughout the world. The pathogenic mechanisms involved in the development of the EPEC diarrhoea, however, remain unclear. Adherence of EPEC strains to the small bowel mucosa as demonstrated by histopathological studies seems to be of prime importance for infection. Using the adherence of EPEC to HeLa- or HEp-2-cells as a model system, different attachment patterns are observed: Localized adherence (LA), where bacteria attach to and form microcolonies in distinct regions, and diffuse adherence (DA), where bacteria adhere evenly to the whole cell surface. The genes responsible for LA have been located on a 60 megadalton plasmid (pMAR2). A fragment of pMAR2 hybridizes to DNA from other EPEC strains expressing LA. However, homology with DA-EPEC strains was not observed (1). Thus, the factors mediating the LA and DA phenotype are genetically different. Recently a chromosomal DNA-fragment coding for a fimbrial adhesin mediating DA was isolated (2).

Here we present the cloning and initial characterization of plasmid encoded factors responsible for the diffuse adherence (DA) pattern exhibited by the EPEC strain 2787 (3).

Results

For our investigations we used EPEC strain 2787. The strain was isolated from a case of infantile diarrhoea. This strain harbours three plasmids, one of about 3 kb and two of about 100 kb. Since the genes coding for expression of LA have been localized on a large plasmid, we screened the extrachromosomal elements of strain 2787 for genes responsible for expression of DA. Plasmid DNA was partially digested with EcoRI and the resulting fragments were ligated to pBR322. After transformation of E.coli C600 amp-resistent transformants were sreened for their ability to adhere to HeLa-cells. One clone expressed the DA phenotype. It contained a plasmid with a 11 kb insert in pBR322 denoted pIB6. Plasmid pIB6 conferred the ability to adhere to HeLa-cells to other E.coli K-12 strains. Southern blot analysis showed homology between pIB6 and one of the large plasmids, thus indicating that this plasmid is responsible for the DA phenotype. To define the smallest DNA fragment mediating the DA phenotype we constructed deletion mutants of pIB6. The fragment length sufficient to encode the factor(s) necessary for the diffuse adherence could be reduced to about 6.0 kb. Plasmid pIB264 carries the smallest insert mediating adherence.

To identify the protein(s) involved in adherence antiserum was raised in rabbits against whole E.coli C600 cells harbouring the pIB4 plasmid (mediates DA phenotype, 9.2 kb insert). Preadsorption with C600(pBR322) resulted in pIB4 specific antiserum which reacted with a 100 kd protein of a total cell extract of C600(pIB4) and all other plasmid containing strains showing DA as well as wildtype strain 2787. The surface associated 100 kd protein could be isolated by mild heat-treatment (20 min, 60°C). Two of the deletion mutants not showing the DA phenotype contain a membrane associated protein of about 80 kd crossreacting with the pIB4 specific antiserum.

Antiserum obtained against strain C600(pIB4) recognized large proteins in other EPEC strains exhibiting a diffuse adherence pattern. This indicates that the 100 kd protein is serologically conserved even if the recognized proteins are not identical. In Southern blotting experiments, however, DNA homology between different DA-EPEC strains could not be demonstrated.

To verify the 100 kd protein to be the adhesin partially purified 100 kd protein was incubated with fixed HeLa-cells. Bound protein was detected by ELISA employing specific antiserum to pIB4 encoded proteins. The results indicate that the 100 kd protein binds to HeLa-cells and may thus be the adhesin involved in diffuse adherence (AIDA-I) of EPEC strains.

For further evidence that we have isolated the adhesin we try to inhibit adherence of bacteria to HeLa-cells by pIB4 specific antiserum. The extent of inhibition was determined by measurement of ß-galactosidase activity of the attached bacteria. It was shown that antiserum as well as Fab-fragments inhibit the attachment of 2787 as well as of C600(pIB6). Attachment was better inhibited by Fab-fragments than by antiserum.

The identification and sequencing of the gene(s) encoding the expression of the AIDA-I protein is in progress.

References

1. J.P. Nataro, I.C.A. Scaletsky, J.B. Kaper, M.M Levine, and L.R. Trabulsi, Plasmid-mediated factors conferring diffuse and localized adherence of enteropathogenic Escherichia coli, Infect. Immun., 48: 378-383 (1985)
2. S.S. Bilge, C.R. Clausen, W. Lau, and S.L. Moseley, Molecular characterization of diarrhea-associated Escherichia coli to HEp-2 cells, J. Bact., 171: 4281-4289 (1989)
3. I. Benz, and M.A. Schmidt, Cloning and expression of an adhesin (AIDA-I) involved in diffuse adherence of enteropathogenic Escherichia coli, Infect. Immun. 57: 1506-1511 (1989)

INTRACELLULAR MECHANISMS REGULATING INTESTINAL SECRETION

Hugo R. de Jonge

Department of Biochemistry I, Medical Faculty
Erasmus University
P.O.Box 1738, Rotterdam 3000 DR, The Netherlands

In intestinal epithelium at least three independent signal transduction pathways have been implicated in the action of neurohumoral agents and microbial toxins on transepithelial transport of electrolytes and water (see Fig. 1):[1-3]

(1) activation of the stimulatory GTP-binding protein G_s through interaction with hormone receptors at the basolateral membrane (VIP, PGE_2) or through covalent modification (i.e. ADP-ribosylation) of its α subunit catalyzed by the A_1 subunit of choleratoxin (CT) or E.coli heat-labile toxin (LT); G_s stimulation of adenylate cyclase in the basolateral membrane; elevation of cyclic AMP (cAMP) levels; activation of cAMP-dependent protein kinases (PK-A type I and II); phosphorylation of specific substrate proteins localized at least in part in the apical membrane; sustained activation of apical Cl^- channels and basolateral K^+ channels resulting in net Cl^- secretion; inhibition of apical Na^+-Cl^- cotransporters (presumably composed of Na^+/H^+ and Cl^-/HCO_3^- exchangers coupled through circular proton movements) in the absorptive villous cells. Inhibitors of this pathway include activators of the inhibitory G-protein G_i (somatostatin, clonidine, AlF_4^-), adenylate cyclase inhibitors (2',3'-dideoxyadenosine), PK-A inhibitors (Rp-cAMPS; Walsh inhibitor, PK-I), K^+ channel blockers (Ba^{2+}; probenicid; phorbolesters acting through proteins kinase $C\alpha$) and Cl^- channel blockers (indanoyloxyacetic acid, IAA;[4] cis-unsaturated fatty acids;[5] CFTR inhibitors; see below); only a few of them have been exploited as anti-diarrhoeal agents (somatostatin; clonidine);

(2) heat-stable E.coli toxin (ST_A)-triggered activation of a unique atriopeptin- and NO-insensitive isoform of guanylate cyclase (GC) in the apical membrane; elevation of cyclic GMP (cGMP) levels; activation of an intestine-specific membrane-bound isotype of cGMP-dependent protein kinase (PK-G II); phosphorylation of substrate proteins including a 25.000 M_R apical proteolipid; this protein serves as a cosubstrate for PK-A and is immunologically related to the band 3 anion exchanger in erythrocytes;[2] submaximal activation of apical Cl^- and basolateral K^+ channels; full inhibition of apical Na^+-Cl^- cotransport. Inhibitors of this pathway include GC blockers (thiol blocking- or disulphide-reducing agents)[6] and PK-G II inhibitors (Rp-cGMPS; the isoquinolinesulfonamide H8; staurosporine);

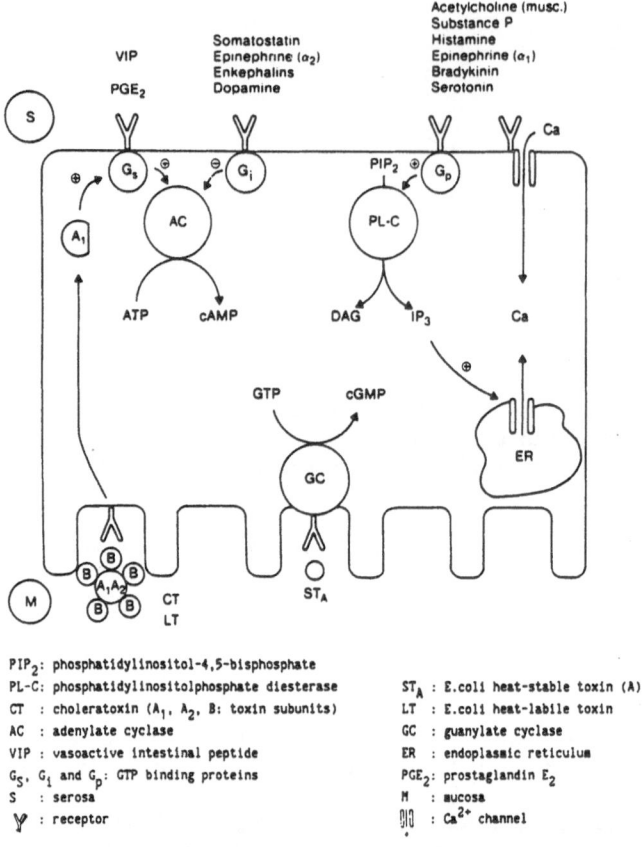

Fig. 1. Transmembrane signalling of hormones, neurotransmitters and toxins
capable of modulating transepithelial transport of salt and water.

(3)   activation of a phosphatidylinositolbisphosphate (PIP$_2$)-specific
      isoform of phospholipase C through interaction with hormone/
      neurotransmitter receptors at the basolateral membrane (e.g.
      acetylcholine, bradykinin); generation of inositoltrisphosphate
      (IP$_3$) and diacylglycerol (DAG) signals; IP$_3$-triggered mobilization
      of Ca^{2+} stores in the endoplasmic reticulum; opening of receptor-
      operated Ca^{2+} channels in the basolateral membrane; Ca^{2+}/DAG-
      activation of protein kinase C (PK-C); Ca^{2+}-provoked and transient
      activation of cAMP/cGMP-insensitive apical and basolateral Cl$^-$
      channels and basolateral K$^+$ channels, presumably mediated through
      a Ca^{2+}-calmodulin-dependent protein kinase; PK-C activation of
      (cAMP-sensitive?) apical Cl$^-$ channels (cf. Fig. 2); Ca^{2+}/PK-C
      inhibition of the apical Na$^+$-H$^+$ exchanger.[7] Much less information
      is yet available about the physiological role of Gp-triggered PIP$_2$
      breakdown recently identified in isolated intestinal brush
      borders[8] and about the involvement of the phospholipase D pathway
      (generating phosphatidic acid, PA; lysoPA; arachidonic acid, AA:
      and DAG)[9] in intestinal secretion.

Additional complexities of stimulus-secretion coupling mechanisms in
the intestine, emerging from our recent biochemical and electrophysio-
logical studies on Cl$^-$-secreting human colon carcinoma cell lines, rat
enterocytes and human ileal and rectal mucosa will be discussed briefly
in the following sections.

Role of the cystic fibrosis gene-encoded protein (CFTR) in intestinal Cl⁻ secretion

Cystic fibrosis (CF), the most frequent autosomal recessive disorder among Caucasians, is a genetic disease characterized clinically by viscous airway secretion and chronic pulmonary infections, pancreatic exocrine insufficiency, intestinal obstruction (meconium ileus) and salt waisting of the sweat glands. Ion transport studies in sweat gland and airway epithelial have identified a defect in the ß-adrenergic (cAMP-mediated) but not in the cholinergic (Ca^{2+}-mediated) activation of apical Cl⁻ channels as the most plausible primary lesion in CF.[10,11] Recent Ussing chamber measurements of short-circuit currents (Isc) across stripped ileal mucosa and rectal biopsies from control and CF patients have revealed the expression of a similar defect in cAMP-provoked Cl⁻ secretion in CF intestine.[12-14] In this tissue, however, in contrast to airway and sweat gland coil, an additional defect was observed in cGMP- and Ca^{2+}-provoked Cl⁻ secretion (Table 1).

Although the molecular basis of the different phenotypic expression of the CF mutation in various epithelia is as yet unknown, the complete loss of the Cl⁻ secretory response to all activating signals in CF intestine has important implications:

(i)   CF patients are apparently resistant to the dehydrating action of microbial toxins and protected against secretory diarrhoea. A proportionally reduced fluid secretion may occur in CF heterozygotes and could represent a main selective advantage for the CF allele;

(ii) the putative CF gene-encoded protein, designated CFTR,[15,16] plays a crucial role in cAMP-, cGMP- and Ca^{2+}-provoked Cl⁻ channel activation;

(iii) inhibitors of CFTR function or expression, e.g. CFTR-directed antibodies, antisense DNA oligomers or newly designed pharmaceutical agents, may mimic the defective Cl- secretion provoked by mutated CFTR and are potentially applicable as a novel class of anti-diarrhoeal agents.

As predicted from the cDNA sequence, CFTR is a highly conserved 150-170 kDa membrane protein (1480 aminoacids) sharing considerable organisational and sequence similarities with a family of ATP-dependent transport systems including the 170 kDa multidrug resistance proteins (MDR, P-glycoprotein).[15,16] In addition, its structure includes a large and rather unique regulatory domain (R) containing potential phosphorylation sites for PK-A, PK-C and PK-G. Using anti-peptide antibodies raised against putative extracellular and ATP binding domains of CFTR, we have recently identified two CFTR candidate proteins (155 and 195 kDa) expressed in a variety of epithelial cells including T-84 and HT 29.cl.19A colonocytes.[17] Exceptionally high levels were detected immunocytochemically in the RER of the Goblet cells, suggesting an additional role of CFTR in the processing and secretion of intestinal mucins.[17] At this stage, at least two models of CFTR function remain plausible; (i) CFTR functions itself as a Cl⁻ channel; (ii) CFTR functions as a MDR-like pump protein; the mutated protein either fails to remove a Cl⁻ channel inhibitor or is unable to translocate a channel-activating factor to its functional compartment. In the second model it remains unclear whether CFTR, the Cl⁻ channel complex, or both are phosphorylated and regulated by the various kinases. Recent speculations that arachidonic acid metabolites, e.g. leukotriene LTC4 or prostaglandin D2 conjugates with glutathione and cysteine, could be the regulator of the epithelial Cl⁻ channel and serve as a CFTR substrate[18] has stimulated our interest in a possible role of eicosanoids in cAMP-provoked intestinal secretion.

Table 1. Short-circuit current response of stripped ileal mucosa from control and CF patients mounted in Ussing chambers

Addition/concentration	Side of addition	I_{SC} (μA/cm^2)*	
		Control(n=8)	CF (n=8)
8-Br-cAMP (10^{-3} M)	Serosa	$+55\pm13$	-10 ± 3
Choleratoxin (10 μg/ml)	Mucosa	$+29\pm6$	$- 6\pm2$
8-Br-cGMP (10^{-4} M)	Serosa	$+28\pm5$	$- 5\pm2$
Heat-stable E.coli toxin (ST$_A$;50 U/ml)	Mucosa	$+32\pm6$	$- 6\pm3$
Carbachol (10^{-5} M)	Serosa	$+61\pm16$	-25 ± 8
Ca^{2+}-ionophore (A23187, 10^{-5} M)	Serosa	$+15\pm5$	$- 1\pm1$
Ca^{2+}-ionophore + phorbolester (PMA, 10^{-7} M)	Serosa	$+40\pm7$	$+ 2\pm2$
Glucose (10^{-2} M)		26 ± 5	29 ± 4

*Data represent maximal changes in I_{SC} provoked by each secretagogue in indomethacin (2.10^{-5} M)-pretreated tissue; + indicates active anion secretion; - indicates active cation (presumably K$^+$) secretion. All patients in this table carried the same, most common, CF mutation (ΔF508).

Electrophysiological and biochemical characterization of intestinal Cl channels; role of apical G-proteins and arachidonic acid metabolites

Filter-grown monolayers of the human colon carcinoma cell line HT 29.cl.19A constitute a highly uniform secretagogue-responsive and Cl$^-$-secreting intestinal epithelium and were therefore selected as a model system in our transport studies. Combined patchclamp-, micro-electrode-, Ussing chamber-, ^{125}I$^-$/^{36}Cl$^-$-isotope efflux-, and vesicle-transport experiments carried out on this cell line have sofar led to the identification of multiple types of Cl$^-$ channels showing different modes of activation and subcellular distribution (Fig. 2).

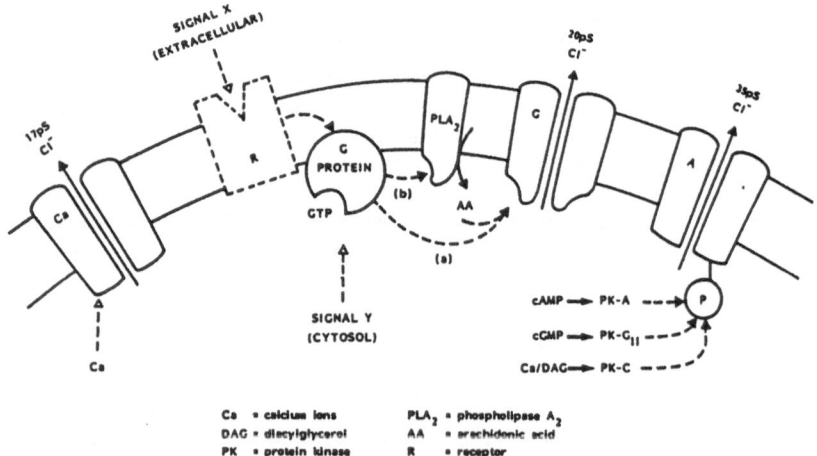

Fig. 2. Chloride channel subtypes expressed in the apical membrane of HT29.cl19A colonocytes.

(i) a 35 pS outwardly rectifying Cl channel observed only in excised patches of apical membranes that is activated by cAMP and protein kinase A (PK-A)/ATP and by depolarizing voltages[10]

(ii) a carbachol and Ca^{2+}-ionophore-activated 17pS non-rectifying channel observed only in cell-attached patches of apical membranes. Isotope efflux and microelectrode experiments suggest that the cAMP-activated anion channels are located exclusively in the apical membrane and prefer ^{36}Cl to ^{125}I whereas the Ca^{2+}-activated anion channels are distributed symmetrically between apical and basolateral membrane and conduct I better than Cl

(iii) a GTPγS/G-protein-activated inwardly rectifying channel (20 pS; "G" channel) detected sofar only in excised patches of apical membranes.[10,19,20] The effect of GTPγS was counteracted by GDPßS and appeared independent of cytosolic messengers including ATP, cAMP and Ca^{2+}, suggesting that protein phosphorylation and/or phospholipase C activation is not involved. A Cl conductance with similar regulatory properties was also detectable in isolated membrane vesicles obtained from HT-29.cl.19A colonocytes and rat enterocytes by measurements of ^{125}I uptake and Cl influx rates following entrapment of the fluorescent "Cl reporter" dye SPQ.[19,20]

Immunoblotting studies using specific G-protein subunit-directed antibodies have provided evidence for a localization of ß, α_s and a novel, pertussis toxin-insensitive subtype of α_i (39 kDa), but not of α_{i1-3}, in the brushborder membrane of rat intestinal villus cells (N. van den Berghe and H.R. de Jonge, manuscript in preparation). This result implies that G_s or the novel G_i protein, but not one of the conventional G_i's, are candidate-regulators of the "G" channel. In this respect the intestinal Cl channel differs clearly from the GTP-activated Na^+ and Cl channels identified recently in renal apical membranes for which α_{i3} was identified as the channel activating G-protein.[21,22] The preliminary outcome of Cl flux measurements in HT-29 and rat intestinal membrane vesicles additionally suggests that, similar to the renal Na^+ channel,[23] arachidonic acid metabolites, presumably leukotrienes, play an intermediary role in the coupling mechanism between the G-protein and the Cl channel (Table 2).

Table 2. Chloride conductance in rat intestinal brush border membrane vesicles.

Loading conditions	Initial rate of Cl induced SPQ quenching (% of control); n=6		
	-GTPγS	+GTPγS $(10^{-5}$ M)	Δ GTPγS
BSA (0.1%)	100±5	230±9	130
Arachidonic acid (50 µg/ml)/ BSA (0.1%)	235±8	243±10	8
Mepacrine $(2.10^{-5}$ M)	121±7	138±12	17
NDGA $(2.10^{-5}$ M)	153±12	164±10	11

For experimental details of vesicle isolation and SPQ-fluorescence techniques see ref. 20. Arachidonic acid, mepacrine, and nordihydroguaiaretic acod (NDGA) were trapped at the vesicle interior prior to SPQ fluorescence measurements.

This observation leads to the interesting speculation that choleratoxin, through ADP-ribosylation of G_s in the apical membrane, might trigger the activation of a Ca^{2+}-insensitive form of phospholipase A_2

and the release of arachidonic acid, and that at least part of the toxin-provoked Cl⁻ secretion might result from a leukotriene-provoked opening of apical "G" channels (Fig. 3). Recent I_{SC} measurements on HT-29.cl.19A monolayers in Ussing chambers indicate that 40-60% of the I_{SC} response to choleratoxin is inhibited by relatively low concentrations of the phospholipase A_2 inhibitor mepacrine (2-5.10⁻⁵ M) and the lipoxygenase inhibitor NDGA (2.10⁻⁵ M), suggesting that a substantial part of the toxin-provoked Cl⁻ secretion takes place through the G-channel. However, since the 8-Br-cAMP- and forskolin-provoked Cl⁻ secretion was inhibited to a similar extent, cAMP rather than apical G_s might at least in part be responsible for the activation of the G channel by the toxin, implying that cAMP itself would be able to trigger the arachidonic acid cascade (Fig. 3). Further studies are clearly needed to evaluate this model and to characterize other aspects of the channel e.g. its physiological importance and activators, its relationship to CFTR, and its sensitivity to the CF mutation.

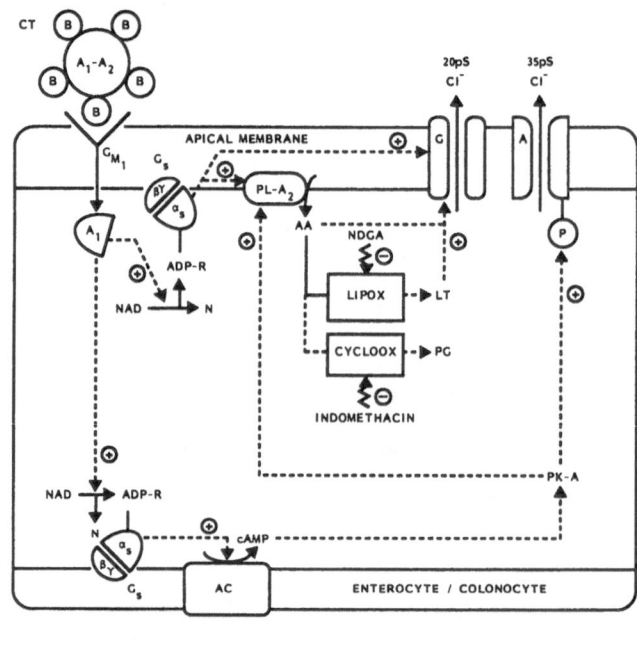

CT	= choleratoxin (A,B subunits)	PL-A₂	= phospholipase A₂
Gₛ	= GTP-binding protein (stimulatory)	AA	= arachidonic acid
ADP-R	= ADP-ribosyl	LIPOX	= lipoxygenase
AC	= adenylate cyclase	CYCLOOX	= cyclooxygenase
PK-A	= cAMP-dependent protein kinase	LT	= leukotrienes
PG	= prostaglandins	pS	= picoSiemens

Fig. 3.

REFERENCES

1. H.R. de Jonge and S.M. Lohmann, Mechanisms by which cyclic nucleotides and other intracellular mediators regulate secretion. Ciba Found. Symp. 112:116 (1985).
2. H.R. de Jonge and M.C. Rao, Role of cAMP- and cGMP-dependent protein kinase in epithelial transport, in: "Pathophysiology of Secretory Diarrhoea", M. Duffey and E. Lebenthal, eds., Raven Press, New York, in the press (1990).

3. M. Field, M.C. Rao and E.B. Chang, Intestinal electrolyte trans-
    port and diarrheal disease (parts I and II) N. Engl. J. Med.
    321:800 and 879 (1989).

4. D.W. Landry, M.H. Ekabas, C. Redhead, A. Edelman, E.J. Cragoe,
    and Q. Al-Awqati, Purification and reconstitution of chloride
    channels from kidney and trachea, Science 244:1469 (1989).

5. T.C. Hwang, S.E. Guggino and W.B. Guggino, Direct modulation of
    secretory chloride channels by arachidonic and other cis
    unsaturated fatty acids, Proc. Natl. Acad. Sci. U.S.A. 87:5706
    (1990).

6. H.R. de Jonge, A.G.M. Bot and A.B. Vaandrager, Mechanism of
    action of E.coli heat-stable enterotoxin, in: "Bacterial
    Protein Toxins", Zentralblatt f. Bact. Microbiol. u. Hygiene
    Suppl. 15:34 (1986).

7. M. Donowitz, M.E. Cohen, M. Gould and G.W.G. Sharp, Elevated
    intracellular Ca acts through protein kinase C to regulate
    rabbit leal NaCl absorption, J. Clin. Invest. 83:1953 (1989).

8. A.B. Vaandrager, M.C. Ploemacher and H.R. de Jonge, Polyphospho-
    inositide metabolism in intestinal brush borders: stimulation
    of IP_3 formation by guanine nucleotides and Ca^{2+}, Am. J.
    Physiol. 259, in the press (1990).

9. M.M. Billah and J.C. Anthes, The regulation and cellular
    functions of phosphatidylcholine hydrolysis, Biochem. J.
    269:281 (1990).

10. H.R. de Jonge, N. van den Berghe, B.C. Tilly, M. Kansen and J.
    Bijman, (Dys)regulation of epithelial chloride channels.
    Biochem. Soc. Transact. 17:816 (1989).

11. N.J. Willumsen and R.C. Boucher, Activation of an apical Cl⁻
    conductance by Ca^{2+}-ionophore in cystic fibrosis airway epithe-
    lia, Am. J. Physiol. 256:C226 (1989).

12. H.R. de Jonge, J. Bijman and M. Sinaasappel, Relation of regula-
    tory enzyme levels to chloride transport in intestinal cells,
    Ped. Pulmonol. (Suppl.) 1:54 (1987).

13. C.J. Taylor, P.S. Baxter, J. Hardcastle and P.T. Hardcastle,
    Failure to induce secretion in jejunal biopsies from children
    with cystic fibrosis, Gut 29:957 (1989).

14. H.M. Berschneider, M.R. Knowles, R.G. Azizkhan, R.C. Boucher,
    N.A. Tobey, R.C. Orlando and D.W. Powell, Altered intestinal
    chloride transport in cystic fibrosis, FASEB J. 2:2625 (1988).

15. J.M. Rommens et al., Identification of the cystic fibrosis gene:
    chromosome walking and jumping, Science 245:1059 (1989).

16. J.R. Riordan et al., Identification of the cystic fibrosis gene:
    cloning and characterization of the complemantary DNA, Science
    245:1066 (1989).

17. A.T. Hoogeveen, J. Keulemans, R. Willemsen, B.J. Scholte, J.
    Bijman, H. Galjaard, M.J. Edixhoven and H.R. de Jonge,
    Immunological identification of 155 and 195 kDa proteins as
    candidates for the products of the cystic fibrosis gene,
    Nature, submitted.

18. D. Ringe and G.A. Petsko, A transport problem? Nature 346:312
    (1990).

19. B.C. Tilly, M. Kansen, P.G.M. van Gageldonk, N. van den Berghe,
    J. Bijman and H.R. de Jonge, Activation of intestinal chloride
    channels by GTP-binding regulatory proteins, in: "Advances in
    Second Messengers and Phosphorylation Research", Y. Nishizuka,
    ed., 24:95 (1990).

20. B.C. Tilly, M. Kansen, P.G.M. van Gageldonk, N. van den Berghe
    and H.R. de Jonge, Activation of intestinal chloride channels
    by GTP-binding G-proteins mediate intestinal chloride channel

activation, <u>J. Biol. Chem.</u>, submitted.

21. D.B. Light, D.A. Ausiello and B.A. Stanton, Guanine nucleotide-binding protein, α^{*}_{1-3}, directly activates a cation channel in rat renal inner medullary collecting duct cells. <u>J. Clin. Invest.</u> 84:352 (1989).

22. E.M. Schwiebert, D.B. Light, G. Fejes-Toth, A. Naray-Fejes-Toth and B.A. Stanton, A GTP-binding protein activates chloride channels in a renal epithelium, <u>J. Biol. Chem.</u> 265:7725 (1990).

23. D.A. Ausiello, C.R. Patenande and H.F. Cantiello, G-protein activation of epithelial Na^{+} channels is mediated via phospholipid metabolites, <u>Proc. Int. Workshop on the CF gene</u>, Sestri Levante, April 9-11, abstract (1990).

CHOLERA TOXIN: ASSEMBLY, SECRETION AND IN VIVO EXPRESSION

J. Holmgren[1], S.J.S. Hardy[2], T.R. Hirst[3],

S. Johansson[1], G. Jonson[1], J. Sanchez[1], and

A.-M. Svennerholm[1]

[1]Department of Medical Microbiology and Immunology, University of Göteborg, S-413 46 Göteborg, Sweden; [2]Department of Biology, University of York, York YO1 5DD, Great Britain; and [3]Department of Genetics, University of Leicester, Leicester LE1 7RH, Great Britain

INTRODUCTION

Cholera toxin (CT), the enterotoxin elaborated from *Vibrio cholerae* during cholera infection and which is responsible for the profuse diarrheal secretion in cholera patients, is the prototype for a family of structurally and functionally related multisubunit enterotoxins that may cause diarrhea. Another prominent member of this family is the heat-labile enterotoxin(s) (LT) from *E. coli*. The molecular properties of CT and LT have been well characterized at both the protein and DNA level. Likewise, their pathogenic action on the intestinal epithelium leading to intracellular cyclic AMP accumulation and fluid secretion is also understood in considerable molecular detail[1-3]. Figure 1 summarizes the role of CT in the pathogenesis of cholera. An analogous picture may be drawn for LT in the pathogenesis of *E. coli* diarrhea.

In contrast, knowledge has been much more limited as to the pathways through which these complex enterotoxins are formed in *V. cholerae* and *E. coli*, respectively; however, recent studies to be described here have shed light on the molecular aspects of toxin assembly and secretion. Another area where knowledge is still scarce and where only preliminary studies have so far been undertaken concerns the mechanisms regulating the expression of enterotoxins (and other virulence factors) *in vivo* during infection. These different topics are the subjects of the studies summarized here.

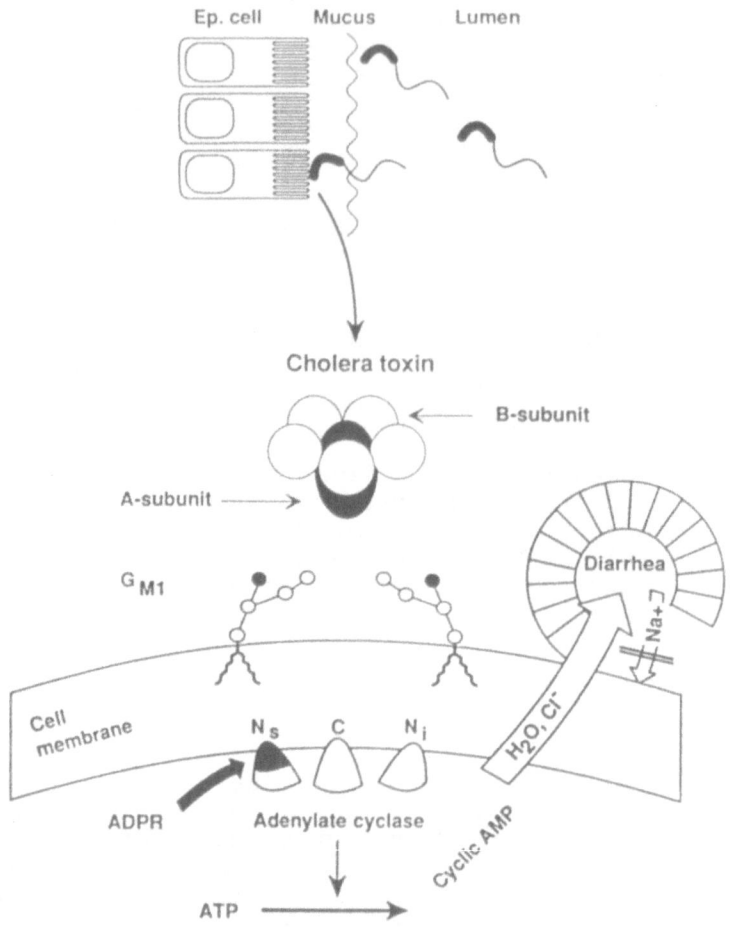

Fig. 1.　　Sequential steps in the pathogenesis of cholera:

(1)　Colonization and multiplication of *V. cholerae* O1 bacteria on the small intestinal epithelium with production of cholera toxin (subunit structure $A-B_5$);

(2)　Binding of toxin to GM1 ganglioside receptors via the B subunits, translocation of the A subunit across the brush-border membrane (probably with release of the A_1 fragment), and activation of adenylate cyclase through ADP-ribosylation of the N_S (G_S) component of the cyclase system;

(3)　Increased formation of cyclic AMP by enterocytes associated with chloride (and bicarbonate) secretion from crypt cells and inhibition of sodium chloride absorption by villous cells;

(4)　These changes in electrolyte transport processes are accompanied by similar changes in water movements resulting in fluid secretion and diarrheal fluid losses of up to 1-2 liters per hour.

In contrast to *E. coli* which retains the assembled LT holotoxin in a cell-associated, periplasmic location[4], *V. cholerae* completely secretes CT extracellularly into the surrounding milieu[5]. When plasmids encoding LT were transferred into *V. cholerae*, LT was likewise completely secreted into the medium[6,7]. This indicates that *V. cholerae* has a secretion pathway which enables either CT or LT to traverse its cell envelope. Since the secretion of complex oligomeric proteins such as CT or LT by *V. cholerae* poses several conceptual problems which have fundamental cell-biological interest beside their cholera-specific relevance, we have examined in some detail the assembly and secretion pathway of CT/LT from both wild-type and recombinant *V. cholerae* strains[7-11]. Our results point to a highly coordinated process that ensures correct assembly as well as secretion of the holotoxin complex, see Fig. 2. The main findings of these studies forming the basis for this model are summarized in the following:

(1) The A and B subunits, which are initially formed as precursors, transiently enter the periplasm. Pulse-chase experiments with [^{35}S]-methionine labeling of a *V. cholerae* strain (TRH7000) from which CT had been genetically deleted but which instead had been made to harbour an LT-encoding plasmid, revealed that radiolabeled LT A and B subunits entered the periplasm rapidly, within 30 sec, which was followed by their much slower efflux ($t_{\frac{1}{2}}$ = 13 min) into the extracellular medium. The rate of efflux from the periplasm was for the B subunit calculated to be ca 170 B monomers per min per cell, corresponding to 34 assembled holotoxin molecules per min per cell. This value was found to closely correspond to the progressive rate of increase in extracellular enterotoxin concentration measured immunologically during exponential cell growth[8].

(2) Oligomerization of toxin subunits occurs in the periplasm prior to their secretion across the outer membrane (OM). The flux of subunits through the periplasm and their presence as monomers or oligomes were monitored by pulse-labeling cells with [^{35}S]-methionine and then isolating periplasmic fractions at different time intervals after the cessation of radiolabel uptake (by "chase" addition of an excess amount of unlabeled methionine). As tested with an SDS-PAGE technique which allowed independent quantitation of total B subunits as well as of B monomers and pentamers, the B subunits were found to be released into the periplasm as monomers which had a relatively rapid turnover corresponding to a half-life of ca 1 min. Concomitantly, assembled toxin could be detected within the periplasm using both this SDS-PAGE method and a GM1-ELISA method showing the presence of associated A and B subunits (i.e. holotoxin) in the periplasm. Since the half-time for efflux of labeled total B subunits from the periplasm was slow ($t_{\frac{1}{2}}$ = 13 min) compared with the rate of B monomer turnover ($t_{\frac{1}{2}}$ = 1 min) it can be concluded that all of

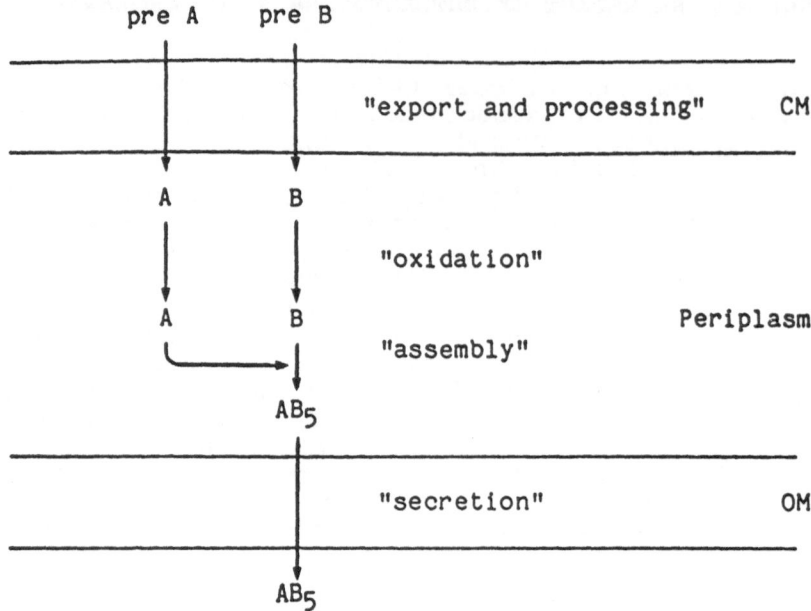

Fig. 2.   Pathway of enterotoxin secretion in *Vibrio cholerae*
          (for explanation see text).

the B subunits that enter the periplasm both assemble and
associate with A subunit to form holotoxin prior to their
secretion across the outer membrane[9].

**(3) The periplasm is an ideal location for toxin assembly
since it ensures the high concentrations of monomers needed
for oligomerization to occur.** Since the major toxin pools in
the periplasm were found to be A and B subunit monomers and
fully assembled holotoxin, we calculated the concentration of
monomers and assembled toxin in the periplasm of an average
V. cholerae cell from the amount of toxin detected by a GM1-
ELISA and from the rates of subunit turnover. The
concentration of A and B monomers in the periplasm was
estimated to be ca 20 µg/ml, and the concentration of
assembled toxin was estimated to be 260 µg/ml. These high
concentrations could clearly favour the spontaneous formation
of holotoxin from subunit monomers. Direct evidence that
toxin assembly is dependent on subunit concentration was
obtained by analysing the *in vitro* dissociation and
reassociation of B subunits of purified CT on acidification
and neutralization, respectively. Reassociation of
dissociated CT took place efficiently at concentrations at or
above ca 20 µg/ml, while it was negligible below a
concentration of 10 µg/ml. A rationale for enterotoxin
assembly taking place in the periplasm of *V. cholerae* before
translocation across the outer membrane would thus appear to
be that the periplasm represents a very concentrated mileu in
comparison to the extracellular environment and that such a
mileu is necessary for native holotoxin to form. It is

conceivable that the high concentration of fully assembled toxin in the periplasm is also important in facilitating movement of the molecule across the outer membrane, since we found an almost 2000-fold difference in toxin concentration between the periplasmic and external sides of the outer membrane[9].

(4) Oxidation of cysteines in the B subunits is a prerequisite step for in vivo formation of B pentamers. Addition of the reducing agent DTT to acid-denatured CT or LT B subunits (monomers) prevented their subsequent reassembly into pentamers at neutral pH. Similarly, addition of DTT to a culture of bacteria harbouring a plasmid encoding for LTB, 0.25 minutes after an [^{35}S]-methionine pulse and chase, inhibited further assembly of radiolabelled B subunit monomers into pentamers but had no effect on the amount of B pentamers that had already assembled[10].

(5) Formation of holotoxin proceeds via A subunit association with B monomers or small oligomers, e g dimers. The ability of B subunits to assemble into pentamers in the absence of A subunits, both *in vivo* and *in vitro*[7,9], could be taken to suggest that the A subunit has no role in B subunit pentamerization. However, subsequent experiments have revealed a subtle but significant role for the A subunit in toxin assembly[10]. Thus, it was found that mixtures of purified A subunit and assembled B pentamers of CT, at a molar ratio of 5 B subunits to 1 A subunit, do not spontaneously associate *in vitro* to form holotoxin. Instead, it was found that the A subunit would only associate with B subunits that were in the process of assembling. We therefore conclude that the pathway of holotoxin assembly involves association of the A subunit, not with the fully assembled B subunit pentamer, but with either B monomers or a B subunit assembly intermediate.

(6) A subunits accelerate the rate of B subunit pentamerization implying that A subunits play a coordinating role in holotoxin assembly. The A subunit was found to influence the kinetics of B subunit assembly *in vivo*. This was shown by pulse labeling bacteria harbouring plasmids encoding either both LT A and B subunits or only the B subunits. It was found that the formation of B pentamers *in vivo* was ca 3 times more rapid in the strain expressing A subunit than in the strain that did not[10]. We conclude that the A subunit can accelerate the rate of B subunit pentamerization, probably by stabilizing an assembly intermediate. The mechanism by which this occurs remains to be defined. One possibility is that the initial association between A and B_n may induce conformational changes in B_n that favour nt interactions with B monomers. Alternatively, domains of the A subunit may form stable salt bridges with B, thereby neutralizing electrostatic charges on the B subunit that might otherwise destabilize subsequent B subunit interactions[10].

(7) Toxin translocation across the *V. cholerae* outer membrane may involve interaction between specific structural domains

in the B pentamers and the outer membrane. When derivatives of CT-deleted *V. cholerae* TRH7000 were constructed which harboured either plasmid pWD605 (encoding the A subunit of LT) or plasmid pWD615 (encoding the B subunit of LT), it was found that approximately 90% of the B subunit were secreted from *V. cholerae* TRH7000 (pWD615) whereas none of the A subunits expressed by TRH7000 (pWD605) were secreted[7]. This shows that the B subunits of the toxin contains structurally information which enables the molecule to translocate across the *V. cholerae* cell envelope and that comparable information is absent from the A subunit; we would tentatively propose that this information may reside in conformational "epitopes" specific for assembled B pentamers.

(8) Hybrid enterotoxin-derived proteins may also be secreted by *V. cholerae*. Not only the native CT and LT holotoxins and their B pentamers but also certain chimeric proteins, obtained by way of gene fusion, have been found to be secreted by *V. cholerae* cells. The CT and LT enterotoxins are in themselves hybrid proteins between A and B subunits, and as mentioned the A subunit is translocated across the outer membrane only by way of its association with the B subunits[7]. Our studies have shown that also A subunit with certain genetically fused peptide extensions (e.g. peptides corresponding to the heat-stable enterotoxin of *E. coli*) can associate with the B subunit in the periplasm of *V. cholerae* and be secreted in the form of a odified holotoxin complex. Likewise, defined gene fusion extensions to either the amino- or carboxy-ends of the B subunits permit B oligomer assembly in the periplasm followed by secretion extracellularly[11,12].

EXPRESSION OF CT FROM CLASSICAL AND EL TOR CHOLERA VIBRIOS IN VIVO AND IN VITRO

In the pathogenesis of cholera, the most important event is the production and secretion of the cholera enterotoxin in the small intestine of the infected individual. We wished to determine in which amounts CT is being produced by strains of *V. cholerae* O1 belonging to different biotypes and serotypes during infection in the intestine as compared with the levels achieved during *in vitro* growth conditions. Bacteria harvested directly from ligated small intestinal loops of rabbits infected with holera vibrios, and the corresponding intestinal fluid after removal of microorganisms were used as *in vivo* specimens for these studies. Comparisons were made with cells and culture-supernatants of bacteria grown *in vitro* during conditions found to be optimal for CT production.

It has been shown that CT expression in *V. cholerae* of both classical and El Tor biotypes is under regulatory control of a transmembrane DNA binding protein, ToxR, which under the influence of environmental factors such as low temperature, slightly acid pH, high salt concentration and certain amino acids can up-regulate the transcription of the ctx cistron manifold[3]. Even so, optimal production of CT in vitro requires different growth conditions for the two biotypes of *V. cholerae* O1. Although CT production is up-regulated at low temperature (27°-30°C versus 37°C) in both biotypes, we confirmed the findings of Iwanaga et al.[13] that while shake-culture in Syncase edium is optimal for CT production by classical strains, CT production by El Tor strains is dramatically improved by using a sequence of low-aeration "stand still" culture followed by maximal-aeration shake cultures and a different medium, "AKI". Thus these "AKI conditions" at 30°C yielded on average 300 times more toxin from our El Tor strains than "classical condition" growth in Syncase (geometric mean of 54 strains 210 ng/ml versus 0.7 ng/ml)[14]. The CT production *in vivo*, as measured with GM1-ELISA using a monoclonal antibody to the B-subunit of the toxin, was strikingly different in the two biotypes (Fig 3). Thus, while the classical strains produced significantly higher levels of toxin *in vivo* than *in vitro* (P<0.001), the *in vivo* production of toxin by the El Tor strains was instead drastically reduced (P<0.001). We conclude that regulation of CT production is complex and probably not identical in classical and El Tor strains.

GENERAL DISCUSSION AND PERSPECTIVES

Our results indicating that toxin secretion across the outer membrane of *V. cholerae* involves translocation of the molecule after it has folded and adopted a stable conformation as holotoxin (or in the absence of A subunit as B pentamer) challenge the paradigm that polypeptides traverse biological membranes as unfolded chains which lack native tertiary structures[15]. This suggests that the physico-chemical principles governing translocation across the outer membrane of *V. cholerae* differ markedly from the export mechanisms envisaged for transfer of proteins across e.g., the endoplasmic reticulum and other membranous organelles in mammalian cells and the cytoplasmic membranes of bacteria. Indeed, we have proposed that proteins, which are synthesized with amino-terminal signal sequences, and which are subsequently secreted from Gram-negative bacteria (e.g., oligomeric proteins such as cholera toxin and pertussis toxin, complex surface adhesins such as the components of P fimbriae, and numerous enzymes including proteases and cellulases) fold into tertiary or quaternay conformations before they are translocated across the outer membrane.

Likewise, our data which reveal that the subtle effect of A subunit on the rate of B subunit pentamerization has a considerable impact on the coordination of toxin assembly and secretion may have broader significance. Thus, we propose that similar kinetic control mechanisms might operate to facilitate coordination of assembly of other secreted oligomeric and multimeric proteins, such as pertussis toxin from *Bordetella pertussis*, and perhaps also in the formation of complex surface adhesins such as type-1 and P fimbriae. To define the extent to which these analogies hold true and to identify the precise mechanism of translocation, first of monomers into the periplasm and of assembled toxin across the outer membrane are important future research problems: e.g.

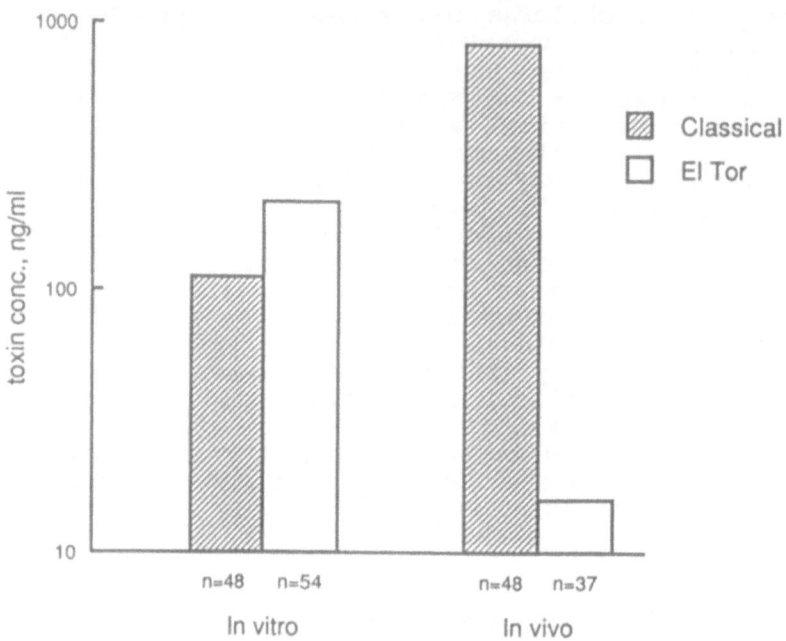

Fig. 3. Production of cholera toxin from *Vibrio cholerae* 01 strains of classical and El Tor biotypes after growth *in vitro* and in rabbit intestine (ligated loops) *in vivo* (for details see ref[17]).

are specific periplasmic "chaperon" or outer membrane "translocator" proteins involved?

Our findings that not only the native CT and LT and their B pentamers but also certain gene fusion proteins based on these proteins can be both overexpressed and secreted by *V. cholerae*, also open up interesting biotechnological applications. CT-deleted *V. cholerae* harbouring overexpression plasmids encoding for hybrid CTB- or LTB-derived hybrid proteins carrying various putative vaccine peptide antigens have already proved useful for production and purification of this novel class of possible recombinant protein vaccines[16]. Other biotechnological applications along similar lines are also feasible.

Finally, our findings of significant differences in the expression of CT by *V. cholerae in vivo* and *in vitro*, and between strains of classical and El Tor biotypes, suggest that a more precise definition of the role of different postulated virulence factors in the processes of infection and immunity should include *in vivo* studies as outlined here and elsewhere[17]. The different regulation of CT expression in classical and El Tor cholera remains to be defined in molecula terms. It also remains to be determined, for instance, whether only a dead ETEC bacterium can release its LT to produce disease or whether *E. coli* specifically during *in vivo* conditions associated with intestinal infection might evolve enterotoxin secretory mechanisms analogous to those exhibited both *in vivo* and *in vitro* by *V. cholerae*.

ACKNOWLEDGEMENTS: The studies were supported by grants from the Medical and Technical Research Councils of Sweden, the World Health Organization and the Wellcome Trust.

REFERENCES

1. J. Holmgren, Cholera toxin, *Nature* 292:413-417 (1981).
2. J. Holmgren, Toxins affecting intestinal transport processes, *in*: "The virulence of *Escherichia coli*", M. Sussman Ed., Academic Press, London (1985).
3. M.J. Betley, V.L. Miller, and J.J. Mekalanos, Genetics of bacterial enterotoxins, *Ann. Rev. Microbiol.* 40:577-605 (1986).
4. T.R. Hirst, L.L. Randall, and S.J.S. Hardy, Cellular location of heat-labile enterotoxin in *Escherichia coli*, *J Bacteriol* 157:637-.642 (1984).
5. S.N. De, Enterotoxicity of bacteria-free culture filtrate of *Vibrio cholerae*, *Nature* 183:1533-1534 (1959).
6. R.J. Neill, B.E. Ivins, and R.K. Holmes, Synthesis and secretion of the plasmid-coded heat-labile enterotoxin of *Escherichia coli* in *Vibrio cholerae*, *Science* 221:289-291 (1983).
7. T.R. Hirst, J. Sanchez, J.B. Kaper, S.J.S. Hardy, and J. Holmgren, Mechanisms of toxin secretion by *Vibrio cholerae* investigated in strains harboring plasmids that encode heat-labile enterotoxins of *Escherichia coli*, *Proc Natl Acad Sci USA* 81:7752-7756 (1984).
8. T.R. Hirst, and J. Holmgren, Transient entry of enterotoxin subunits into the periplasm during their secretion from *Vibrio cholerae*, *J. Bacteriol* 169:1037-1045 (1987)
9. T.R. Hirst, and J. Holmgren, Conformation of protein secreted across bacterial outer membranes: A study of enterotoxin translocation from *Vibrio cholerae*, *Proc Natl Acad Sci, USA* 84:7418-7422 (1987).
10. S.J.S. Hardy, J. Holmgren, S. Johansson, J. Sanchez, and TR Hirst, Coordinated assembly of multisubunit proteins: oligomerization of bacterial enterotoxins *in vivo* and *in vitro*, *Proc Natl Acad Sci USA* 85:7109-7113 (1988).
11. J. Sanchez, and J. Holmgren, Recombinant system for overexpression of cholera toxin B subunit in *V. cholerae* as a basis for vaccine development. *Proc. Natl Acad Sci, USA* 86:481-485 (1989).
12. J. Sanchez, S. Johansson, B. Löwenadler, A-M. Svennerholm, and J. Holmgren, Recombinant cholera toxin B-subunit and gene fusin proteins for oral vaccination, *Res Microbiol*, in press (1990).
13. M. Iwanaga, K Yamamoto, N. Higa, Y. Ichinose, N. Nakasone, and M. Tanage, Culture conditions for stimulating cholera toxin production by *Vibrio cholerae* O1 El Tor, *Microbiol Immunol* 30:1075-1083 (1986).

14. G. Jonson, A.-M. Svennerholm, and J. Holmgren, Expression of virulence factors by classical and El Tor *Vibrio cholerae in vivo* and *in vitro*,

15. L.L. Randall, S.J.S. Hardy, and J.R. Thom, Export of protein, a biochemical view, *Ann Rev. Microbiol* 41:507-541 (1987).

16. J. Holmgren, J. Clemens, D. Sack, J. Sanchez, and A.-M. Svennerholm, Development of oral vaccines with special reference to cholera, *in* "Topics in pharmaceutical sciences" D.D. Breimer, D.J.A. Crommelin, and K.J. Midha, eds., Springer, Basel (1989).

17. G. Jonson, A.-M. Svennerholm, and J. Holmgren, *Vibrio cholerae* expresses cell surface antigens during intestinal infection which are not expressed during *in vitro* culture, *Infect Immun* 57:1809-1815 (1989).

HEAT-STABLE ENTEROTOXINS PRODUCED BY ENTERIC BACTERIA

Yoshifumi Takeda[1], Shinji Yamasaki[1],
Toshiya Hirayama[2] and Yasutsugu Shimonishi[3]

Department of Microbiology, Facutly of Medicine
Kyoto University, Kyoto[1]
The Institute of Medical Science
The University of Tokyo, Tokyo[2] and
Institute for Protein Research
Osaka University, Osaka, Japan[3]

INTRODUCTION

Enterotoxigenic <u>Escherichia</u> <u>coli</u> produces two types of enterotoxin that cause diarrhea in humans and cattles. One is a high molecular weight heat-labile enterotoxin (LT) and the other is a low molecular weight heat-stable enterotoxin (ST). Two different kinds of ST have been described. One is STa or STI that is methanol soluble and active in suckling mice and piglets, but inactive in weaned pigs. The other is STb or STII that is methanol insoluble and active in weaned pigs, but inactive in suckling mice. At least two genetically distinct STa's, that is STIa and STIb, have been reported. STIa and STIb also have been called STp and STh, respectively, because STIa and STIb were originally reported in <u>E. coli</u> isolated from pigs (and cows) and humans, respectively, although human isolates produce both STh and STp.

It is known that diarrhea induced by LT is due to continuous stimulation of membrane-bound adenylate cyclase of intestinal brush border membrane by ADP-ribosyl transferase activity of LT. On the other hand, ST stimulates guanylate cyclase and increases the intracellular concentration of cGMP. It has been reported that the increase in cGMP results in the activation of cGMP-dependent protein kinase[1] followed by phosphorylation of brush-border membrane proteins[2]. However, further detailed mechanism to cause diarrhea by ST is still obscure. To elucidate the detailed mechanism by which ST cause diarrhea, we studied the structure-function relationship of ST produced by enteric bacteria.

PRIMARY STRUCTURE AND MODE OF DISULFIDE BOND FORMATION OF ST'S

STh and STp from enterotoxigenic E. coli have been purified from the culture supernatant of human and porcine strains and their primary structures with 19 and 18 amino acid residues have been determined[3,4] (Fig. 1). Similar ST's are produced by several enteric bacteria, such as Yersinia enterocolitica, Vibrio cholerae non-01 and Vibrio mimicus and their primary structures have also been determined[5-7] (Fig. 1). All the ST's share the highly homologous sequence of 13 amino acid residues boxed by a line in Fig. 1. It is also characteristic that there are 6 cysteine residues at the same positions in all ST's.

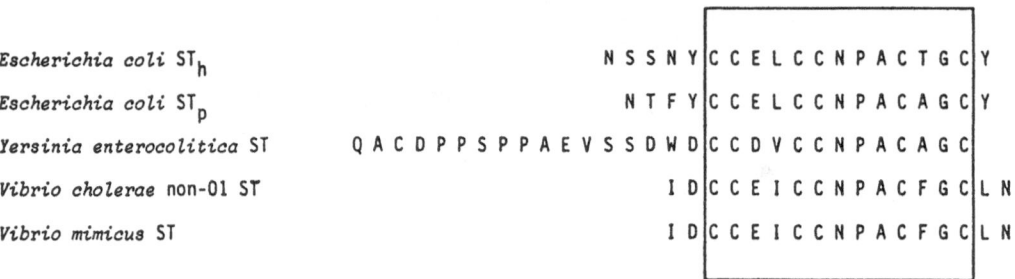

Escherichia coli STh	N S S N Y	C C E L C C N P A C T G C Y
Escherichia coli STp	N T F Y	C C E L C C N P A C A G C Y
Yersinia enterocolitica ST	Q A C D P P S P P A E V S S D W D	C C D V C C N P A C A G C
Vibrio cholerae non-01 ST	I D	C C E I C C N P A C F G C L N
Vibrio mimicus ST	I D	C C E I C C N P A C F G C L N

Fig. 1. Amino acid sequences of heat-stable enterotoxins produced by various enteric bacteria

Experimental work to examine the mode of disulfide bond formation in STh and STp showed that the mode of three disulfide bonds formation in both ST's was the same (Fig. 2) although the primary structures of these two ST's in the common 13 amino acid residues differ each other[8,9]. Moreover, it has been shown that the mode of disulfide bond formation in other ST's from Yersinia enterocolitica and Vibrio cholerae non-01 is also the same as that of E. coli ST's[10] (Fig. 2).

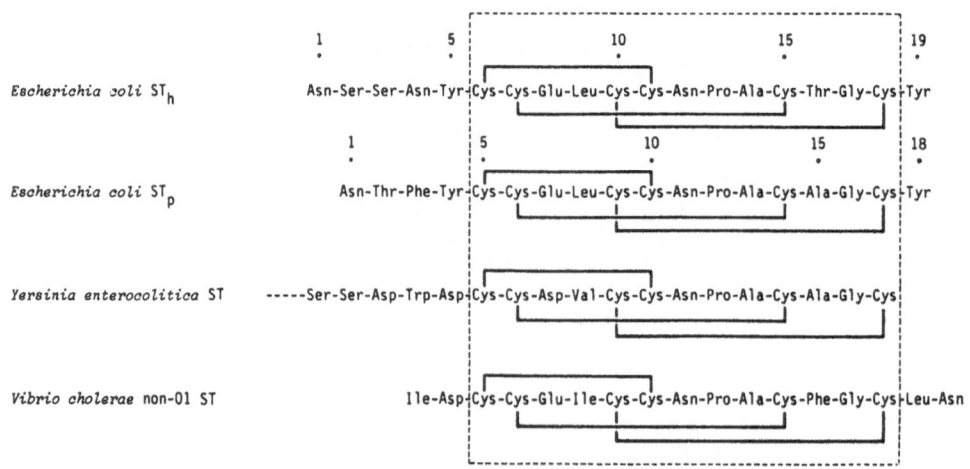

Fig. 2. Disulfide bond structures of ST's produced by various enteric bacteria

126

CHEMICAL SYNTHESIS OF ST'S

To study the structure and function of ST's, we chemically synthesized STh, STp and their analogues and examined biological and immunological characters of the synthesized ST's and their analogues[11-16]. Table 1 shows a list of synthesized ST's and their analogues. To confirm that chemically synthesized ST has the same tertiary structure as native ST, the 500 MHz ^{1}H-NMR spectra of synthesized and native ST were measured. Fig. 3 shows the results of the NMR spectra of synthesized and native STh. As shown in Fig. 3, the proton chemical shifts of synthesized STh were superimposable on those of native STh, indicating that the synthesized STh had the same three dimensional structure as native toxin. Similarly, the synthesized STp was shown to have the same tertiary structure as native STp, including the positions of disulfide linkages.

The biological activity of synthesized STh, STp and their analogues was examined by the fluid accumulation test in suckling mice[17]. The minimum effective doses of all synthesized peptides are summarized in Table 2. As shown in Table 2, not only synthesized STh(1-19) and STp(1-18), but also synthesized shorter analogues of STh and STp gave similar values of the minimum effective dose. Moreover, when these synthesized peptides were treated with an antiserum raised against purified native STh, biological activities of all peptides were neutralized in a similar manner as native STh and STp.

From these results, it is concluded that the sequence with the 13 amino acid residues, from the cysteine residue near the N-terminus to the cysteine residue near the C-terminus, is essential for the biological properties of both STh and STp.

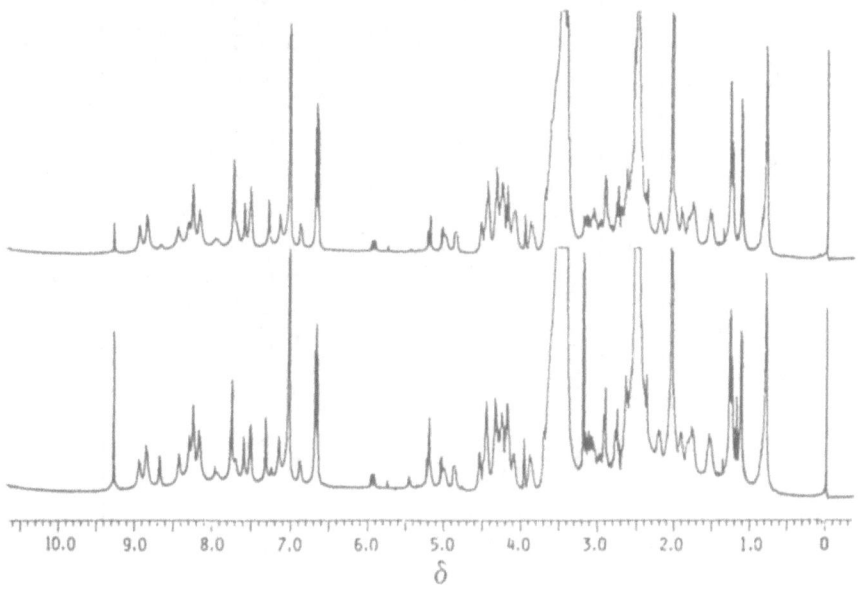

Fig. 3. 500 MHz ^{1}H-NMR spectra of synthesized STh (upper panel) and native STh (lower panel)

Table 1. Chemically synthesized STh, STp and their analogues

Sequence	Abbreviation
Asn-Ser-Ser-Asn-Tyr-Cys-Cys-Glu-Leu-Cys-Cys-Asn-Pro-Ala-Cys-Thr-Gly-Cys-Tyr	STh(1-19)
Asn-Ser-Ser-Asn-Tyr-Cys-Cys-Glu-Leu-Cys-Cys-Asn-Pro-Ala-Cys-Thr-Gly-Cys	STh(1-18)
Ser-Ser-Asn-Tyr-Cys-Cys-Glu-Leu-Cys-Cys-Asn-Pro-Ala-Cys-Thr-Gly-Cys-Tyr	STh(2-19)
Ser-Asn-Tyr-Cys-Cys-Glu-Leu-Cys-Cys-Asn-Pro-Ala-Cys-Thr-Gly-Cys-Tyr	STh(3-19)
Asn-Tyr-Cys-Cys-Glu-Leu-Cys-Cys-Asn-Pro-Ala-Cys-Thr-Gly-Cys-Tyr	STh(4-19)
Tyr-Cys-Cys-Glu-Leu-Cys-Cys-Asn-Pro-Ala-Cys-Thr-Gly-Cys-Tyr	STh(5-19)
Tyr-Cys-Cys-Glu-Leu-Cys-Cys-Asn-Pro-Ala-Cys-Thr-Gly-Cys	STh(5-18)
Cys-Cys-Glu-Leu-Cys-Cys-Asn-Pro-Ala-Cys-Thr-Gly-Cys-Tyr	STh(6-19)
Cys-Cys-Glu-Leu-Cys-Cys-Asn-Pro-Ala-Cys-Thr-Gly-Cys	STh(6-18)
Asn-Thr-Phe-Tyr-Cys-Cys-Glu-Leu-Cys-Cys-Asn-Pro-Ala-Cys-Ala-Gly-Cys-Tyr	STp(1-18)
Asn-Thr-Phe-Tyr-Cys-Cys-Glu-Leu-Cys-Cys-Asn-Pro-Ala-Cys-Ala-Gly-Cys	STp(1-17)
Thr-Phe-Tyr-Cys-Cys-Glu-Leu-Cys-Cys-Asn-Pro-Ala-Cys-Ala-Gly-Cys-Syr	STp(2-18)
Phe-Tyr-Cys-Cys-Glu-Leu-Cys-Cys-Asn-Pro-Ala-Cys-Ala-Gly-Cys-Tyr	STp(3-18)
Tyr-Cys-Cys-Glu-Leu-Cys-Cys-Asn-Pro-Ala-Cys-Ala-Gly-Cys-Tyr	STp(4-18)
Tyr-Cys-Cys-Glu-Leu-Cys-Cys-Asn-Pro-Ala-Cys-Ala-Gly-Cys	STp(4-17)
Cys-Cys-Glu-Leu-Cys-Cys-Asn-Pro-Ala-Cys-Ala-Gly-Cys-Tyr	STp(5-18)
Cys-Cys-Glu-Leu-Cys-Cys-Asn-Pro-Ala-Cys-Ala-Gly-Cys	STp(5-17)

128

Table 2. Biological properties of synthesized STh, STp and
        their analogues

		Minimum effective dose (ng/100ul)	Neutralization by anti-native STh
Native STh		1.0	
Synthesized STh	STh(1-19)	0.8	+
	STh(1-18)	0.4	+
	STh(2-19)	0.8	+
	STh(3-19)	1.3	+
	STh(4-19)	1.1	+
	STh(5-19)	0.8	+
	STh(5-18)	0.5	+
	STh(6-19)	0.6	+
	STh(6-18)	0.6	+
Native STp		1.0	+
Synthesized STp	STp(1-18)	1.0	+
	STp(1-17)	1.3	+
	STp(2-18)	0.5-2	+
	STp(3-18)	1.5-2	+
	STp(4-18)	0.8-1.0	+
	STp(4-17)	1.2	+
	STp(5-18)	0.8-1.0	+
	STp(5-17)	0.7	+

IDENTIFICATION OF AMINO ACID RESIDUES ESSENTIAL FOR THE TOXIN
ACTIVITY

To study which amino acid(s) are important for the toxic
activity of ST, we synthesized the analogues of STh(6-19) that
has a single amino acid replacement at position 12, 13 and 14 and
toxic activities of the synthesized analogues were examined[18,19]

Table 3 shows the results of suckling mouse assay for
various analogues of STh(6-19) with an amino acid replacement at
position 12 (asparagine in the native ST). Among 8 analogues,
replacement by valine showed the lowest minimum effective dose
(MED) value and it was very similar to the MED of native ST.
Replacement by neutral amino acids, such as alanine,
phenylalanine, serine and glycine, influenced toxic activity very
little while replacement by acidic amino acid, such as glutamic
acid, and basic amino acids, such as arginine and lysine, greatly
reduced toxic activity. Particularly, replacement by gltamic
acid reduced the activity by the factor of about 400 although the
size of the side chains of asparagine and glutamic acid are not
significantly different. These results indicate that the amino
acid residue at position 12 should not be charged, especially
should not be basic, to maintain toxic activity.

The results obtained with the analogues replacing proline at
position 13 are summarized in Table 4. Replacement by valine had
a minimum effect. Replacement by neutral amino acids, such as
alanine, glutamine and serine, and by basic amino acids, such as

Table 3. Biological activities of various analogues of STh(6-19) with an amino acid replacement at position 12

Amino Acid Sequence	MED(pmol)
6 12 19	
Cys-Cys-Glu-Leu-Cys-Cys-Asn-Pro-Ala-Cys-Thr-Gly-Cys-Tyr	0.4
- - - - - - Val - - - - - - -	0.5
- - - - - - Ala - - - - - - -	4.5
- - - - - - Phe - - - - - - -	7.8
- - - - - - Ser - - - - - - -	11
- - - - - - Gly - - - - - - -	32
- - - - - - Glu - - - - - - -	150
- - - - - - Arg - - - - - - -	340
- - - - - - Lys - - - - - - -	1,300

MED: minimum effective dose

lysine and arginine, reduced toxic activity only little. Replacement by glutamic acid and phenylalanine influenced toxic activity significantly. Glutamine, which has amide group at the terminal carboxylic acid of glutamic acid was 7 times more toxic than glutamic acid. Therefore, it is assumed that the amino acid residue at position 13 should not be acidic or should not have bulky and hydrophobic side chains to show the biological activity.

Table 5 summarizes the results obtained with the analogues replacing alanine at position 14. Replacement by glycine or serine had minimum effect. Replacement by aspartic acid, glutamic acid or glutamine reduced activity considerably.

Table 4. Biological activities of various analogues of STh(6-19) with an amino acid replacement at position 13

Amino Acid Sequence	MED(pmol)
6 13 19	
Cys-Cys-Glu-Leu-Cys-Cys-Asn-Pro-Ala-Cys-Thr-Gly-Cys-Tyr	0.4
- - - - - - - Val - - - - - -	1.4
- - - - - - - Ala - - - - - -	6.9
- - - - - - - Gln - - - - - -	13
- - - - - - - Ser - - - - - -	15
- - - - - - - Lys - - - - - -	20
- - - - - - - Arg - - - - - -	20
- - - - - - - Glu - - - - - -	93
- - - - - - - Phe - - - - - -	1,200

MED: minimum effective dose

Table 5. Biological activities of various analogues of STh(6-19) with an amino acid replacement at position 14

Amino Acid Sequence			MED(pmol)
6	14	19	
Cys-Cys-Glu-Leu-Cys-Cys-Asn-Pro-Ala-Cys-Thr-Gly-Cys-Tyr			0.4
- - - - - - - - Gly - - - - -			10
- - - - - - - - Ser - - - - -			14
- - - - - - - - Asp - - - - -			506
- - - - - - - - Glu - - - - -			540
- - - - - - - - Gln - - - - -			2090
- - - - - - - - Val - - - - -			>6600
- - - - - - - - Leu - - - - -			>6600
- - - - - - - - Phe - - - - -			>6400
- - - - - - - - Lys - - - - -			>6400
- - - - - - - - Arg - - - - -			>6500

MED: minimum effective dose

Replacement by valine, leucine, phenylalanine, lysine or arginine totally abolished the toxic activity. The MED values of these analogues were at least 17,000 times more than that of native toxin. Replacement by glycine and serine showed significant toxic activity while valine and leucine with a little bigger side chains did not show toxic activity at all. It seems therefore that the size of the side chain of the 14th amino acid residue have great effect upon the activity of the toxin.

When we compare the MED values of all the analogues shown in Table 3-5, it is tempting to speculate that, out of the 3 amino acid residues, alanine at position 14 plays most important role in demonstrating activity of the toxin.

IMPORTANCE OF DISULFIDE BONDS FOR THE TOXIC ACTIVITY

To study the importance of the three disulfide bonds on the toxic activity, we synthesized peptides with various combinations of two disulfide bonds[10,20], as shown in Table 6. In peptides 1-9, two cystein residues were protected by acetamide thus the disulfide bond between these cystein residues was not formed. In peptides 10-15, cystein residues were replaced by alanine, thus the two disulfide bond between these cysteines were not formed.

The biological activities of all the synthetic peptides were examined and the results are shown in Table 6. It was found that peptides 1 and 9 had weak, but significant toxic activity, while the other peptides showed no toxic activity at a dose of less than 1 μg.

Table 6. Synthetic analogues of STh with various combinations of two disulfide bonds and their toxic activities

No.	Amino acid Sequence													Toxic activity
	6	7	8	9	10	11	12	13	14	15	16	17	18	

```
          6    7    8    9   10   11   12   13   14   15   16   17   18
        Acm                      Acm
 1      Cys-Cys-Glu-Leu-Cys-Cys-Asn-Pro-Ala-Cys-Thr-Gly-Cys          +

        Acm                                     Acm
 2      Cys-Cys-Glu-Leu-Cys-Cys-Asn-Pro-Ala-Cys-Thr-Gly-Cys          -

        Acm                                               Acm
 3      Cys-Cys-Glu-Leu-Cys-Cys-Asn-Pro-Ala-Cys-Thr-Gly-Cys          -

             Acm            Acm
 4      Cys-Cys-Glu-Leu-Cys-Cys-Asn-Pro-Ala-Cys-Thr-Gly-Cys          -

             Acm                 Acm
 5      Cys-Cys-Glu-Leu-Cys-Cys-Asn-Pro-Ala-Cys-Thr-Gly-Cys          -

             Acm                                Acm
 6      Cys-Cys-Glu-Leu-Cys-Cys-Asn-Pro-Ala-Cys-Thr-Gly-Cys          -

             Acm                                          Acm
 7      Cys-Cys-Glu-Leu-Cys-Cys-Asn-Pro-Ala-Cys-Thr-Gly-Cys          -

                            Acm                 Acm
 8      Cys-Cys-Glu-Leu-Cys-Cys-Asn-Pro-Ala-Cys-Thr-Gly-Cys          -

                            Acm                           Acm
 9      Cys-Cys-Glu-Leu-Cys-Cys-Asn-Pro-Ala-Cys-Thr-Gly-Cys          +

10      Ala-Ala-Glu-Leu-Cys-Cys-Asn-Pro-Ala-Cys-Thr-Gly-Cys          -

11      Ala-Cys-Glu-Leu-Ala-Cys-Asn-Pro-Ala-Cys-Thr-Gly-Cys          -

12      Cys-Cys-Glu-Leu-Ala-Ala-Asn-Pro-Ala-Cys-Thr-Gly-Cys          -

13      Cys-Cys-Glu-Leu-Cys-Ala-Asn-Pro-Ala-Ala-Thr-Gly-Cys          -

14      Cys-Cys-Glu-Leu-Cys-Ala-Asn-Pro-Ala-Cys-Thr-Gly-Ala          -

15      Cys-Cys-Glu-Leu-Cys-Cys-Asn-Pro-Ala-Ala-Thr-Gly-Ala          -
```

From the above findings we assume that one or two of these three disulfide bonds are closely related to formation of the spatial structure of the STh molecule, which is necessary for expression of its toxic activity. To examine this hypothesis, we synthesized peptides 16-18 (Table 7) with two disulfide bonds at specific positions between four cystines. Peptides 16 and 18 correspond to peptides 1 and 9, respectively, in Table 6. As shown in Table 7, peptides 16 and 18 which had two disulfide bonds between residues 7 and 15, and 6 and 11 or 10 and 18 were toxic, while peptide 17 with no disulfide bond at positions 7 and 15 was inactive. Digestion of peptide 16 with aminopeptidase M, and of peptide 18 with carboxypeptidase A gave peptides 19 and 20, respectively, which lack their N-terminal amino acid residue and C-terminal three amino acid residues. These peptides were about twice as toxic as their original peptides. These results indicates that a disulfide bond between Cys^7 and Cys^{15} is necessary for the toxic activity of STh and activity center of STh is located in a peptide from Cys^7 to Cys^{15}.

Table 7. Synthetic analogues of STh with one or two disulfide bonds and their toxic activities

No.	Peptide	MED(pmol)
	Cys-Cys-Glu-Leu-Cys-Cys-Asn-Pro-Ala-Cys-Thr-Gly-Cys	0.4
16	Ala-Cys-Glu-Leu-Cys-Ala-Asn-Pro-Ala-Cys-Thr-Gly-Cys	380
17	Cys-Ala-Glu-Leu-Cys-Cys-Asn-Pro-Ala-Ala-Thr-Gly-Cys	Inactive
18	Cys-Cys-Glu-Leu-Ala-Cys-Asn-Pro-Ala-Cys-Thr-Gly-Ala	290
19	Cys-Glu-Leu-Cys-Ala-Asn-Pro-Ala-Cys-Thr-Gly-Cys	150
20	Cys-Cys-Glu-Leu-Ala-Cys-Asn-Pro-Ala-Cys	110
21	Cys-Glu-Leu-Ala-Cys-Asn-Pro-Ala-Cys-Thr-Gly-Ala (Cys)	Inactive
22	Cys-Cys-Glu-Leu-Ala-Cys-Asn-Pro-Ala-Cys-Thr-Gly-Ala	Inactive
23	Cys-Cys-Glu-Leu-Ala-Cys-Asn-Pro-Ala-Cys-Thr-Gly-Ala	Inactive
24	Cys-Ala-Glu-Leu-Cys-Cys-Asn-Pro-Ala-Ala-Thr-Gly-Cys	Inactive
25	Cys-Glu-Leu-Ala-Cys-Asn-Pro-Ala-Cys-Thr-Gly-Ala	Inactive

MED: minimum effective dose

We further syntehsized a peptide with two disulfide bonds between Cys^6 and Cys^{11} and Cys^7 and Cys^{15}, but without the peptide bond between Cys^6 and Cys^7 (peptide 21) and three kinds of peptides with only one disulfide bond between Cys^6 and Cys^{11}, Cys^7 and Cys^{15} or Cys^{10} and Cys^{18} (peptides 22-24)[11,20]. As shown in Table 7, all these peptides were inactive. Therefore, the spatial structure of STh with two disulfide linkages between Cys^7 and Cys^{15}, and Cys^6 and Cys^{11} or Cys^{10} and Cys^{18} is essential for the toxic activity of STh. In other words, the spatial structure of STh cannot be maintained by only one disulfide bond between Cys^7 and Cys^{15}, and Cys^6 and Cys^{11} or Cys^{10} and Cys^{18} is essential for the toxic activity of STh.

A TWISTED "8"-SHAPE OF ST

A CPK-model of a short analogue of STh with 13 amino acid residues was constructed from its three disulfide linkages[9] and nuclear Overhauser effects (NOE) data measured by two-dimensional ^{1}H-NMR spectroscopy[21] (Fig. 4). The main chain of the peptide viewed from the front in Fig. 4 is seemingly folded in a twisted "8"-shape and there are two loop structures consisting of two turns between residues 9 to 11 and 12 to 15. The three disulfide linkages are located in the hinge region binding these two loops and presumably stabilize the spatial structure of STh. From the results shown in Table 6 and 7, it is assumed that cleavage of the disulfide bond between Cys^6 and cys^{11} or Cys^{10} and Cys^{18} does not appreciably disrupt the spatial structure necessary for expression of toxicity, and the twisted "8"-structure shown in Fig. 4. is still maintained, at least partially. However, cleavage of the disulfide bond between Cys^7 and Cys^{15} may completely loosen the spatial conformation of STh, because peptide 17 in Table 7 was completely inactive. Cleavage of any two of the disulfide bonds shown in Fig. 4 may completely disrupt the structure fixed by the hinge region. Therefore, a twisted "8"-structure of the main chain seems important for the toxic activity of STh or for stabilization of its spatial structure of the peptide chain from Cys^7 to Cys^{15} may be necessary for expression of the toxic activity of STh.

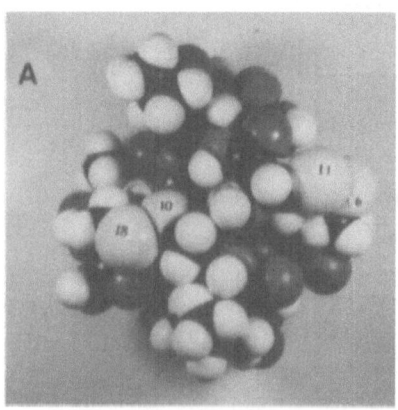 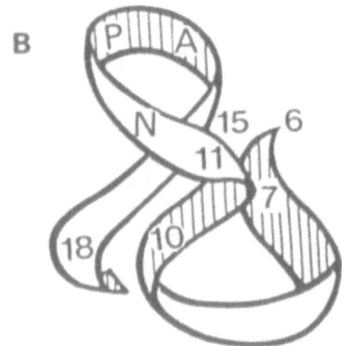

Fig. 4. A CPK-model (A) and schematic ribbon drawing (B) of a STh analogue STh(6-18).

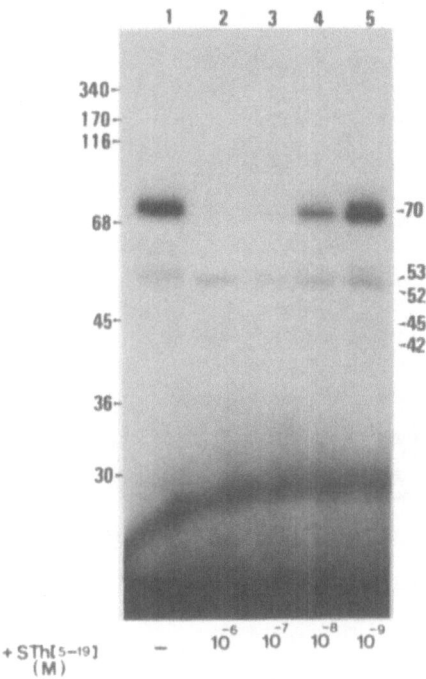

Fig. 5. Effect of unlabeled STh(5-19) on the binding of ^{125}I-ANB-STh(5-19) to 70 kDa protein.

70 kDa PROTEIN OF RAT INTESTINAL BRUSH-BORDER MEMBRANE AS A RECEPTOR FOR ST

To identify a receptor(s) for ST, radioiodinated N-5-azido-nitrobenzoyl-STh(5-19) (^{125}I-ANB-STh(5-19)) was incubated with rat intestinal brush-border membrane and was followed by photolysis. As shown in Fig. 5, radioactivity was mainly incorporated into 70 kDa protein. Although other proteins with 53 kDa, 45 kDa and 42 kDa were also labeled, radiolabelings of these proteins were much lower than that of 70 kDa, Fig 5 shows a specificity of 70 kDa protein to bind ST. When an excess STh(5-19) at concentration of 10^{-8} M or more was added during the initial incubation before photolysis (lane 2-4 in Fig. 5), radiolabeling of 70 kDa protein was remarkably reduced. These results suggest that the photoaffinity labeling of the 70 kDa protein with ^{125}I-ANB-STh(5-19) involved a specific receptor-ligand interaction and that the 70 kDa protein is unique receptor of STh.

BINDING OF STh ANALOGUES TO THE 70 kDa PROTEIN, A RECEPTOR OF ST

Binding of various synthetic STh analogues with a single amino acid replacement at position 14 to the 70 kDa protein was

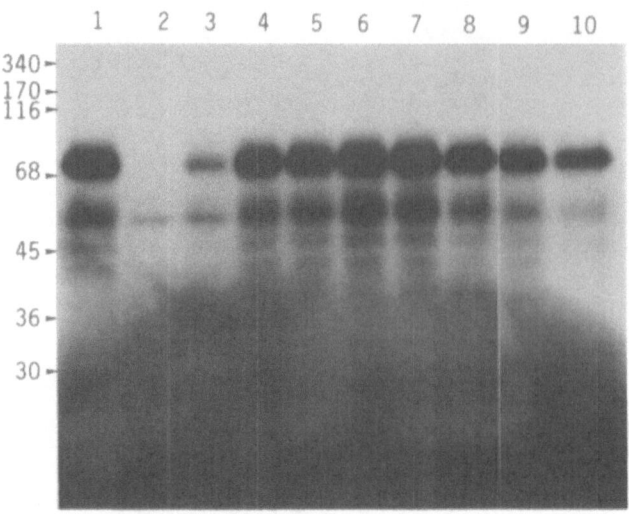

Fig. 6. Effect of various STh analogues with a single amino acid replacement on the binding of ^{125}I-ANB-STh(5-19) to 70 kDa protein

studied by competitive binding of the synthetic STh analogues with the binding of ^{125}I-ANB-STh(5-19) to the 70 kDa protein. As shown in Fig. 6, the binding of ^{125}I-ANB-STh(5-19) to the 70 kDa protein was completely inhibited by STh(6-19) (lane 2 in Fig. 6) and partly by the STh analogues with a replacement by glycine at position 14 (lane 3 in Fig. 6), which showed about one-twentieth of the activity of the original toxin (Table 5). At the same molar level as that of STh(6-19) other analogues with a single amino acid replacement by glutamic acid, glutamine, phenylalanine, arginine, valine, lysine and leucine did not inhibit the binding of the ^{125}I-ANB-STh(6-19) to the 70 kDa protein. These results indicate that the binding abilities of synthetic peptides to the 70 kDa protein are correlated well with their toxic activities.

REFERENCES

1. T. Hirayama, H. Ito and Y. Takeda, Inhibition by the protein kinase inhibitors, isoquinolinesulfonamides, of fluid accumulation induced by Escherichia coli heat-stable enterotoxin, 8-bromo-cGMP and 8-bromo-cAMP in suckling mice, Microb. Pathog., 7: 255-261 (1989).
2. T. Hirayama, M. Noda, H. Ito and Y. Takeda, Stimulation of phosphorylation of rat brush-border membrane proteins by Escherichia coli heat-stable enterotoxin, cholera enterotoxin and cyclic nucleotides, and its inhibition by protein kinase inhibitors, isoquinolinesulfonamides, Microb. Pathog., 8: 421-431 (1990).
3. S. Aimoto, T. Takao, Y. Shimonishi, S. Hara, T. Takeda, Y. Takeda and T. Miwatani, Amino-acid sequence of a heat-stable enterotoxin produced by human enterotoxigenic Escherichia coli, Eur. J. Biochem., 129: 257-263 (1982).
4. T. Takao, T. Hitouji, S. Aimoto, Y. Shimonishi, S. Hara, T.

Takeda, Y. Takeda and T. Miwatani, Amino acid sequence of a heat-stable enterotoxin isolated from enterotoxigenic Escherihcia coli strain 18D, FEBS Lett., 152: 1-5 (1983).

5. T. Takao, N. Tominaga, Y. Shimonishi, S. Hara, T. Inoue and A. Miyama, Primary structure of heat-stable enterotoxin produced by Yersinia enterocolitica, Biochem. Biophys. Res. Commun., 125: 845-851 (1984).

6. T. Takao, N. Tominaga, S. Yoshimura, Y. Shimonishi, S. Hara, T. Inoue and A. Miyama, Isolation, primary structure and synthesis of heat-stable enterotixn produced by Yersinia enterocolitica, Eur. J. Biochem., 152: 199-206 (1985).

7. T. Takao, Y. Shimonishi, M. Kobayashi, O. Nishimura, M. Arita, T. Takeda, T. Honda and T. Miwatani, Amino acid sequence of heat-stable enterotoxin produced by Vibrio cholerae non-01, FEBS Lett., 193: 250-254 (1985).

8. Y. Shimonishi, Y. Hidaka, M. Koizumi, M. Hane, S. Aimoto, T. Takeda, T. Miwatani and Y. Takeda, Mode of disulfide bond formation of a heat-stable enterotoxin (STh) produced by a human strain of enterotoxigenic Escherichia coli, FEBS Lett., 215: 165-170 (1987).

9. Y. Hidaka, H. Kubota, S. Yoshimura, H. Ito, Y. Takeda and Y. Shimonishi, Disulfide linkages in a heat-stable enterotoxin (STp) produced by a porcine strain of enteriotoxigenic Escherichia coli, Bull. Chem. Soc. Jpn., 61: 1265-1271 (1988).

10. Y. Hidaka, H. Kubota, S. Yamasaki, H. Ito, Y. Takeda and Y. Shimonishi, Structural requirements for expression of the toxicity of heat-stable enterotoxins produced by enteric bacteria, Adv. Research Cholera Related Diarrhea, 7: 113-123 (1990).

11. S. Aimoto, H. Watanabe, H. Ikemura, Y. Shimonishi, T. Takeda, Y. Takeda and T. Miwatani, Chemical synthesis of a highly potent and heat-stable analog of an enterotoxin produced by a human strain of enterotoxigenic Escherichia coli, Biochem. Biophys. Res. Commun., 112: 320-326 (1983).

12. H. Ikemura, S. Yoshimura, S. Aimoto, Y. Shimonishi, S. Hara, T. Takeda, Y. Takeda and T. Miwatani, Synthesis of a heat-stable enterotoxin (STh) produced by a human strain SK-1 of enterotoxigenic Escherichia coli, Bull. Chem. Soc. Jpn., 57: 1381-1387 (1984).

13. H. Ikemura, H. Watanabe, S. Aimoto, Y. Shimonishi, S. Hara, T. Takeda, Y. Takeda and T. Miwatani, Heat-stable enterotoxin (STh) of human enterotoxigenic Escherichia coli (strain SK-1). Structure-activity Relationship, Bull. Chem. Soc. Jpn, 57: 2550-2556 (1984).

14. S. Yoshimura, M. Miki, H. Ikemura, S. Aimoto, Y. Shimonishi, T. Takeda, Y. Takeda and T. Miwatani, Chemical synthesis of a heat-stable enterotoxin produced by enterotoxigenic Escheirchia coli strain 18D, Bull. Chem. Soc. Jpn., 57: 125-133 (1984).

15. S. Yoshimura, T. Takao, H. Ikemura, S. Aimoto, Y. Shimonishi, S. Hara, T. Takeda, Y. Takeda and T. Miwatani, Chemical synthesis of fully active and heat-stable enterotoxin of enterotoxigenic Escherichia coli strain 18D. Bull. Chem. Soc. Jpn, 57: 2543-2549 (1984).

16. S. Yoshimura, T. Takao, Y. Shimonishi, S. Hara, M. Arita, T, Takeda, H. Imaishi, T. Honda and T. Miwatani, A heat-stable enterotoxin of Vibrio cholerae non-01: Chemical synthesis, and biological and physicochemical properties, Biopholymers, 25: 69-83 (1986).

17. Y. Takeda, T. Takeda, T. Yano, K. Yamamoto and T. Miwatani, Purification and partial characterization of heat-stable

enterotoxin of enterotoxigenic Escherichia coli, Infect. Immun., 25: 978-985 (1979).

18. S. Yamasaki, T. Sato, Y. Hidaka, H. Ozaki, H. Ito, T. Hirayama, Y. Takeda, T. Sugimura, A, Tai and Y. Shimonishi, Structure-activity relationship of Escherichia coli heat-stable enterotoxin: Role of Ala residue at position 14 in toxin-receptor interaction, Bull. Chem. Soc. Jpn., 63: 2063-2070 (1990).

19. S. Yamasaki, H. Ito, T. Hirayama, Y. Takeda and Y. Shimonishi, Effects on the activity of amino acids replacements at position 12, 13 and 14 of heat-stable enterotoxin (STh) by chemical synthesis, Adv. Research Cholera Related Diarrhea, 8: in press.

20. S. Yamasaki, Y. Hidaka, H. Ito, Y. Takeda and Y. Shimonishi, Structural requirements for the spatial structure and toxicity of heat-stable enterotoxin (STh) of enterotoxigenic Escherichia coli, Bull. Chem. Soc. Jpn., 61: 1701-1706 (1988).

21. T. Ohkubo, Y. Kobayashi, Y. Shimonishi and Y. Kyogoku, A conformational study of polypeptides in solution by [1]H-nmr and distance geometry, Biopolymers, 25: 123-134 (1986).

22. T. Hirayama, K. Matsumoto, Y. Shimonishi and Y. Takeda, Glycoprotein receptors for a heat-stable enterotoxin (STh) produced by enterotoxigenic Escheirchia coli, Adv. Research Cholera Related Diarrhea, 9: in press.

MOLECULAR ANALYSIS OF POTENTIAL ADHESIONS

OF *VIBRIO CHOLERAE* O1

Paul A. Manning

Department of Microbiology and Immunology
The University of Adelaide
Adelaide S.A. 5001 AUSTRALIA

INTRODUCTION

   Vibrio cholerae O1 is a non-invasive enteric pathogen of humans responsible for asiatic cholera. Thus, in order for it to cause disease it must adhere to and colonize the small intestinal mucosa where it is then able to efficiently deliver its toxins responsible for inducing the typical watery diarrhoea. The adhesions, colonization factors and toxins all have been the subject of interest because of the implications for vaccine devlopment.

   V. cholerae O1 exists as two biotypes , classical and El Tor, which are defined on the basis of several biological characteristics. Within these biotypes are two major serotypes, Inaba and Ogawa. A third, minor and apparently unstable serotype, Hikojima, has been defined. However, recent genetic evidence from this laboratory casts some doubt on its actual nature.

   Various components of the cell envelope have been implicated as adhesions in *V. cholerae*. These include the flagellum and its sheath, the O-antigen of the lipopolysaccharide (LPS), various outer membrane proteins, haemagglutinins and fimbriae or pili (1,2,7,9,11,15). In most cases each of these antigens represents more than one determinant. For example, Hanne and Finkelstein (7) have defined at least four haemagglutinins which differ in their sugar sensitivity, requirement for divalent cations and phase of growth at which they are expressed. Similarly, Hall et al. (6) have identified three different fimbrial types. Thus, it was clear that a genetic analysis of the various determinants was needed in order to precisely define the roles of these different components.

   This laboratory has been concentrating on the genetic and biochemical analysis of a number of potential adhesions. These studies have begun to clarify the relative importance of these factors, as well as demonstrating both physical and genetic interactions.

THE POTENTIAL ADHESIONS

The three components which have been the subject of detailed genetic analysis are a mannose-fucose resistant haemagglutinin (MFRHA), the O-antigens of the lipopolysaccharides and the toxin coregulated pilus (TCP).

The Mannose-Fucose Resistant Haemagglutinin (MFRHA)

A MFRHA is produced by both biotypes of *V. cholerae* and if involved in adherence could represent a cross-protecting antigen. Consequently, the genetic locus determining biosynthesis of a MFRHA was cloned and identified by its ability to confer the haemagglutinating phenotype on laboratory strain *Escherichia coli* K-12 (4). Extensive genetic analysis and determination of the nucleotide sequence of this locus has defined the genes for several proteins flanked by a series of repeat units of about 130 bp (Fig. 1). The role of these highly conserved repeats is unknown, and there appear to be further copies elsewhere in the chromosome.

Truncation of the MFRHA at the *Xba*I site within its gene resulted in loss of haemagglutinating activity. Consequently this site was used for the insertion of a kanamycin resistance cassette, which served to both inactivate the gene and also to provide a selection for introducing the mutated gene back into the *V. cholerae* chromosome by allelic exchange. Such a mutation was introduced into strain 569B (classical, Inaba) and examined for virulence in the infant mouse cholera model. This mutation increased the LD_{50} at 48h from about 10^5 to greater than 2×10^8 clearly demonstrating its significance. This was not an effect of the antibiotic resistance cartridge since a similar mutation introduced into the extracellular DNAse gene had no effect on virulence.

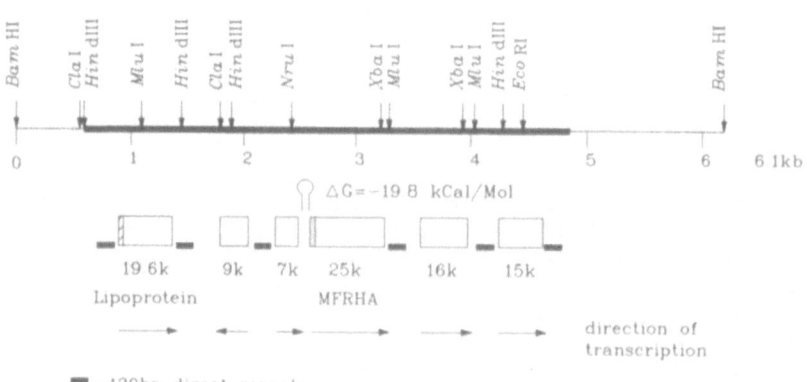

Fig. 1.   Genetic organization of the MFRHA locus. The boxes correspond to the genes for the various proteins encoded within the locus. The direct repeat sequences flanking the genes are indicated as is the direction of transcription of the genes. A terminator-like structure is found between the genes for the 7kDa protein and the MFRHA.

Analysis of the predicted proteins encoded at the MFRHA locus suggests that the MFRHA itself is a cationic outer membrane protein which is held on the cell surface primarily by charge interactions with the LPS. However, a second protein corresponds to a lipoprotein, which is also predicted to be surface located. It is thought that the lipoprotein may in fact correspond to an anchor to hold the MFRHA on the surface.

Analysis of the regulation of the MFRHA suggests that its gene is under the control of iron starvation and that it is subject to repression by a Fur-like protein (8). Northern blot and primer extension analyses identify the first nucleotide of the mRNA as being in the centre of a Fur binding site, prior to the gene for the 7kDa protein immediately preceding the MFRHA. This implies that transcription would be blocked in the presence of Fur protein (Barker, A., Williams, S.G. and Manning, P.A., manuscript in preparation). These genes may form part of a virulence regulon which also includes the El Tor haemolysin, since a Fur binding site has also been identified near the start of transcription of its mRNA (S.G. Williams and P.A. Manning, manuscript in preparation). However, a new gene encoding a positive activator of transcription of the haemolysin structural gene, hlyA, has also been identified and cloned, indicating that its regulation is more complex.

The TCP Colonization Factor

Taylor et al have described a pilus colonization factor, TCP (toxin co-regulated pilus), which is co-regulated with cholera toxin. Because of the requirement of TCP expression for the positive activator protein ToxR, our cloning strategy involved constructing a gene bank in E. coli K-12 using a mobilizable cosmid vector, which then permitted the clones to be conjugally mobilized into V. cholerae strain O17(El Tor, Ogawa). This strain was chosen because it appeared unable to express TCP (13). Following this protocol, we cloned the gene cluster encoding TCP biosynthesis (12).

Introduction of the cosmid clone pPM2103 into strain O17 resulted in the production of large bundles of TCP fimbriae as visualized by immuno-gold electron microscopy. Thus by comparing strain O17 and O17[pPM2103] we could ask a number of questions about the role of TCP in colonization and protection in the infant mouse cholera model.

The expression of TCP, mediated by plasmid pPM2103, decreased the LD_{50} of strain O17 from 3×10^5 to 5.7×10^3 demonstrating that TCP enhanced its virulence. O17[pPM2103] was also capable of absorbing out the non-LPS protective activity against a TCP^+ strain whereas O17 harbouring a control cosmid (strain DS9) had no effect on the protective titre of the serum. This indicated that TCP was a protective antigen. In order to confirm this O17[pPM2103] and DS9 were used to immunize mice and the resulting antisera were absorbed with O17 and assayed for protective activity against O17 and another TCP^+ strain Z17561. Only the antiserum raised against O17[pPM2103] was protective and only against Z17561. Thus, TCP appeared to be both a protective antigen and immunogenic.

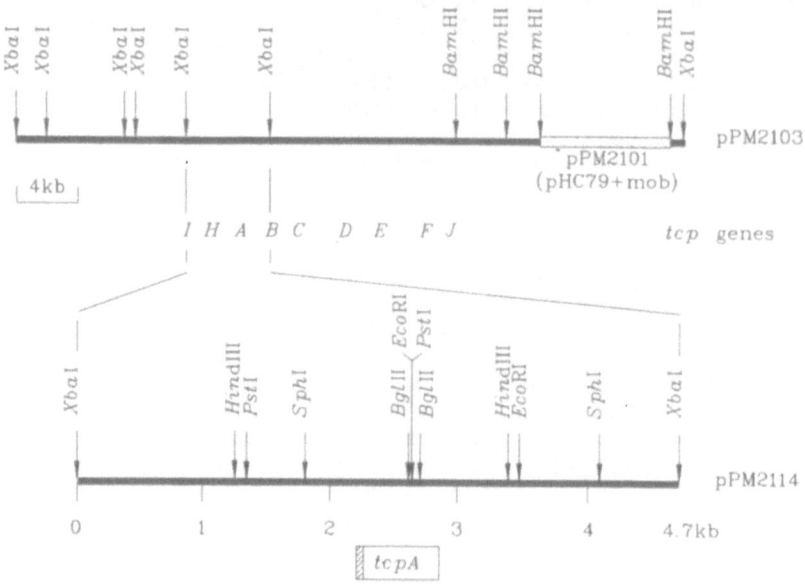

Fig. 2.   Localization of the *tcpA* structural gene.   The
          nucleotide sequence for much of the *tcp* region has
          been determined.   In particular, the gene, *tcpA*, for
          the major structural subunit has been localized
          within a 4.7kb *Xba*I fragment as shown.   Plasmid
          pPM2103 is a cosmid clone containing the entire *tcp*
          region and pPM2114 is the 4.7kb *Xba*I fragment cloned
          in pK18 (3).

     Southern hybridization analysis of pPM2103 using a
synthetic oligonucleotide probe predicted from the NH$_2$
terminal sequence of TcpA allowed the localization of the
structural gene for the major structural subunit (Fig. 2).
Further analyses have localized other genes associated with
biosynthesis, processing and assembly of the fimbriae within
the *tcp* region.   Nucleotide sequence analysis of the gene,
tcpA, (3) has greatly facilitated the construction of specific
mutants.   The mutated *tcpA* gene was constructed by deleting
the DNA between the *Bgl*II sites within *tcpA* and inserting a
kanamycin resistance cassette.

     In order to assess the importance of TCP in colonization,
infant mice were fed sub-LD$_{50}$ doses of either strain 569B or
its *tcpA* mutant and at various time intervals the mice were
sacrificed and the intestines removed, homogenized and the
number of *V. cholerae* enumerated (Fig. 3).   With a low
inoculum initial colonization of the *tcpA* mutant was observed,
however, the organisms steadily decreased at the higher dose
possibly due to the much higher inherent levels of
diahrroeagenic toxins.   These results demonstrated that the
tcpA mutant could not persist in the gut, implying a defect in
colonization.   However, the initial colonization observed at
lower doses is suggestive of an additional adherence factor,
which we believe to be the MFRHA described above.

The O-antigens of the Lipopolysaccharides (LPS)

Protection against *V. cholerae* is at least in part serotype
specific, which implicates the serotype specific determinant,
the O-antigen of the LPS.   The genes, *rfb*, encoding O-antigen

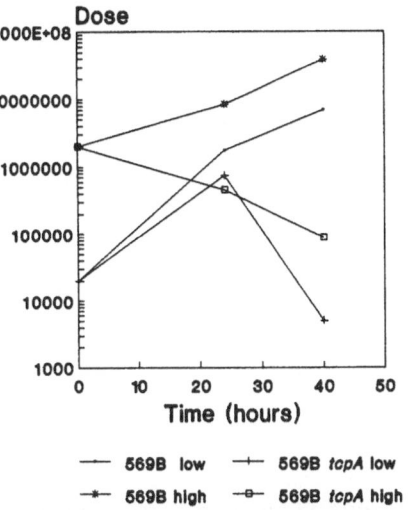

Fig. 3. Colonization of the infant mouse gut by 569B and its *tcpA* mutant. Infant mice were fed either a low (2 x 10^4) or high (2 x 10^6) dose and at 24h and 48h the intestines were removed, homogenized and the number of organisms counted.

biosynthesis for both the Inaba and Ogawa serotypes have been cloned, localized and mapped (10,16,17). These studies also confirmed that antibodies to the O-antigen alone were protective in the infant mouse cholera model. Nucleotide sequence analysis of the DNA fragment encoding the genes for O-antigen biosynthesis has enabled a precise map of the genes to be constructed. Comparisons between Inaba and Ogawa strains has enabled the gene determining the Ogawa specificity to be localized (Fig. 4). Bacteriophage CP-T1 uses O-antigen as its receptor (5), and this provided an excellent selection for isolating transposon insertion mutants in the *rfb* genes (17).

Several such mutants have been isolated and all map within the 5' region of the *rfb* operon resulting in a total block in O-antigen biosynthesis. These have been assessed for virulence in the infant mouse model and are clearly attenuated (Table 1).

Table 1 . Virulence of *rfb* mutants in the infant mouse cholera model.

Strain	Dose	
	10^8	10^7
569B	0/8	1/8
rfb-2	3/8	8/8
rfb-6	3/8	7/8
rfb-7	8/8	8/8
rfb-8	8/8	8/8
rfb-9	8/8	8/8
rfb-10	7/8	8/8

Survival of the infant mice was measured at 48h. The LD_{50} of 569B given at the same time was 5 x 10^5.

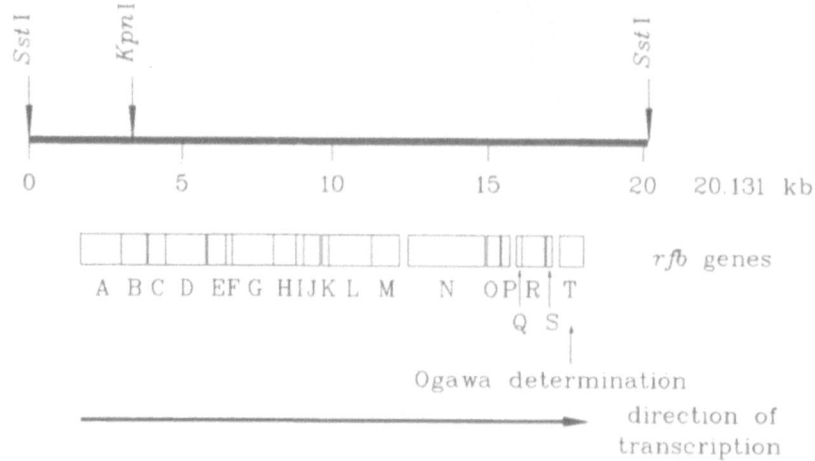

Fig. 4.   Genetic organization of the *rfb* locus.   The
         nucleotide sequence of the entire *Sst*I (*Sac*I)
         fragment containing the *rfb* region has been
         determined from both Ogawa and Inaba serotypes
         (Karageorgos, L.E., Brown, M.H., Stroeher, U.H.,
         Morona, R. and Manning, P.A., manuscript in
         preparation).   This and studies with specific
         subclones has enabled *rfbT* to be defined as the gene
         responsible for determining the Ogawa serotype
         specificity (Stroeher, U.H., Karageorgos, L.E.,
         Morona, R. and Manning, P.A., manuscript in
         preparation).

     It has been shown that the LPS is not only present on the
cell surface but also extends along the sheath of the
flagellum.   Consequently we have examined the cells electron
microscopically.   These studies demonstrated that the *rfb*
mutants produced large amounts of flagellar sheath material
and very few normal flagella.   This was reflected in the poor
motility of the cells as measured by a lack of swarming on
supersoft agar.   Furthermore, the electron micrographs showed
greatly reduced levels of TCP, and Western blot analysis
revealed that a pool of unassembled pilin was being retained
within whole cells.   Thus, an intact LPS is essential for the
correct assembly of both the TCP fimbriae and the flagellum.

CONCLUDING COMMENTS

These studies implicate both the MFRHA and TCP as important
colonization factors of *V. cholerae* O1, whereas the role of
the O-antigen of the LPS is clearly different.   It would seem
that the O-antigen is required for the export and/or assembly
of functional flagella and also TCP, and is probably also
essential for the MFRHA to localize correctly on the cell
surface.   The interactions between these molecules and their
relative significance are the subject of ongoing studies.   The
recognition domains and the receptors which they recognize are
also being defined.

ACKNOWLEDGEMENTS

The author would especially like to acknowledge the numerous co-workers who have been involved in the various stages of the analyses reported here. Work in the author's laboratory is supported by the National Health and Medical Research Council of Australia, the Australian Research Committee and the Clive and Vera Ramaciotti Foundations.

REFERENCES

1. Attridge, S.R. and Rowley, D., 1983, The role of the flagellum in the adherence of *Vibrio cholerae*. J. Infect. dis., 147:864.
2. Chitnis, D.S., Sharma, K.D. and Kamat, R.S., 1982, Role of somatic antigen of *Vibrio cholerae* in adhesion to intestinal mucosa. J. Med. Microbiol., 5:52.
3. Faast, R., Ogierman, M.A., Stroeher, U.H. and Manning, P.A., 1989, Nucleotide sequence of the structural gene, *tcpA*, for a major pilin subunit of *Vibrio cholerae*. Gene, 85:227.
4. Franzon, V.L. and Manning, P.A., 1986, Molecular cloning and expression in *Escherichia coli* K-12 of the gene for a hemagglutinin from *Vibrio cholerae*. Infect. Immun., 52:279.
5. Guidolin, A. and Manning, P.A., 1985, Bacteriophage CP-T1 of *Vibrio cholerae*: identification of the cell surface receptor. Eur. J. Biochem., 153:89.
6. Hall, R.H., Vial, P.A., Kaper, J.B., Mekalanos, J.J. and Levine, M.M., 1988, Morphological studies on fimbriae expressed by *Vibrio cholerae* 01. Microbial Pathogenesis, 4:257.
7. Hanne, L.F. and Finkelstein, R.A., 1982, Characterization and distribution of the hemagglutinins produced by *Vibrio cholerae*. Infect. Immun., 36:209.
8. Hantke, K., 1982, Negative control of iron uptake systems in *Escherichia coli*. FEMS Microbiol. Lett., 15:83.
9. Kabir, S. and Showkat, A., 1983, Characterization of surface properties of *Vibrio cholerae*. Infect. Immun., 39:1048.
10. Manning, P.A., Heuzenroeder, M.W., Yeadon, J., Leavesley, D.I., Reeves, P.R. and Rowley, D., 1986, Molecular cloning and expression in *Escherichia coli*K-122 of the O-antigens of the Inaba and Ogawa serotypes of the *Vibrio cholerae* 01 lipopolysaccharides and their potential for vaccine development. Infect. Immun., 53:272.
11. Neoh, S.H. and Rowley, D., 1970, The antigens of *Vibrio cholerae* involved in vibriocidal action of antibody and complement. J. Infect. Dis., 121:505.
12. Sharma, D.P., Stroeher, U.H., Thomas, C.J., Manning, P.A. and Attridge, S.R., 1989a, The toxin-coregulated pilus (TCP) of *Vibrio cholerae*: molecular cloning of genes involved in pilus biosynthesis and evaluation of TCP as a protective antigen in the infant mouse model. Microbial Pathogenesis, 7:437.

13. Sharma, D.P., Thomas, C.J., Hall, R.H., Levine, M.M. and Attridge, S.R., 1989b, Significance of toxin-coregulated pili as protective antigens of *Vibrio cholerae* in the infant mouse model. Vaccine, 7:451.

14. Taylor, R.K., Miller, V.L., Furlong, D.B. and Mekalanos, J.J., 1987, Use of *phoA* fusions to identify a pilus colonization factor coordinately regulated with cholera toxin. Proc. Natl. Acad. Sci. USA, 84:2833.

15. Tweedy, J.M., Park, R.W.A. and Hodgkiss, W., 1968, Evidence for the presence of fimbriae (pili) on vibrio species. J. Gen. Microbiol., 51:235.

16. Ward, H.M., Morelli, G., Kamke, M., Morona, R., Yeadon, J., Hackett, J.A. and Manning, P.A., 1987, A physical map of the chromosomal region determining O-antigen biosynthesis in *Vibrio cholerae* O1. Gene, 55:197.

17. Ward, H.M. and Manning, P.A., 1989, Mapping of chromosomal loci associated with lipopolysaccharide synthesis and serotype specificity in *Vibrio cholerae* O1 by transposon mutagenesis using Tn*5* and Tn*2680*. Mol. Gen. Genet., 218:367.

SHIGELLA TOXIN AND RELATED PROTEINS - TRANSLOCATION

TO THE CYTOSOL AND MECHANISM OF ACTION

Sjur Olsnes[*], Kirsten Sandvig[*] and Bo van Deurs[+]

[*]Institute for Cancer Research
The Norwegian Radium Hospital
Montebello, Oslo, Norway
and
[+]Structural Cell Biology Unit
Department of Anatomy
The Panum Institute
University of Copenhagen
Copenhagen, Denmark

INTRODUCTION

Shigella species and certain E. coli strains produce toxins that are exceedingly toxic to many mammalian cells (van Heyningen and Gladstone, 1953; Olsnes and Eiklid, 1980; Karmali et al., 1985; O'Brian and Holmes, 1987). Thus, as little as 0.1 pg/ml Shigella toxin is enough to kill a culture of sensitive HeLa cells. Shigella toxin and the related Shiga-like toxins have been cloned and sequenced (Calderwood et al., 1987; Kozlov et al., 1987; Strockbine et al., 1988). The toxins act by inactivating the ribosomes and thereby block protein synthesis (Reisbig, Olsnes and Eiklid, 1981). A necessary step in their mechanism of action is to translocate to the cytosol an enzymatically active polypeptide chain.

STRUCTURE AND MECHANISM OF ACTION

Shigella toxin and Shiga-like toxins have the same general structure, consisting of an A-chain which is linked by non-covalent interactions to 5 copies of the B-chain (Olsnes, Reisbig and Eiklid, 1981; O'Brian and Holmes, 1987). The B-subunit binds the toxin to cell surface glycolipids containing Gal 1-4Galß (Lindberg et al., 1987; Lingwood et al., 1987). The A-chain has a trypsin-sensitive region bridged by a disulfide. Upon proteo-lytic cleavage, and reduction of the disulfide, the enzymatically active A_1-fragment is released. This fragment inactivates eukaryotic ribosomes by removal of a particular adenine residue from 28 S ribosomal RNA without cleaving the RNA backbone. As a result, the elongation factors are not able to bind properly to the ribosomes and protein synthesis is blocked (Endo et al., 1988).

Certain plant toxins, such as abrin, ricin, modeccin and others have the same enzymatic activity as Shigella toxin (Endo et al., 1988). Furthermore, the enzymatically active A-chain of the plant toxins is structurally closely related to Shigella toxin A-chain. On the other hand, the B-chain of the plant toxins, which binds to carbohydrates with terminal galactose is structurally unrelated to Shigella toxin.

There are two kinds of Shiga-like toxins produced by certain enteropathogenic strains of E. coli, viz. Shiga-like toxin 1 and 2 (O'Brian and Holmes, 1987). Whereas Shiga-like toxin 1 differs insignificantly from Shigella toxin, Shiga-like toxin 2 differs considerably. All three toxins have, however, the same general structure and mechanism of action.

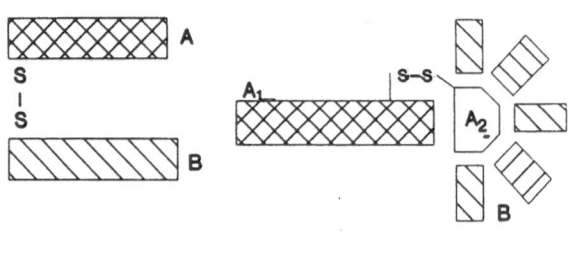

Fig. 1. Schematic structure of ricin and Shigella toxin.

TOXIN ENTRY

The process least well understood in the mechanism of action of toxins in this group is how they penetrate through the cellular membrane to reach their target in the cytosol, the ribosomes.

In all cases studied, it appears that the process is initiated by endocytic uptake of the surface-bound toxins (Olsnes and Sandvig, 1985). Two different kinds of endocytic uptake may be involved in this process. Shigella toxin is bound to receptors that are scattered over the surface of the cell, but after binding of the toxin the complexes aggregate in coated pits and the toxin is subsequently endocytosed rapidly by the formation of coated vesicles (Sandvig et al., 1989).

Ricin on the other hand is endocytosed mainly by an alternative pathway that does not appear to involve coated pits and which may represent a more slow internalization from any location at the surface membrane (Sandvig et al., 1987, 1988). In accordance with this, the endocytic uptake of ricin occurs much more slowly than that of Shigella toxin.

The translocation to the cytosol is best understood in the case of diphtheria toxin where it appears to occur from the

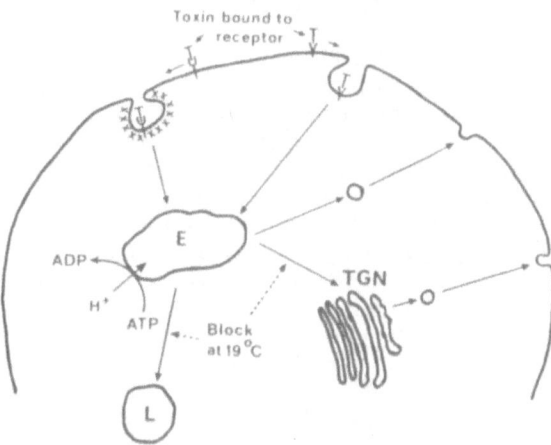

Fig. 2. Schematic presentation of uptake and intracellular routing of toxins. Shigella toxin and ricin bind to sites at the cell surface and are then internalized. Whereas the internalization of Shigella toxin occurs from coated pits, ricin is internalized to a large extent by an other mechanism, probably not involving coated pits. Both toxins are subsequently transported via endosomes (E) to the trans-Golgi network (TGN) by a process that is blocked at 19°C. L, lysosome. (Modified from Olsnes et al., 1989.

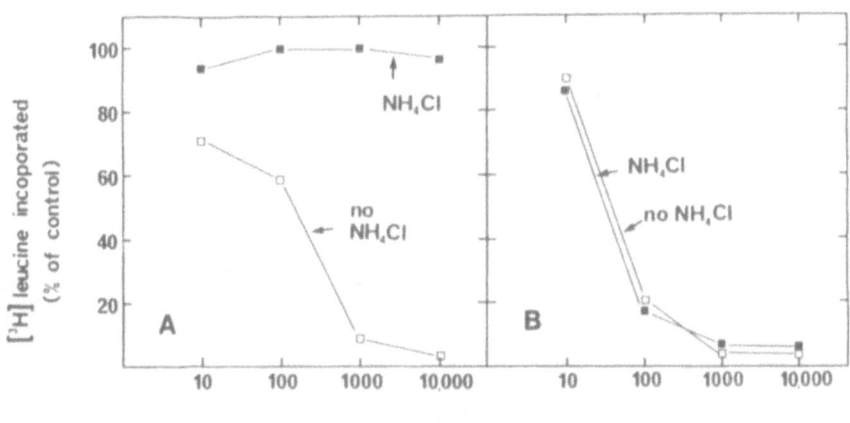

Fig. 3. Inhibition of intoxication when acidified cells are exposed to Shigella toxin. Cells were preincubated with ammonium chloride which was later removed. This induced intracellular acidification. The cells were kept under conditions were they were unable to correct the intracellular pH and then they were exposed to toxin. After 15 min the cells were washed and treated with anti-Shigella toxin serum to remove any toxin remaining at the outside. Finally, the cells were transferred to normal medium and incubated overnight to allow internalized toxin time to express its toxic effect. The next day the ability of the cells to incorporate [^{3}H]leucine was measured.

endosomes (Olsnes et al.,1988). In the case of the other toxins in this group, it appears that transport to another compartment is necessary.

In the case of ricin, modeccin and Shigella toxin there is a temperature-sensitive step in the translocation process which is blocked at temperatures below 20°. This indicates that the toxin must be transported to a compartment beyond the endosomes before translocation can take place.

Ultrastructural studies with Shigella toxin and ricin (Fig. 4) have shown that part of the endocytosed toxin molecules are transported to the trans-Golgi network. In the case of ricin, approximately 5% of the endocytosed toxin is transported to this compartment (van Deurs et al., 1986; 1988).

Transport to the trans-Golgi network appears to be required for toxic effect, at least in the case of ricin. Thus, it was found that myeloma cells producing monoclonal anti-ricin antibodies were highly resistant to ricin even under conditions -

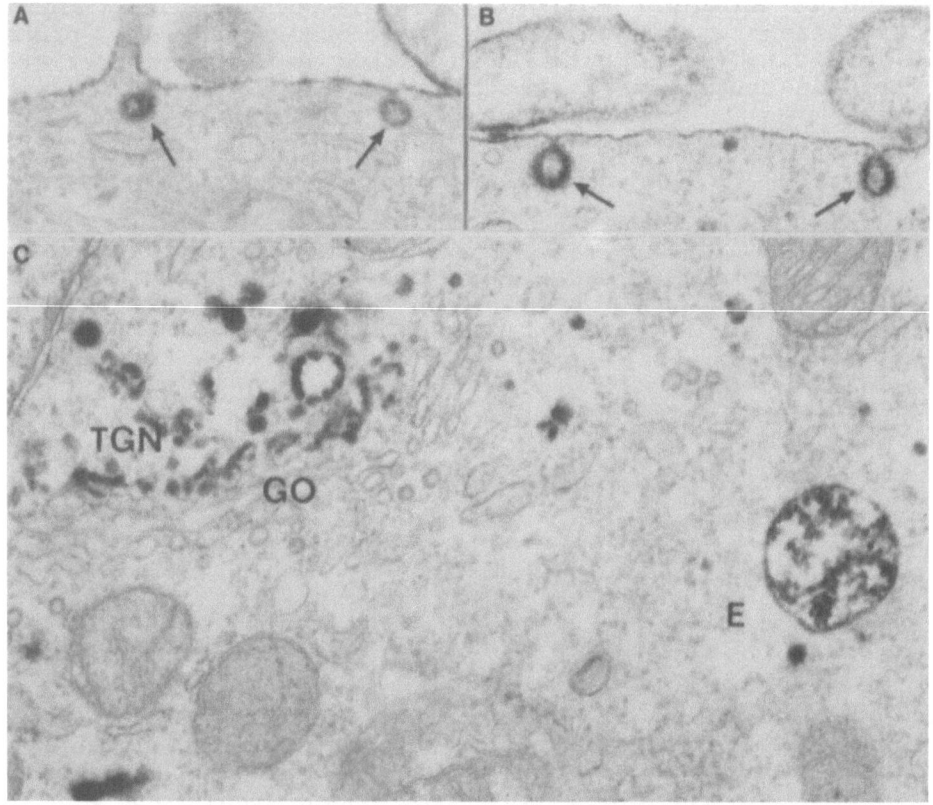

Fig. 4. Ultrastructural studies of the internalization of Shigella toxin in Vero cells. Panel A and B show Shigella toxin-horse radish peroxidase conjugate being internalized from coated pits (arrows). Panel C shows the presence of toxin-horse radish peroxidase in endosome (E) and in the TGN of a Golgi complex (Go).

where the antibodies could not react with the toxin at the cell surface (Youle and Colombatti, 1987). Apparently, incoming ricin that reached the <u>trans</u>-Golgi network was inactivated by antibodies passing through this organelle on its way out of the cell. In contrast to ricin and Shigella toxin that require temperatures above 20° for toxic effect consistant with the possibility that they are translocated from the <u>trans</u>-Golgi, diphtheria toxin appears to be translocated from the endosomes as soon as the pH of the vesicles reaches values below pH 5.3.

Diphtheria toxin, which like Shigella toxin and ricin acts on intracellular targets, appear to enter the cytosol from endosomes, the low pH of these organelles being the trigger for penetration (Olsnes et al.,1988). In accordance with this, diphtheria toxin entry is not strongly inhibited when the temperature is lowered to 20°C, as the transport to the endosomes is not inhibited at this temperature. Brefeldin A inhibits transport of material from the endoplasmic reticulum to the Golgi apparatus. As a result, the Golgi apparatus disintegrates (Lippincott-Schwartz et al., 1989). To test if this interferes with the intoxication process, we tested if brefeldin A treatment altered the sensitivity of Vero cells to Shigella toxin and diphtheria toxin. As shown in Fig. 5, the drug strongly protected the cells against Shigella toxin, but not against diphtheria toxin. This further supports the conclusion that Shigella toxin must be transported to the <u>trans</u>-Golgi compartment for intoxication to take place, whereas this is not necessary in the case of diphtheria toxin.

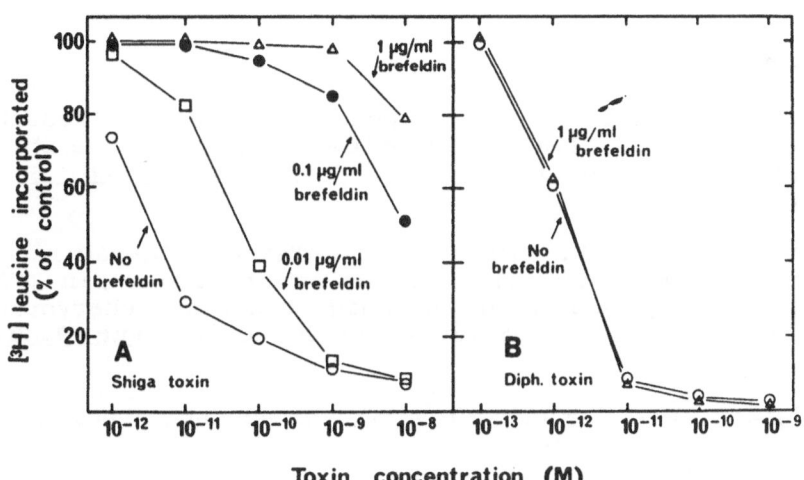

Fig. 5. Ability of brefeldin A to inhibit in Vero cells the toxic effect of Shigella toxin, but not that of diphtheria toxin. Cells were incubated with and without brefeldin A for 30 min and then increasing concentrations of toxin were added. After further incubation for 3 h, the ability of the cells to incorporate [^{3}H]leucine was measured.

THERAPEUTIC POTENTIAL

In recent years much effort has been made to improve the specificity of cytostatic drugs used in cancer treatment and in immunosuppression. Often antibodies against surface antigens have been used as a delivery system to target toxic components to the cells in question (Frankel,1988; Olsnes et al., 1989). True tumor-specific antigens are rare, but the reexpression of developmental antigens on many tumor cells may provide the required selectivity. The amounts of antibodies that reach and bind to the target cells are often small and the antibodies must therefore be armed with very active compounds. Cytocidal toxins of the group here described may fulfill this requirement.

In humans immunotoxins have proved valuable for in vitro purging of bone marrow to remove T-lymphocytes before transplantation to avoid graft-versus-host reaction (see reviews in Frankel, 1988). Moreover, clinical trials with patients treated systemically with immunotoxins have been initiated, and encouraging results have been reported in a number of studies.

Immunotoxins were first considered as anticancer drugs, but more recent research indicates that their potential applications reaches much further. Anti T-cell immunotoxins may be used in immunosuppression as already mentioned, and conjugates with anti-idiotype antibodies to the acetylcholine receptor may eradicate B-cells producing antibodies to the acetylcholine receptor in myasthenia gravis. Allergic conditions might be treated with IgE-toxin conjugates that bind to and destroy mast cells and basophile leukocytes. It should also be noted that the protein synthesis machinery of eukaryotic parasites is sensitive to the toxins, and the possibility of making immunotoxins against such organisms is currently being explored.

REFERENCES

Calderwood, S.B., Auclair, F., Donohue-Rolfe, A., Keusch, G.T. and Mekalanos, J.J. 1987. Nucleotide sequence of the Shiga-like toxin genes of Escherichia coli. Proc. Natl. Acad. Sci. USA, 84:4364-4368.

Endo, Y., Tsurugi, K., Yutsudo, T., Takeda, Ogasawara, Y. T., and Igarashi, K. 1988. Site of action of Vero toxin (V2) from Escherichia coli 0157:H7 and of Shiga toxin on eukaryotic ribosomes. RNA N-glycosidase activity of the toxins. Eur. J. Biochem. 171:45-50.

Frankel, A.E. 1988.Immunotoxins. Kluver Academic Publishers, pp. 39-73.

Karmali, M.A., Petric, M., Lim, C., Fleming, P.C., Arbus, G.S., and Lior, H. 1985. The association between idiopathic hemolytic uremic syndrome and infection by vero toxin-producing Escherichia coli. J. Infect. Dis. 151:775-782.

Kozlov, Yu.V., Kabishev, A.A., Fedchenko, V.I. and Bayev, A.A. 1987. Cloning and sequencing of shiga-toxin structural genes. Proc. Acad. Sci.USSR, 295:740-744.

Lindberg, A. A., Brown, J. E., Strömberg, N., Westling-Ryd, M., Schultz, J. E. and Karlsson, K. A. 1987. Identification of the carbohydrate receptor for Shiga toxin produced by Shigella dysenteriae Type 1. J. Biol. Chem. 262:1779-1785.

Lingwood, C.A., Law, H., Richardson, S., Petric, M., Brunton, J.L., De Grandis, S. and Karmali, M. 1987. Glycolipid binding of purified and recombinant Escherichia coli produced Verotoxin in vitro J. Biol. Chem. 262:8834-8839.

Lippincott-Schwartz, J., Yuan, L.C., Bonifacino, J.S., and Klausner, R.D. 1989. Rapid redistribution of Golgi proteins into the ER in cells treated with brefeldin A: Evidence for membrane cycling from Golgi to ER. Cell, 56:801-813.
O'Brien, A., and Holmes, R.K. 1987. Shiga and Shiga-like toxins. Microbiol. Rev. 51:206-220.

Olsnes, S., and Eiklid, K. 1980. Isolation and characteri-zation of Shigella shigae cytotoxin. J. Biol. Chem. 255:284-289.

Olsnes, S., Moskaug, J.Ø., Stenmark, H. and Sandvig, K. 1988. Diphtheria toxin entry - protein translocation in the reverse direction. Trends Biochem. Sci., 13: 348-351.

Olsnes, S., Reisbig, R., and Eiklid, K. 1981. Olsnes, S., Reis-big, R. and Eiklid, K.: Subunit structure of Shigella toxin. J. Biol. Chem. 256: 8732-8738.

Olsnes, S., Moskaug, J.Ø., Stenmark, H. and Sandvig, K. (1988) Diphtheria toxin entry - protein translocation in the reverse direction. Trends Biochem. Sci., 13:348-351.

Olsnes, S. and Sandvig, K. 1988. How protein toxins enter and kill cells. In Immunotoxins. A.E. Frankel, editor. Kluver Academic Publishers, pp. 39-73.

Olsnes, S., Sandvig, K., Petersen, O.W. and van Deurs, B. 1989. Immunotoxins - entry into cells and mechanism of action. Immunol. Today, 10:291-295.

Reisbig, R., Olsnes, S., and Eiklid, K. 1981. The cytolytic ac-tivity of Shigella toxin. Evidence for catalytic inactivation of the 60S ribosomal subunit. J. Biol. Chem. 256:8739-8744.

Sandvig, K., Olsnes, S., Petersen, O. W., and van Deurs, B. 1987. Acidification of the cytosol inhibits endocytosis from coated pits. J. Cell Biol. 105:679-689.

Sandvig, K., Olsnes, S., Brown, J.E., Petersen, O.W. and van Deurs, B. 1989. Endocytosis from coated pits og Shiga toxin: A glycolipid-binding protein from Shigella dysenteriae 1. J. Cell Biol., 108:1331-1343.

Strockbine, N.A., Jackson, M.P., Sung, L.M., Holmes, R.K. and O'Brian, A.D. 1988. Cloning and sequencing of the genes for Shigella dysenteriae type 1. J. Bacteriol. 170:1116-1122.

van Deurs, B., Sandvig, K., Petersen, O. W., Olsnes, S., Simons, K., and Griffiths, G. 1988. Estimation of the amount of internalized ricin that reaches the <u>trans</u> Golgi network. <u>J. Cell Biol</u>. 106:253-267.

van Deurs, B., Sandvig, K., Petersen, O.W. and Olsnes, S.: Endocytosis and intracellular sorting of ricin. **CRC Review: Trafficking of bacterial toxins**. (C.B. Saelinger, Ed.) CRC Press (1990) pp. 91-119.

van Deurs, B., Tønnessen, T. I., Petersen, O. W., Sandvig, K. and Olsnes, S. 1986. Routing of internalized ricin and ricin conjugates to the Golgi complex. <u>J. Cell Biol</u>. 102:37-47.

van Heyningen, W. E., and Gladstone, G. P. 1953. The neurotoxin of <u>Shigella shigae</u>. I. Production, purification and properties of the toxin. <u>Br. J. Pathol</u>. 34:202-216.

Youle, R.J., and Colombatti, M. (1987) Hybridoma cells containing intracellular anti-ricin antibodies show ricin meets secretory antibody before entering the cytosol. <u>J Biol Chem</u> 262:4676-4682.

VERO CYTOTOXINS (SHIGA-LIKE TOXINS) OF *ESCHERICHIA COLI*

Sylvia M. Scotland

Division of Enteric Pathogens
Central Public Health Laboratory
61 Colindale Avenue, London NW9 5HT, UK

INTRODUCTION

Shiga toxin produced by strains of *Shigella dysenteriae* type 1 and Vero cytotoxins (Shiga-like toxins) produced by strains of *Escherichia coli* belong to a family of related toxins. Two major forms of Vero cytotoxin (VT) are recognised. Polyclonal antiserum to Shiga toxin neutralises the activity of VT1 (SLTI) but not VT2 (SLTII). Strains of *E. coli* produce VT1 only, VT2 only or both toxins. Some strains of animal origin, but none as yet of human origin, also produce heat labile enterotoxin or heat stable enterotoxin. By definition, VT is detected by its cytotoxic effect on a monolayer of Vero cells grown in tissue culture, with serum neutralisation tests to determine which toxins are present[1]. For routine testing, bacteria are grown in a medium such as trypticase soy broth and the toxin present in filtered culture supernatants is assayed. Cell bound toxin can be released by polymixin treatment or sonication to obtain maximum yields. Production of VT1, but not VT2, is increased in an iron-depleted medium[2]. HeLa cells have also been used to test for VT, but this cell line can no longer be recommended as there are variant forms of VT2, termed VT2v, that have litle or no effect on HeLa cells although they are cytotoxic for Vero cells. The first variant toxin to be identified, produced by strains causing porcine (o)edema disease, has been termed VT2vp[3] or VT2e[4]. VT2vh[5] and VT2va[6] are variant toxins from strains of human origin. All the VT2 forms are neutralised at least to some extent by polyclonal antisera raised against one of the other VT2 forms.

BIOLOGICAL ACTIVITY

VT-producing *E. coli* (VTEC) are a cause of diarrhoea in man and animals. In particular in human infections they are associated with bloody diarrhoea and the haemolytic uraemic syndrome in which major features are kidney failure and microangiopathic anaemia. In addition there may be neurological disturbances which also occur in porcine oedema disease caused by VTEC. The role of cytotoxins, and other properties, in the pathogenicity of VTEC is still under investigation[7]. VTEC are not invasive in the way that Shigellae or enteroinvasive *E. coli* are, but many have other properties in addition to VT production that may contribute to their virulence. These include the ability to cause attaching and effacing lesions[8] and the production of an entero-haemolysin[9].

Shiga toxin, VT1 and VT2 are all composed of one A and several B protein subunits. All three cause fluid accumulation when injected into tied loops of the rabbit ileum. They cause hind leg paralysis and sub-sequently death when injected intraperitoneally into mice. All are cyto-toxic in picogram amounts for certain cell lines, such as Vero cells, growing in tissue culture. A single monoclonal antibody (MAb) that bound to the B subunit in Western blots neutralised all three activities confirming that all were produced by a single toxin[10].

These biological activities of VT1, VT2 and Shiga toxin are the result of their inhibition of protein synthesis. They inhibit elongation factor 1-dependent binding of aminoacyl-tRNA to the eukaryotic ribosome. At the molecular level the toxins cleave a glycosidic bond at site 4324 in the 28S ribosomal RNA of the 60S ribosomal subunit[11]. The same residue is cleaved by ricin, a toxin from the castor bean plant, which has two regions of significant amino acid homology with the enzymatically active A subunits of VT1 and VT2[12]. There is speculation as to whether the similarity in structure and function of these toxins is due to a common origin of the genes or due to convergent evolution.

Subunit B is involved in toxin binding. Globotriosyl ceramide (Gb_3) has been shown to be the receptor for VT1, VT2 and Shiga toxin. A terminal galα1-4 gal residue is essential. Samuel et al.[3] have recently shown that VT2vp and VT2vh bind to additional glycolipids, and that the binding affinities of the various VT2 toxins with glycolipids from different cells are complex. There is now much interest in determining the presence of these VT receptors in different tissues and in a variety of animals, and in showing whether changes occur as animals mature. An association of VTEC with haemolytic uraemic syndrome, which is a major cause of renal failure especially in children, was first shown in Canada[7]. The association has been confirmed in a number of countries, and as part of a British study the P1 blood group status of children with the haemolytic uraemic syndrome was determined. Gb_3 is a component of this blood group and the aim was to see whether children with the P1 receptor might be at greater risk of developing the syndrome[13]. However, to the contrary, a statistically significant number of children failed to express this antigen or expressed it weakly, including 6 of 7 children with a particularly poor outcome. Possession of the P1 antigen appeared to be a protective factor and it was proposed that red blood cells with the P1 receptor may absorb circulating toxin in a relatively harmless way before it has a chance to reach more vulnerable nucleated endothelial cells.

GENETICS AND RELATEDNESS OF THE DIFFERENT VT FORMS

VT genes have been shown to be phage-encoded in several strains of human origin and in two strains of bovine origin[14]. Several groups have sought such phages in strains of porcine origin but without success. VT-encoding phages have been isolated from two strains, both of serotype O157:H7, that produced both VT1 and VT2[14]. Each yielded two phages separately carrying the VT1 and VT2 genes. These phages were morpho-logically indistinguishable and gave very similar DNA restriction digest patterns. In contrast a phage encoding VT1 genes from an O26 strain was different morphologically from the O157 phages and the DNA gave different digest patterns although the VT1 genes were on a conserved fragment.

Table. 1. Comparison of nucleotide homology:
percentage relatedness of gene sequences for A and B subunits

Toxin (E. coli serogroup)	VT1		VT2		VT2vp	
	A	B	A	B	A	B
Shiga	99	100			60	64
VT2 (0157)	57	60	.	.	94	79
VT2va (0128)	58.8	64.3	69.5	78.1	70.6	98.0
VT2vh (091)			98.6	95.5	94.5	82.8
VT2vp (0139)	60	64	94	79	.	.

Data obtained from references 5, 6, 15, 16, 17 and 18.

The nucleotide sequences of the genes for Shiga toxin and several VTs have been determined and a comparison of sequence homologies is shown in the table. The sequences of VT1 and Shiga toxin are virtually identical. The nucleotide sequence for a VT1 toxin gene from an O26 strain differs in 3 bases from that of a Shiga toxin gene but this results in only one amino acid residue difference at position 45 from the N-terminus of the A subunit where serine in VT1 replaces threonine in Shiga toxin[15]. Shiga toxin and this VT1 have identical sequences for signal peptides, for ribosome binding sites, and for putative promoter regions and iron regulated operator sequences. The nucleotide sequence for a VT1 gene from an O157 strain also differed slightly from that of a Shiga toxin gene but it gave a predicted amino acid sequence identical to that of Shiga toxin. This sequence was confirmed when purified VT1 was analysed by Edman degradation[16].

Although VT1 and VT2 have many biological properties in common their gene sequences have diverged considerably with homology of <65% in both the A and B subunits (table). Nevertheless VT1 and VT2 have similar secondary structures and hydropathy plots[18]. The table also shows the relationship of gene sequences of the different VT2 variants to those of VT2. VT2vh is more closely related to VT2 than is VT2va. VT2vp shows considerable divergence from VT2 in the B subunit but not in the A subunit. The gene sequence of the B subunit of VT2va, but not of the A subunit, is closely related to that of VT2vp. Ito et al.[5] have determined that the nucleotide sequences for presumptive promoters and ribosome binding sites were identical for VT2, VT2vp and VT2vh.

Clearly both the A and B subunit structures have been highly conserved in Shiga toxin and those VT1 toxins studied to date. In contrast a variety of VT2 forms have already been identified and there may be others as toxins from additional serogroups are examined. We have examined 19 strains of human origin producing only VT2 and belonging to 9 serogroups other than O157; 16 produced variant forms with much less activity on HeLa cells. There even appear to be differences in those VT2 toxins produced by strains of O157 as shown by serum neutralisation studies and determinations of isoelectric points, which for four strains ranged from 4.1 to 6.5[19].

IMMUNOLOGICAL TESTS FOR VT

Several immunological assays, including enzyme-linked immunosorbent assays (ELISAs) and colony blot assays[10], have been described for VT. In general the ELISAs can be divided into those using an anti-VT monoclonal antiserum to bind the toxin[20] and those using the glycolipid receptor[21]. Bound toxin is then detected with monoclonal or polyclonal anti-VT antiserum followed by the appropriate enzyme-labelled immunoglobulin and enzyme substrate. Most of these assays although simple to perform have a sensitivity 70 to 2000 fold less than Vero cells.

More recently Basta et al.[22] have described an ELISA for VT1 in which the receptor was lyso-Gb3, a more polar form of Gb3 obtained by de-N-acylation. With this modified receptor the sensitivity of the test was at least equal to the Vero cell test. Because of the differences in the receptor specificities of the VT2 variants the choice of receptor for such tests will be very important.

The specificity of any MAb used in an immunological test will also need to be determined in detail as, for example, some raised against VT2 recognise the variant forms and others do not. Each can be useful depending on whether a test is required for a wide or narrow range of toxins. A MAb has been described[23] that recognised both VT1 and VT2 in an ELISA and also neutralised both toxins. Polyclonal antisera reacting with both toxins have not been reported and such a MAb with a wide spectrum of activity should be very useful for the primary screening of strains for any VT.

DNA HYBRIDISATION TESTS TO DETECT VT GENES

VT gene probes, comprising sequences of DNA lying within the VT1 or VT2 genes, have been developed from the VT-encoding phages[24]. They are a 750 base pair HincII fragment from an O26 strain for VT1 and an 850 base pair AvaI-PstI fragment from an O157 strain for VT2. The 2 probes show no cross hybridisation when used under stringent conditions and therefore are also convenient to determine whether a strain has genes for VT1 or VT2 or both. We have examined over 600 strains with the probes, including strains producing VT1, VT2, VT2vp and VT2vh. Only two strains hybridised (with the VT1 probe) but failed to produce VT in the Vero cell test. When they were tested with an alternative VT1 gene segment they failed to hybridise and it was concluded that both had incomplete gene sequences[25]. Only one strain, of serogroup O128, produced a toxin (neutralised by an antiserum to VT2) but did not hybridise with either VT probe and this strain is being investigated further[25].

DNA sequencing has revealed areas of low and high homology within the various toxin genes. The regions of high homology are likely to have been conserved because they are essential for the activity of the toxin. Several groups have now described the use of synthetic oligonucleotide probes for the detection of strains producing either VT1 or VT2. Sequences from both the A and B subunit genes have been tested. Because of the variation in degrees of DNA homology care has to be taken in the choice of oligonucleotide sequence. If the aim is, for example, to detect all VT2- and VT2v-producing strains in a clinical specimen probes from conserved regions are preferable. If the aim is to distinguish variant forms of VT2, for example to show that a VT2-producer in a food is of porcine origin, poorly conserved regions with high specificity should be chosen.

Karch and Meyer[26] have used a polymerase chain reaction technique that combined the use of regions of low and high homology to advantage. The primers were based on regions conserved between VT1, VT2 and VT2v but synthesised with some changed bases to minimise the variation further. Amplified sequences were then tested with oligonucleotide probes of high specificity to determine which VT gene was present.

In the near future, the general availability of such probe or immunological tests as an alternative to the time consuming tissue culture assays should prove very valuable for detecting VTEC in clinical specimens and foods. Additionally they will enable progress to be made in understanding the evolution of this complex family of toxins.

REFERENCES

1. S.M. Scotland, G.A. Willshaw, H.R. Smith, and B. Rowe, Properties of strains of *Escherichia coli* belonging to serogroup 0157 with special reference to production of Vero cytotoxins VT1 and VT2, *Epidemiol Infect*. 99: 613 (1987).

2. H. Chart, S.M. Scotland, and B. Rowe, Production of Vero cytotoxin by *Escherichia coli* and Shiga toxin by *Shigella dysenteriae* 1 as related to the growth medium and availability of iron, *Zentralbl. Bakteriol*. 272: 1 (1989).

3. J.E. Samuel, L.P. Perera, S. Ward, A.D. O'Brien, V. Ginsburg, and H.C. Krivan, Comparison of the glycolipid receptor specificities of Shiga-like toxin type II and Shiga-like toxin type II variants, *Infect Immun*. 58: 611 (1990).

4. D.R. Pollard, W.M. Johnson, H. Lior, S.D. Tyler, and K.R. Rozee, Rapid and specific detection of Verotoxin genes in *Escherichia coli* by the polymerase chain reaction, *J Clin Microbiol*. 28: 540 (1990).

5. H. Ito, A. Terai, H. Kurazono, Y. Takeda, and M. Nishibuchi, Cloning and nucleotide sequencing of Vero toxin 2 variant genes from *Escherichia coli* 091:H21 isolated from a patient with the hemolytic uremic syndrome, *Microb Pathog*. 8: 47 (1990).

6. V.P.J. Gannon, C. Teerling, S.A. Masri, and C.L. Gyles, Molecular cloning and nucleotide sequence of another variant of the *Escherichia coli* Shiga-like toxin II family, *J Gen Microbiol*. 136: 1125 (1990).

7. M.A. Karmali, Infection by Verocytotoxin-producing *Escherichia coli*, *Clin Microbiol Rev*. 2: 15 (1989).

8. S. Knutton, T. Baldwin, P.H. Williams, and A.S. McNeish, Actin accumulation at sites of bacterial adhesion to tissue culture cells; basis of a new diagnostic test for enteropathogenic and enterohemorrhagic *Escherichia coli*, *Infect Immun*. 57: 1290 (1989).

9. L. Beutin, M.A. Montenegro, I. Ørskov, F. Ørskov, J. Prado, S. Zimmerman, and R. Stephan, Close association of verotoxin (Shiga-like toxin) production with enterohemolysin production in strains of *Escherichia coli*, *J Clin Microbiol*. 27: 2559 (1989).

10. N.A. Strockbine, L.R.M. Marques, R.K. Holmes, and A.D. O'Brien, Characterization of monoclonal antibodies against Shiga-like toxin from *Escherichia coli*, *Infect Immun*. 50: 695 (1985).

11. Y. Endo, K. Tsurugi, T. Yutsudo, Y. Takeda, T. Ogasawara, and K. Igarashi, Site of action of a Vero toxin (VT2) from *Escherichia coli* 0157:H7 and of Shiga toxin on eukaryotic ribosomes, *Eur J Biochem*. 171: 45 (1988).

12. C.J. Hovde, S.B. Calderwood, J.J. Mekalanos, and R.J. Collier, Evidence that glutamic acid 167 is an active-site residue of Shiga-like toxin I, *Proc Natl Acad Sci USA*. 85: 2568 (1988).

13. C.M. Taylor, D.V. Milford, P.E. Rose, T.C.F. Roy, and B. Rowe, The expression of blood group P1 in post-enteropathic haemolytic uraemic syndrome, *Pediatr Nephrol*. 4: 59 (1990).

14. P.J.G.M. Rietra, G.A. Willshaw, H.R. Smith, A.M. Field, S.M. Scotland, and B. Rowe, Comparison of Vero-cytotoxin-encoding phages from *Escherichia coli* of human and bovine origin, *J Gen Microbiol*. 135: 2307 (1989).

15. N.A. Strockbine, M.P. Jackson, L.M. Sung, R.K. Holmes, and A.D. O'Brien, Cloning and sequencing of the genes for Shiga toxin from *Shigella dysenteriae* type 1, *J Bacteriol*. 170: 1116 (1988).

16. T. Takao, T. Tanabe, Y-M. Hong, Y. Shimonishi, H. Kurazono, T. Yutsudo, C. Sasakawa, M. Yoshikawa, and Y. Takeda, Identity of molecular structure of Shiga-like toxin I (VT1) from *Escherichia coli* O157:H7 with that of Shiga toxin, *Microb Pathog*. 5: 357 (1988).

17. D.L. Weinstein, M.P. Jackson, J.E. Samuel, R.K. Holmes, and A.D. O'Brien, Cloning and sequencing of a Shiga-like toxin type II variant from an *Escherichia coli* strain responsible for edema disease of swine, *J Bacteriol*. 170: 4223 (1988).

18. M.P. Jackson, R.J. Neill, A.D. O'Brien, R.K. Holmes, and J.W. Newland, Nucleotide sequence analysis and comparison of the structural genes for Shiga-like toxin I and Shiga-like toxin II encoded by bacteriophages from *Escherichia coli* 933, *FEMS Microbiol Lett*. 44: 109 (1987).

19. N. Dickie, J.I. Speirs, M. Akhtar, W.M. Johnson, and R.A. Szabo, Purification of an *Escherichia coli* serogroup O157:H7 Verotoxin and its detection in North American hemorrhagic colitis isolates, *J Clin Microbiol*. 27: 1973 (1989).

20. F.P. Downes, J.H. Green, K. Greene, N. Strockbine, J.G. Wells, and I.K. Wachsmuth, Development and evaluation of enzyme-linked immunosorbent assays for detection of Shiga-like toxin I and Shiga-like toxin II, *J Clin Microbiol*. 27: 1292 (1989).

21. S. Ashkenazi, and T.G. Cleary, Rapid method to detect Shiga toxin and Shiga-like toxin I based on binding to globotriosyl ceramide (Gb$_3$), their natural receptor, *J Clin Microbiol*. 27: 1145 (1989).

22. M. Basta, M. Karmali, and C. Lingwood, Sensitive receptor-specified enzyme-linked immunosorbent assay for *Escherichia coli* Verocytotoxin, *J Clin Microbiol*. 27: 1617 (1989).

23. A. Donohue-Rolfe, D.W.K. Acheson, A.V. Kane, and G.T. Keusch. Purification of Shiga toxin and Shiga-like toxins I and II by receptor analog affinity chromatography with immobilized P1 glycoprotein and production of cross-reactive monoclonal antibodies, *Infect Immun*. 57: 3888 (1989).

24. G.A. Willshaw, H.R.Smith, S.M. Scotland, A.M. Field, and B. Rowe, Heterogeneity of *Escherichia coli* phages encoding Vero cytotoxins: comparison of cloned sequences determining VT1 and VT2 and development of specific gene probes, *J Gen Microbiol*. 133: 1309 (1987).

25. S.M. Scotland, H.R. Smith, G.A. Willshaw, H. Chart, and B. Rowe, Production of Vero cytotoxins by strains of *Escherichia coli* isolated from human infections and correlation with hybridisation with gene probes for VT1 and VT2, p395, *in*: "Bacterial protein toxins", Fehrenbach et al., ed., Gustav Fischer, Stuttgart (1988).

26. H. Karch, and T. Meyer, Single primer pair for amplifying segments of distinct Shiga-like-toxin genes by polymerase chain reaction, *J Clin Microbiol*. 27: 2751 (1989).

STRUCTURE AND FUNCTION OF *CLOSTRIDIUM DIFFICILE* TOXINS

S. P. Borriello

Microbial Pathogenicity Research Group
Clinical Research Centre, Watford Road,
Harrow, Middlesex, HA1 3UJ, UK

INTRODUCTION

To understand the structure function relationship of a protein toxin it is essential to first have a clear understanding of the functional activities. For the purpose of this brief overview on the structure-function relationship of *Clostridium difficile* toxins A and B an attempt will be made to concentrate on known or proposed functions as opposed to effects. This can be best exemplified by comparing the haemagglutinating activity of toxin A which is mediated by binding to a specific tri-saccharide (a known function), to the ability of both toxins to cause cell rounding by disruption of the cell cytoskeleton (an effect exerted by unknown mechanisms and for which particular functions have not been identified). This brief overview will attempt to identify known functions of the toxin molecules, to highlight what is known of secondary structure and to comment on identified structure-function relationships.

PHYSICOCHEMICAL PROPERTIES

One of the most striking properties of the toxins is their extremely large size. Native toxin A has an Mr of between 400,000 and 600,000 and that of toxin B is between 360,000 and 500,000; under denaturing conditions both toxins have an Mr in the region of 250,000[1]. Although there has been some lack of agreement of the molecular weights of these toxins, their major denatured form, and consequently of the ultimate sub-unit composition of the native form (see reviews by Lyerly *et al*[1] and Borriello *et al*[2]) the molecular mass of 308,103 Da for toxin A and 269,696 Da for toxin B deduced from the recently published deoxyribo-nulceic acid sequences[3,4] accord well with the larger size estimates quoted above and directly contradict the smaller size estimates reported by some for the toxins[5,6]. This point is critical, as will be highlighted below, when trying to assess what significance to place on some of the reported functions of the toxins. The evidence to date indicates that the native toxins exist as dimers.

Both toxins are susceptible to a variety of proteases[7] and extremes of pH, though toxin A is least susceptible[7,8], and both toxins can be inactivated by oxidizing agents[9]. The inability of a variety of reducing agents to inactivate the toxins indicates that sulfhydryl groups are not involved in binding or cytotoxicity of the toxins[9]. However, di-sulphide bonds may be involved in aspects of the secondary and/or tertiary structure of the toxins. For toxin A it is known that dissociation into the major c. 250 kDa polypeptide can occur in the presence of sodium dodecyl sulphate in the absence of a reducing agent, but that numerous larger protein bands can be seen on a gel which migrate to the c. 250 kDa position following reduction with β-mercapto-ethanol[10]. It would appear that dissociation in the absence of reduction is incomplete.

FUNCTIONS OF TOXINS A AND B

The functions of the two toxins are shown in Table 1. Each of these functions will be discussed in turn.

Table 1   Functions of *C. difficile* Toxins A and B

Function	Toxin A	Toxin B	Reference
Binds nucleic acid	...	+	Meador and Tweten[11]; Bisseret et al[12]; Borriello and Stewart (unpublished).
Binds phosphorylated nucleotides	+	+	Florin and Thelestam[13]; Lobban and Borriello (unpublished).
Non specifically binds monoclonal antibodies	+	+	Lyerly et al[14]; Borriello and Vale (unpublished).
Binds troponin	−	+	Martig et al[15].
Binds Galα1-3Galβ1-4GlucNAc	+	+	Krivan et al[16].
Has enolase activity	...	?	Knoop et al[17].

INTERACTION WITH NUCLEIC ACID

There are two publications which contain within them some evidence that there is an interaction between toxin B and nucleic acid[11, 12]. In the first report[11] it was noted that approximately 50% of the total cytotoxic activity of toxin B was lost during the concentration and dialysis of the culture supernatant, and that this apparent loss in activity appeared to be due to inhibition by DNA fragments predominantly of greater than 20 MDa in mass. Addition of this DNA to toxin B could partially inhibit toxin B induced cytotoxicity. However, it must be clearly stated that there is no evidence in this report that toxin B and DNA are bound to each other. The conclusion from their finding is that possible interaction between toxin B and DNA decreases its cytotoxic activity. It is possible however that the DNA interacted with the target cells and not the toxin. In the second report[12] nucleic acids were recovered from toxin B purified by anion-exchange chromatography and shown to consist of both ribonucleic and deoxyribonucleic acid material of about 20 nucleotides in length. Whether this is nucleic acid bound to toxin B or simply co-eluting nucleic acid still needs to be demonstrated. However, an effect on the cytotoxicity of toxin B was indicated from their finding that growth of *C. difficile* in the presence of nucleases resulted in a decreased cytotoxin titre and yielded toxin B with a three-fold lower specific activity. Although the toxin B reported in these experiments is of a lower molecular weight, in keeping with a co-purified contaminant[11], the contaminants are unlikely to be cytotoxic so the effects on cytotoxicity by nucleases are probably valid but need to be interpreted cautiously.

More importantly the two reports appear to be contradictory in that one maintains that the presence of nucleic acids decreases cytotoxic activity and that their removal enhances cytotoxicity[11] while the other states that their presence enhances cytotoxicity and that removal of nucleic acids is detrimental to the cytotoxic activity of toxin B[12].

We have taken a slightly different approach which has been to add nucleases or nucleic acid to purified toxin and to monitor its effect on cytotoxicity to African Green Monkey kidney (Vero) cells (Borriello and Stewart, unpublished). Our findings show a clear and consistant eight-fold reduction in the cytotoxin titre of toxin B following treatment with nucleases (both RNA'se and DNA'se), but no such effect with toxin A. These findings would support the observations of Bisseret et al[12]. In addition, contrary to the findings of Meador and Tweten[11], addition of C. difficile DNA to toxin B did not reduce its cytotoxicity. Although there are many possible explanations of this apparent interaction between nucleic acid and toxin B, one that we find intriguing is the possibility that toxin B can directly bind nucleic acid and that this ability could be a candidate for its mechanism of action ie that toxin B turns of an essential gene involved in cytoskeleton control by directly binding to host cell DNA. This is purely hypothetical, and preliminary findings show that Vero DNA does not inhibit its cytotoxicity for Vero cells. However the possibility of this potential mechanism of action merits further investigation.

BINDING OF PHOSPHORYLATED NUCLEOTIDES

There is very good evidence that toxin B has a phosphate binding site and that occupation of that site(s) delays cytotoxicity though it does not prevent it[13]. A series of nuclear di-, tri-, and tetraphosphates, inorganic polyphosphates and polyphosphorylated sugars all caused a dose-dependent delay in cytotoxicity. This delay was not due to irreversible inactivation of the toxin or interference with its binding to tissue culture cells. Direct binding studies with ATP showed that between 7% and 20% of a given toxin B preparation bound it. Although the polyphosphate binding site shares a number of characteristics with diphtheria toxin, the activity of toxin B, unlike diphtheria toxin, was also affected by inositol hexasulphate, implying that toxin B may interact with any molecule containing highly anionic regions. Although a common binding site is infered for the polyphosphate compounds and inositol hexasulphate this was not examined directly and there may in fact be a polyphosphate specific site(s) and one or more that react with highly anionic regions on molecules.

We have looked at the interaction of toxin A with ATP by following the change in intrinsic fluorescence at an excitation wavel length of 285nm and emission at 333nm ie monitoring free tryptophan residues (Lobban and Borriello, unpublished). The titration curve and scatchard plot analysis of the data indicates that there is specific binding over the concentration range of 0.1 to 100µM of ATP. What effect this has on the biological activity of toxin A has not yet been examined.

NON-SPECIFIC BINDING OF MONOCLONAL ANTIBODIES

In experiments to raise murine monoclonal antibodies (M-Mabs) to C. difficile toxin A we routinely included in the screening procedure a negative control of an M-Mab to Herpes simplex-1 virus. To our surprise it consistently gave positive results with toxin A. The appearance a few months later of a paper by Lyerly and colleagues[14] showed not only that non-specific interaction with M-Mabs had been observed by others but that it had been done some considerable time before us. Both toxins A and B are capable of interacting non-specifically (presumably) with a variety of commercially available M-Mabs raised to other antigens and in general this type of binding occurs to a higher degree with toxin A than toxin B[14]. It has also been shown (Borriello and Vale, unpublished) that toxin A bound to four of 11 human paraproteins. Though it is conceivable that the paraproteins are specifically directed against toxin, we think this unlikely. It is unknown whether this binding is mediated though the Fc or Fab portion of the molecule or whether there is a common binding site(s) for this interaction with the toxin molecules. The identification of Mabs from different species that appear to interact non-specifically with toxin A permits us to study the commonality of the binding region in competitive binding assays.

BINDING TO TROPONIN

The known disruption of myofilament bundles in cells in tissue culture by toxin B prompted a study to determine which cytoskeletal protein interacted with toxin B[15]. Co-incubation of cytoskeletal proteins with toxin B showed that troponin caused an 88% reduction in the cytotoxic activity of the toxin. Exposing toxin B to SDS-polyacrylamide separated troponin components showed that the toxin bound to one major and several minor components of the troponin complex. It is not possible from this published abstract to assess the purity of the toxin B preparation used, which may influence interpretation of the experiment designed to see which troponin components were bound. This does not however detract from the observation that troponin inhibits toxin B induced cytotoxicity.

BINDING TO Galα1-3Galβ1-4GlucNAc

The observation that toxin A would haemagglutinate rabbit erythrocytes at 4°C led to the identification of the tri-saccharide Galα1-3Galβ1-4GlucNAc as a receptor for toxin A[16]. Although there is no direct evidence published that the toxin and this tri-saccharide bind to each other, the indirect evidence is very strong. This haemagglutination and presumed specific tri-saccharide binding is the only effect for which a specific binding function has been identified, and further it is the only function for which a component of the toxin molecule has been identified (see below).

DOES TOXIN B HAVE ENOLASE ACTIVITY?

There is little doubt that a protein with enolase activity can be purified from *C. difficile*[17]. The problem of ascribing this activity to toxin B lies in the fact that there is a lack of agreement on the relative molecular mass (Mr) of the native and denatured forms of the toxin (see reviews[1,2]). The report that there was a high similarity between sequenced toxin B peptides and enolases from rat and *Saccharomyces cerevisiae*[12] was related to a protein product with an estimated Mr of 290,000 under non-denaturing conditions and 52,000 under denaturing conditions, which has the general characteristics of a contaminating protein described by Meador and Tweten[11]. More difficult to explain is the finding by Knoop *et al*[16] that toxin B degrades 2-phosphoglycerate (ie exhibits enolase activity). The toxin B preparation in this study[16] had a molecular mass of 163,000 as determined by SDS-polyacrylamide electrophoresis, which does not accord with the deduced molecular weight of 269,696 Da from the sequenced gene[4]. A comparison of the sequence for toxin B with that of enolase should resolve this controversy.

STRUCTURE OF THE TOXINS

Since the recent discovery of the deoxyribonucleoside sequence of both toxins[3,4] it has been possible to deduce the primary structure and to infer that both toxins are produced as a single large polypeptide which aggregate into dimers (unless dimerisation is a consequence of purification procedures). From the primary structure one can model likely higher structure configurations and/or identify likely locations of folds. Direct analysis of toxin A by circular dichroism indicates that it is composed of 8% α-helices, 65% β-sheets with the remainder at 27% (unpublished).

There is some evidence that toxin A may be produced in a protoxin form that is cytotoxic but lacks haemagglutinating activity[2,19] (and therefore presumably is non-enterotoxic). As the native molecular mass and breakdown profile under denaturing conditions with and without reduction of the protoxin form was the same as fully active toxin A it was deduced that the change from protoxin to toxin was essentially one of changes in tertiary structure, and that the protoxin form was less in a less relaxed state. This has to some extent been borne out by the observation that the protoxin form consists of 17% α-helices (ie twice the degree of fully active toxin A), and 62% β-sheets with the remainder at 21%.

STRUCTURE FUNCTION RELATIONSHIP

The only known structure function relationship is that for cold haemagglutination of rabbit erythrocytes and the inferred binding to the tri-saccharide Galα1-3Galβ1-4GlucNAc. The evidence clearly shows that the binding site consists of the essentially hydrophilic protein product of a series of repeats at the 3' end of the toxin A gene[20]. What secondary or tertiary structure this part of the polypeptide adopts, or where it resides in relation to the toxin molecule as a whole is unknown.

CONCLUSION

Essentially little is known of function, less of higher structure and next-to-nothing of structure-function relationships for either toxin. However, it is encouraging that some concensus is emerging on functions, but this is going to continue to depend on workers ensuring purity of their toxin preparations.

Initial approaches that may be fruitful in furthering our understanding of structure-function would be a comparison for the degree of homology between the two toxins in an attempt to identify common areas that my code for some of the common effects/functions eg cytotoxicity and polyphosphate binding. Our analyses indicate an overall 60% homology, this being predominantly at the 5' end. In fact from 201 base pairs in from the 5' end of toxin B there is a surprisingly high degree of similarity between toxin A and B over at least 4000 base pairs. A comparison with other known toxins may also yield useful information. For example, we compared the gene sequence for toxin A and B with that of the diphtheria (β) toxin gene and found c. 40% homology.

I have little doubt that over the next five years reports on a combination of cross-linking, intrinsic fluorescence, circular dichroism, site directed mutagenesis and computer modelling studies will greatly further our knowledge on the structure of these large proteins and facilitate our understanding of how the structure relates to function. I would suspect that the goal of those active in this area is to produce crystals of the toxins.

REFERENCES

1. D. M. Lyerly, H. C. Krivan, and T. D. Wilkins, *Clostridium difficile*: its disease and toxins, Clin. Microbiol. Rev. 1:1 (1988).
2. S. P. Borriello, H. A. Davies, S. Kamiya, P. J. Reed, and S. Seddon, Virulence factors of *Clostridium difficile*, Rev. Infect. Dis. 12(S2):185 (1990).
3. C. H. Dove, S-Z. Wang, S. B. Price, C. J. Phelps, D. M. Lyerly, T. D. Wilkins, and J. L. Johnson, Molecular characterization of the *Clostridium difficile* toxin A gene, Infect. Immun. 58:480 (1990).
4. L. A. Barroso, S-Z, Wang, C. J. Phelps, J. L. Johnson, and T. D. Wilkins, Nucleotide sequence of *Clostridium difficile* toxin B gene, Nucl. Acids Res. 18:4004 (1990).
5. C. Pothoulakis, L. M. Barone, R. Ely, B. Faris, M. E. Clark, C. Franzblau, and J. T. LaMont, Purification and properties of *Clostridium difficile* toxin B, J. Biol. Chem. 261:1316 (1986).
6. B. Rihn, J. M. Scheftel, R. Girardot, and H. Monteil, A new purification procedure for *Clostridium difficile* enterotoxin, Biochem. Biophys. Res. Commun. 124:690 (1984).
7. D. M. Lyerly, P. E. Carrig, and T. D. Wilkins, Susceptibility of *Clostridium difficile* toxins A and B to trypsin and chymotrypsin, Microb. Ecol, Hlth. Dis. 2:219 (1989).
8. N. S. Taylor, G. M. Thorne, and J. G. Bartlett, Comparisons of two toxins produced by *Clostridium difficile*, Infect. Immun. 34:1036 (1981).
9. D. M. Lyerly, M. D. Roberts, C. J. Phelps, and T. D. Wilkins, Properties of toxins A and B of *Clostridium difficile*, FEMS Microbiol. Lett. 33:31 (1986).
10. S. Kamiya, P. J. Reed, and S. P. Borriello, Purification and characterisation of *Clostridium difficile* toxin A by bovine thyroglobulin affinity chromatography and dissociation in denaturing conditions with or without reduction, J. Med. Microbiol. 30:69 (1989).
11. J. Meador, and R. K. Tweten, Purification and characterization of toxin B from *Clostridium difficile*, Infect. Immun. 56:1708 (1988).
12. F. Bisseret, G. Keith, B. Rihn, I, Amiri, B. Werneburg, R. Girardot, O. Baldacini, G. Green, V. K. Nguyen, and H. Monteil, *Clostridium difficile* toxin B: characterisation and sequence of three peptides, J. Chromatog. 490:91 (1989).
13. I. Florin, and M. Thelestam, Polyphosphate-mediated protection from cellular intoxication with *Clostridium difficile* toxin B, Biochim. Biophys, Acta. 805:131 (1984).
14. D. M. Lyerly, P.E. Carrig, and T. D. Wilkins, Nonspecific binding of mouse monoclonal antibodies to *Clostridium difficile* toxins A and B, Curr. Microbiol. 19:303 (1989).
15. R. Martig, F. Knoop, and M. Owens, Troponin: a possible receptor for *Clostridium difficile* cytotoxin B, Abstracts of the Annual Meeting of the American Society of Microbiology, Washington DC, p65 (1989).
16. H, C. Krivan, G. F. Clark, D. F. Smith, and T. D. Wilkins, Cell surface binding site for *Clostridium difficile* enterotoxin: evidence for a glycoconjugate containing the sequence Galα1-3Galβ1-4GlcNAc, Infect. Immun. 53:573 (1986).
17. F. Knoop, R. Martig, and M. Owens, Degradation of 2-phosphoglycerate by cytotoxin B of *Clostridium difficile*, FEBS Lett. 267:9 (1990).
18. S. P. Borriello, The pathogenesis of *Clostridium difficile* infection of the gut, J. Med. Microbiol. 33 (In Press).
19. S. P. Borriello, S. Kamiya, and P. J. Reed, Detection and characterisation of a non-haemagglutinating form of *Clostridium difficile* toxin A, In: "Molecular and Medical Aspects of Anaerobes", S. P. Borriello ed. Wrightson Biomedical, Petersfield (1990).

20. S. B. Price, C. J. Phelps, T. D. Wilkins, and J. L. Johnson, Cloning of the carbohydrate-binding portion of the toxin A gene of *Clostridium difficile*, Curr. Microbiol. 16:55 (1987).

ON THE CYTOTOXIC MODES OF ACTION OF *CLOSTRIDIUM DIFFICILE* TOXINS

Monica Thelestam, Mimmi Caspar Shoshan and Carla Fiorentini*

Department of Bacteriology
Karolinska institute, STOCKHOLM, Sweden and

*Department of Ultrastructures
Istituto Superiore di Sanita, ROME Italy

INTRODUCTION

Clostridium difficile, the major aetiological agent of antibiotic associated colitis, produces two high molecular weight protein toxins, designated as toxins A and B, which are involved in the pathogenesis of the disease[1]. Toxin A is an enterotoxin which induces hemorrhagic fluid secretion and tissue necrosis in intestinal loops. Toxin B lacks these activities. Both are cytotoxins which act intracellularly[2,3] although the mechanisms involved in the cytotoxic effects are unknown.

Toxins A and B both cause cell retraction, rounding and cytoskeletal redistributions in all mammalian cells tested[4,5]. The morphological effects induced by both toxins appear identical as viewed in the light microscope. Based on this and other facts we previously proposed that the cytotoxic components of both toxins might be identical[6].

This study was designed to obtain further information about the cytotoxic modes of action of toxins A and B using a biochemical approach. With compounds which selectively interfere with certain signal transduction pathways in cells, we investigated whether these pathways are involved in the cellular intoxications caused by toxins A and B. While the molecular details of the cytotoxic actions are still not clear, our results indicate that certain pathways can be excluded. Another conclusion from this study is that the cytotoxic modes of action of toxins A and B differ from each other, thus refuting our previous hypothesis[6].

MATERIALS AND METHODS

<u>Toxins</u>. Toxin A was purified as previously described[5] and was homogeneous on SDS-PAGE. Toxin B was purified by a method modified from the one described by Meador and Tweten[7] as reported by Caspar Shoshan et al[8]. Toxin B showed one major and one minor band on SDS-PAGE.

<u>Test compounds</u>. All test compounds were from Sigma Chemical Co., St.Louis, Missouri except for amitriptylin (H.Lundbeck & Co., Copenhagen, Denmark), verapamil (ACO, Solna, Sweden), pertussis toxin (a kind gift from Dr Per Askelöf, National Bacteriological

Laboratory, Stockholm, Sweden) and cholera toxin (a kind gift from Prof. J.P.Craig, State Univ. of New York, N.Y.,USA). The test compounds were dissolved either in medium or DMSO, and further dilutions were made in Eagle's Minimal Essential Medium (MEM) with the supplements detailed below. The amount of DMSO never exceeded 1% in the final solutions incubated with cells.

Cells. Human embryonic lung fibroblasts (MRC-5) were cultivated in Eagle's MEM supplemented with 10% newborn bovine serum. Mouse adrenocortical cells (line Y1, ATCC CCL 79) were cultivated in Ham's F10 medium with 10% fetal bovine serum. Rat intestinal crypt cells (line IEC-6, ATCC No. CRL 1592) were cultivated in Dulbecco's modification of Eagle's medium with 5% fetal bovine serum and 0.1 U insulin/ml. All three media were additionally supplemented with 5 mM L-glutamine, 100 U penicillin/ml and 100 ug streptomycin/ml. Y1 and IEC-6 cells were used as targets for toxin A, while Y1 and MRC-5 cells were used for toxin B.

Table 1.   Compounds tested for interference with the CPE caused by *C.difficile* toxins

---

Compounds affecting cAMP and G-proteins
Cholera toxin (elevates cAMP via AC regulatory G-protein, Gs)	370 ug/ml
Pertussis toxin (elevates cAMP via AC regulatory G-protein, Gi)	320· ug/ml
Caffeine (elevates cAMP without involvement of G-protein)	10 ug/ml

Inhibitors of ther phosphatidyl inositol cycle
Neomycin (blocks phosphatidyl inositol specific phospholipase C)	15 mM
LiCl (inhibits inositol phosphate phosphatase)	20 mM

Compounds affecting eicosanoid metabolism
Quinacrine (blocks phospholipase A2)	20 uM
Indomethacin (inhibits cyclooxygenase)	200 uM
Esculetin (inhibits lipoxygenase)	5 uM

Compounds affecting protein kinase C (PKC)
12-O-tetradecanoylphorbol-13-acetate (TPA) (stimulates PKC)	50 ng/ml
4 -phorbol12,13 dibutyrate (PDBu) (inhibits PKC)	200 ng/ml
1-(4-isoquinolynsulfonyl)-2-methylpiperazine(H-7)(inhibitsPKC)	50uM

Cellular intoxication was assessed microscopically as the percentage of cells showing a characteristic cytopathic effect (CPE). The values are averages of duplicate samples in each of which three fields of view were scored, each field chosen to contain roughly 200 cells. All experiments were performed at least three times. To investigate the effects of different test compounds on toxin-induced CPE, cells were pretreated for 5-45 min with the highest concentration of each drug (Table 1), causing no morphological modification of the cells. Toxin A (1.5 ug/ml) or toxin B (0.2-1.0 ng/ml) was added and after 1, 3 and 5 h the CPE was scored. These doses of the toxins caused the rounding up of 100% of the cells in 1-3 h.

Cytosolic free calcium concentration was assayed as previously described[9] using the fluorescent Ca^{2+} indicator fura-2. These analyses were performed with the generous help of Dr Pierluigi Nicotera, Dept Toxicology, Karolinska institute.

RESULTS

Toxin A. None of the compounds listed in Table 1 had any influence on the intoxication induced by toxin A in IEC-6 or Y1 cells.

The involvement of cytosolic Ca^{2+} in toxin A-induced CPE was tested using the Ca^{2+} channel blockers verapamil (160 uM) and $LaCl_3$ (130 uM), the Ca^{2+} ionophore A 23187 (25 uM), the Ca^{2+} chelator Quin 2 (25 uM) and the Ca^{2+} antagonists calmidazolium (10 uM) and amitriptylin (100 uM). Pretreatments of IEC-6 or Y1 cells with these substances had no influence on the cellular intoxication. Moreover the toxin A-induced CPE in these cells was not accompanied by any change in the cytosolic Ca^{2+} level as tested directly after toxin addition and after exposure to toxin for 10, 20, 40 and 60 min. While ATP (10 uM) caused a rapid and marked cytosolic Ca^{2+} increase, even up to 100 ug/ml of toxin A was without effect. However deprivation of extracellular Ca^{2+} during a 10 min toxin binding step caused a 80% and 50% inhibition of the CPE in IEC-6 and Y1 cells respectively, suggesting that extracellular Ca^{2+} promotes the binding of the toxin to the cell surface.

Toxin B. Of the compounds listed in Table 1 only the phorbol ester TPA and the phospholipase A_2 (PLA_2) inhibitor quinacrine were able to affect the cellular intoxication caused by toxin B. In MRC-5 cells TPA at a concentration of 50 ng/ml clearly had a synergistic effect with toxin B while 2 ng/ml were without effect. However, neither of the protein kinase C inhibitors (PDBu and H-7) had any influence on the intoxication.

Quinacrine at a concentration of 20 uM completely prevented the toxin B-induced CPE in both MRC-5 and Y1 cells, while 10 uM strongly delayed the development of the CPE. An involvement of PLA_2 in the toxin B action was further supported by the detection of ^{14}C-arachidonic acid release from prelabelled MRC-5 cells treated with a high dose of toxin B. About 85% of total ^{14}C-arachidonate was released after 5 h of incubation with 1 ug/ml toxin B as compared to 35% from control cells. Lower (but still cytopathogenic) doses of toxin B did not elicit any measurable release of arachidonate.

DISCUSSION

It is increasingly recognized that multiple signal transduction pathways cooperate synergistically in mitogenic stimulation of cells. The test compounds used in this study were selected in order to cover the most important of the known pathways in the target cells. These are: one involving cAMP, the cycle involving phosphatidylinositol and inositol phosphates, the pathway involving phospholipase A_2 and eicosanoid metabolism and finally pathways dependent on protein kinase C.

It appears that none of these pathways are involved in the cytotoxic action of toxin A in either IEC-6 or Y1 cells. This contrasts a previous report that toxin A caused a release of the inflammatory mediators prostaglandin E_2 and leukotriene B_4, both dependent on PLA_2 activity[10]. However, those observations were made after *in vivo* intestinal exposure to toxin A, implying that other intermediary influences, e.g., the enteric nervous system, could be involved.

Certain agents which interfere with the cellular Ca^{2+} metabolism and with calmodulin, have been reported to protect Vero cells against intoxication with crude supernatants of *C.difficile*, containing both toxins A and B[11]. More recently, we found that toxin B required active calmodulin as well as uptake of extracellular calcium in order to intoxicate human lung fibroblasts[12]. Further studies concerning the involvement of Ca^{2+} in toxin B-induced CPE are under way in our laboratory.

For toxin A the present study indicates that while extracellular Ca^{2+} apparently promotes binding of the toxin to both IEC-6 and Y1 cells, the intracellular Ca^{2+} metabolism is not involved. Our observations, using Ca^{2+} antagonists, are in accordance with previous findings[13] that the toxin A-induced CPE in CHO cells was not affected by the calcium channel blocker diltiazem nor by the calmodulin inhibitor trifluoperazine. Moreover, the toxin A induced CPE in IEC-6 and Y1 cells was not accompanied by any change in the cytosolic Ca^{2+} level, even with very high doses of the toxin. Consistent with this neither the intracellular chelator Quin 2 nor the Ca^{2+} ionophore A23187 did influence the CPE.

An earlier investigation indicated a rapid and transient stimulation of intracellular calcium release upon treatment of human granulocytes with toxin A[14]. This effect occurred within less than 3 seconds and required toxin A doses of 30 ug/ml or more. The rapidity of this effect is not consistent with prior endocytotic internalization of the toxin, but instead is suggestive of a transmembrane activation immediately upon contact between toxin A and granulocytes. Thus the increase of cytosolic Ca^{2+} in granulocytes might be due to the peculiar characteristics of this cell type and might not be directly related to the cytotoxicity of toxin A. This assumption is also supported by the fact that much lower concentrations of toxin A than those inducing the Ca^{2+} release in granulocytes do elicit the CPE in other cells.

A phospholipase A_2 activation by toxin B appears plausible in view of the facts that quinacrine inhibited the CPE and that arachidonic acid was released from cells treated with a high dose of the toxin. This latter result is consistent with our observation (manuscript in preparation) that toxin B at high dose levels is cytolytic to human lung fibroblasts. Conceivably this lytic effect could be due to the production of lysophospholipids in the plasma membranes of intoxicated cells. No significant release of arachidonic acid could be detected at lower but still cytopathogenic doses of the toxin. We propose that in this case only a minor activation of PLA_2 takes place, which might not generate enough lysophospholipid to cause a membrane permeabilisation, while it could affect the anchoring of cytoskeletal elements to the plasma membrane. Whether the toxin B-induced PLA_2 activation is a primary or secondary effect of the toxin is not yet known.

The synergism between toxin B and TPA suggests that protein kinase C-mediated reactions may also be involved. A regulation of PKC by lysophospholipids has been suggested[15]. However, the fact that the PKC inhibitors did not affect the intoxication indicates that at least the PKC activity affected by these substances is not required for the toxin B-induced CPE. Further studies to clarify the significance of different isoforms of PKC in the cytotoxic action of toxin B are under way.

Although the molecular modes of action of the *C. difficile* toxins A and B are still not clarified, the results of this study demonstrate that certain key signal transduction pathways can be excluded from involvement in either intoxication. For toxin A neither of five major metabolic pathways appears involved, while for toxin B clearly phospholipase A_2 is of some significance, although we still do not understand how. In any case these studies have made it clear that the biochemistry of the cellular intoxication by toxin A differs from that of toxin B. This conclusion is also supported by our recent observation that toxin B causes multinucleation of transformed B lymphocytes[8] while toxin A blocks not only cytokinesis but also nuclear division (manuscript in preparation). Taken together all these data contradict our previous hypothesis that these two toxins contain one common cytotoxic component[6]. Toxin A in addition to its enterotoxic effect is cytotoxic by a mechanism different from that of the antiproliferative action of toxin B. The contribution of these activities to the intestinal effects of *Clostridium difficile* are yet to be elucidated.

ACKNOWLEDGEMENTS

We thank Kerstin Andreasson and Greger Blomquist for expert technical assistance. These studies were supported by the Swedish Medical Research Council grant No. 16X-05969 and the Italian National Council of Research, grant No. 203.04.15.

REFERENCES

1. D.M. Lyerly, H.C. Krivan, and T.D. Wilkins, *Clostridium difficile* : its disease and toxins, Clin. Microbiol. Rev. 1: 1 (1988).
2. B. Henriques, I. Florin, and M. Thelestam, Cellular internalization of *Clostridium difficile* toxin A, Microb. Pathogen. 2: 455 (1987).
3. I. Florin and M. Thelestam, Lysosomal involvement in cellular intoxication with *Clostridium difficile* toxin B, Microb. Pathogen. 1: 373 (1986).
4. C. Fiorentini, G. Arancia, S. Paradisi, G. Donelli, M. Giuliano, F. Piemonte, and P. Mastrantonio, Effects of *Clostridium difficile* toxins A and B on cytoskeleton organization in HEp-2 cells: a comparative morphological study, Toxicon, 27: 1209 (1989).
5. C. Fiorentini, W. Malorni, S. Paradisi, M. Giuliano, P. Mastrantonio, and G. Donelli, Interaction of *Clostridium difficile* toxin A with cultured cells: cytoskeletal changes and nuclear polarization, Infect. Immun. 58:2329 (1990).
6. M. Thelestam and I. Florin, Studies on cellular intoxication with *Clostridium difficile* toxins, in: "Anaerobes Today", J. M. Hardie and S. P. Boriello, eds, John Wiley & Sons, New York.
7. J. Meador and R.K. Tweten, Purification and characterization of toxin B from *Clostridium difficile*, Infect. Immun. 56: 1708 (1988).
8   M. Caspar Shoshan, P. Åman, S. Skog, I. Florin, and M. Thelestam, Microfilament-disrupting *Clostridium difficile* toxin B causes multinucleation of transformed cells but does not block capping of membrane Ig, Eur. J. Cell Biol. 53, (1990, in press).
9. P. Nicotera, H. Thor, and S. Orrenius, Cytosolic-free Ca^{2+} and cell killing in hepatoma 1c1c7 cells exposed to chemical anoxia, FASEB J. 3: 59 (1989).
10. G. Triadafilopoulos, C. Pothoulakis, R. Weiss, C. Giampaolo, and J. T. LaMont, Comparative study of *Clostridium difficile* toxin A and cholera toxin in rabbit ileum, Gastroenterol., 97:1186 (1989).
11 L.G. Giugliano and B.S. Drasar, The influence of drugs on the response of a cell culture preparation to bacterial toxins, J. Med. Microbiol. 17: 151 (1984).
12. M. Caspar, I. Florin, and M. Thelestam, Calcium and calmodulin in cellular intoxication with *Clostridium difficile* toxin B, J. Cell. Physiol. 132:168 (1987).
13. A.A.M. Lima, D. M. Lyerly, T. D. Wilkins, D. J. Innes, and R. L. Guerrant, Effects of *Clostridium difficile* toxins A and B in rabbit small and large intestine in vivo and on cultured cells in vitro, Infect. Immun. 56: 582 (1988).
14. C. Pothoulakis, R. Sullivan, D. A. Melnick, G. Triadafilopoulos, A.-S. Gadenne, T. Meshulam, and J. T. LaMont, *Clostridium difficile* toxin A stimulates intracellular calcium release and chemotactic response in human granulocytes, J. Clin. Invest. 81:1741 (1988).
15 K. Oishi, R. L. Raynor, P. A. Charp, and J. F. Kuo, Regulation of protein kinase C by lysophospholipids. Potential role in signal transduction, J. Biol. Chem. 263: 6865 (1988).

SALMONELLA AS AN INVASIVE ENTERIC PATHOGEN

P. Helena Mäkelä, Marianne Hovi, Harri
Saxén, Anna Muotiala, Petri Riikonen,
Marjatta Nurminen, Suvi Taira, Soila
Sukupolvi and Mikael Rhen

National Public Health Institute
SF-00300 Helsinki, Finland

NATURAL SALMONELLA INFECTION

Salmonella can infect a surprisingly wide variety of
hosts ranging from reptiles to birds to mammals. Yet the
bacteria isolated are so close to each other in DNA
sequence that they must be considered as one species. The
species name is at this moment still uncertain, because of
the earlier separation of its many serovars as species, and
the familiarity of their names to both microbiologists and
clinicians. A new name, *Salmonella enterica*, has therefore
been proposed to cover the whole species, but this name has
not (yet) been accepted.[1] The use of the serovar names as
trivial names is a practical compromise, also adopted in
this paper.

Indeed, a large variety of serovars exist within the
single *Salmonella* species, based both on its O antigen (the
structure of the side chain of its lipopolysaccharide, LPS)
and the flagellar H antigens. Several of these serovars
have characteristic pathogenic properties either in respect
of host specificity or clinical course of the infection.

Clinically *Salmonella* infections can be described as
of three different types: an invasive generalizd infection,

local enteric infection with diarrhea, and asymptomatic intestinal colonization. In all cases the infection is acquired *via* the alimentary tract from bacteria in the food or drinking water. Some serovars cause an invasive infection often restricted to one host: Typhi in man, Abortus-ovis in sheep, Gallinarum in poultry. Other serovars can infect a wide range of hosts - e.g. Typhimurium, which causes an invasive infection in the mouse, mostly enteric type of disease in humans, and asymptomatic carriage in poultry. The disease picture is further modified by specific features of the host, e.g. age: Typhimurium causes an invasive, lethal disease in very young chicken.[2]

THE COURSE OF INVASIVE INFECTION

The invasive form of salmonellosis has been most extensively studied in experimentally infected mice.[3-8] The mouse is actually very sensitive to the invasive infection mode, and thus strains of several serovars have been used in this model: most often Typhimurium or Enteritidis, but also Dublin and Cholerae-suis. Systemic invasion of tissues takes place through the intestinal epithelium[3] and/or the M cells of Peyer's plaques.[4] These early events are currently studied by several groups and will not be discussed here. Instead, this paper will focus on the fate of the bacteria in the tissues, where they are faced with the full system of parenteral host defenses.

When entering the submucosa, the bacteria are exposed to the tissue fluids and the phagocytic cells patrolling there. An essential question is whether or not the bacteria can avoid being taken up by the macrophages; this point has been studied extensively and will be described shortly. Those bacteria that escape the local macrophages will be transported along with the flow of the tissue fluids, first to the lymphatics and then to the blood stream. Their survival in this environment is the next question for survival, and will be discussed in the following section. The early bacteremia is, however, transient, and the bacteria are soon taken up by the specialized macrophages lining the blood vessels in the liver and spleen. These are the two organs that in the mouse account for most of the *Salmonella* load as well as for the subsequent increase in the numbers of bacteria. This process as a rule takes several days and forms the main phase of invasive salmonellosis.

Once the bacteria in the liver and spleen reach a certain level (appr. 10^9 bacteria/mouse), the final phase of the infection ensues. This is characterized by overwhelming bacteremia, toxic symptoms and death. Apparently endotoxin, and sensitization of the animal to its effects, play a major role in this phase.[6] A more favorable outcome - recovery, clearance of the bacteria and development of immunity - is possible at a lower challenge dose or a lower degree of virulence of the bacteria. These final events will not be discussed in this paper, partially because their molecular level characterization is still very incomplete.

THE FIRST, TISSUE PHASE OF SALMONELLA INFECTION

This phase can be best studied by injecting the bacteria intraperitoneally (ip). The resident macrophages in the peritoneal cavity are likely to be similar to those in the submucosa of the intestine; furthermore, O antigen that is a major determinant of bacterial virulence within the first hours after ip challenge has been shown to have a similar effect after infection by the oral route.[7,8]

Dramatic differences in bacterial survival in the peritoneal cavity can be seen between isogenic derivatives of Typhimurium or Enteritidis that differ in the quality of their O antigen, and these differences are reflected in the final outcome of the infection. Thus, bacteria with O-6,7 LPS will be taken up by the peritoneal macrophages within minutes and killed rapidly thereafter, so that only few percent of the original inoculum remain viable after an hour.[9] At the same time, their O-4,12 sister bacteria are not phagocytized and remain fully viable. The reduction in the number of viable bacteria *de facto* decreases the challenge dose, and thus fewer bacteria reach the subsequent stages of infection.

The differences in phagocytosis have been shown to be due to differences in the deposition of the complement component C3b on the bacterial surface.[9] This takes place *via* activation of the alternative complement pathway.[10] The molecular mechanism of the difference in complement activation has been studied by the groups of Loretta Leive and Keith Joiner.[11] It seems that the O-4,12 and O-6,7 LPS differ in their ability to amplify C3b deposition on the bacteria by modifying the affinity of surface-bound C3b to the accessory factor B. The phagocytosis then takes place *via* CRT, the macrophage receptor for C3b.

Antibodies are not required for this complement acti-vation. However, if they are present and bind to the bacteria, complement is activated *via* the classical pathway, and C3b is again deposited on the bacterial surface. The opsonized bacteria are then phagocytized and killed in the peritoneal cavity quite irrespective of their O antigen type.[12]

THE EARLY BACTEREMIA

Careful *in vitro* studies on the activation of complement by *Salmonellae* or by their LPS (coated onto suitable particles) have shown that the differences mentioned above are differences in the rate of the process, and are most marked at a low concentration of complement. This would accurately reflect the situation in the tissue fluids, and give the results seen *in vivo*. However, at a higher concentration of complement, such as present in serum, its activation takes place quite efficiently by both types of LPS. Thus, when O-6,7 or O-4,12 bacteria are injected intravenously (iv), a difference in their clearance rates (as an indication of phagocytosis by macrophages in the liver and spleen) can be seen in the first minutes,[13] but soon both types have been taken up in the liver and spleen. Thus the numbers of viable bacteria reaching these organs are not affected by the quality of the O antigen when the bacteria are injected iv.

The high concentration of complement in the blood is expected to have a direct killing effect on the bacteria. All virulent *Salmonellae* are, however, very resistant to this effect *in vitro*, and their serum resistance is generally considered a virulence factor. The high serum resistance in many *Salmonella* strains appears to be chromosomally determined, but in others genes of the virulence plasmid (to be discussed shortly) are needed for full resistance.[14] However, when a mutation was introduced in the *traT* gene of the virulence plasmid of Typhimurium it did not reduce the mouse virulence of the bacteria.[14] This apparently clearcut result does not necessarily tell that serum resistance is not a virulence determinant, because mouse complement is defective in the formation of the bactericidal membrane attack complex[15]. Mouse serum does not kill even definitely serum-sensitive bacteria - serum sensitivity being tested e.g. in guinea pig serum. Our preliminary data suggest in fact that serum resistance is important for virulence of *Salmonellae* to the guinea pig, in which bacterial serum-sensitivity leads to a reduced number of bacteria in the liver after iv injection.

MULTIPLICATION OF SALMONELLAE IN THE LIVER AND SPLEEN

The main multiplication of *Salmonellae* during invasive infection takes place in the liver and spleen[5]. This period lasts several days, depending on the challenge dose, the virulence of the bacteria and the susceptibility of the host. After iv injection, the bacteria are cleared from the blood and appear in the liver and spleen in about an hour. During the next few hours, growth of the bacteria and their killing proceed concurrently, resulting in a small net decrease in the total number of bacteria recovered.[16] For the following several days, however, no further killing is observed,[16] while the bacteria grow at a slowly decreasing rate until their net numbers start declining after appr. 7

days. The fact that the bacteria are sheltered from both antibiotics and antibodies during this time strongly speaks for their intracellular localization. The cell type for this intracellular growth phase has not been definitively demonstrated, but the macrophages of these tissues are the most likely cells. The macrophages of the liver, the so called Kupffer cells, have been shown to differ from other macrophages,[17] and one such difference could be their inability to kill intracellular *Salmonellae*. On the other hand, the initial killing of the bacteria suggests that possibly most of the cells that initially phagocytize the *Salmonellae* kill them, whereas bacterial growth can start in only a fraction of them. For an understanding of the virulence requirements of *Salmonellae* it would be important to clarify this question; it is, however, not an easy task.

It would be equally important to know the exact intra-cellular compartment where the *Salmonellae* grow. *In vitro* studies in cultured macrophages are not very helpful since they do not seem to mimick very well the *in vivo* site [17,] and data to be published. The current view that the site is a phagolysosome is supported by several lines of indirect evidence. Two potentially useful pieces of further information are available. One is that γ-interferon is produced during the invasive *Salmonella* infection and acts to reduce the growth rate of bacteria from the second day of infection on.[20] The second is that a mouse gene, *Ity*, also influences the growth rate, with a faster than usual rate in the salmonellosis-sensitive Ity^S mutant mice.[17]

The intracellular growth of the bacteria is clearly an essential parameter of their virulence. Some of its determinants are known, but the molecular mechanisms involved are mostly not clarified. The most clearcut situation is seen with the few types of nutritional *Salmonella* mutants known to reduce virulence. Thus mutations of the *aro* genes deprive the bacteria of essential metabolites, p-aminobenzoic acid required for the synthesis of folic acid, and 2,3-dihydrobenzoic acid required for the synthesis of enterochelin. Such mutants of *Salmonella* are nonvirulent[18] and do not grow in the liver -

they are not killed either, consistent with the above.[17,19] Fields et al. isolated mutants affected in their capacity of survival in cultured macrophages.[20] All these mutants were also avirulent for the mouse, and many were nutritionally defective.

Mutations that result in defective, rough type LPS also influence the intracellular growth phase. Mutants with the complete core do not grow, but are not killed either during this time.[21] "Deep rough" mutants with a very defective LPS core are sensitive to a variety of agents, and do not survive even the uptake phase to the liver; the mechanism of their elimination has not been determined.[22]

A very interesting finding was the realization that a large plasmid is required for this intracellular growth phase. Such a virulence plasmid has now been identified in several serovars, including Typhimurium, Enteritidis, Dublin, Choleraesuis and Gallinarum, and seems to be a regular feature of the isolates.[23] The virulence plasmids of the different serovars vary to some extent in size from appr. 50 to 100 kb, but contain large highly homologous regions. Surprisingly, the amount of DNA required for mouse virulence is much less, 4-6 kb. Four genes have been identified in this region, and they seem to be the same, with very minor sequence alterations, in all strains studied.[24,25,26] So far, nothing is known of how these genes function in promoting the growth of the bacteria. A serious difficulty for studying their effect is their low level of expression at least when the bacteria are grown *in vitro*; therefore reasonable amounts of the corresponding proteins have become available only recently when the genes have been cloned and expressed separately. The low level of expression of the genes of the virulence plasmid suggests that they are under strict control, which would not be unexpected in view of the extensive control systems seen with e.g. the virulence plasmids of *Yersiniae*.

Further regulatory systems of virulence determinants of *Salmonella* have been identified or are suggested by recent observations. These include the *cya*-cyclic AMP control system, the regulatory system under the control of the chromosome genes *phoP/phoQ*, and control by osmolarity.[27-30] Mutants of *phoP* were first identified among those isolated on the basis of their reduced survival in macrophages *in vitro*;[20] they were subsequently found to be sensitive to microbicidal cationic peptides and proteins of phagocytes.[28] *PhoP* is part of a regulatory system controlling the expression of a number of genes; several other genes affecting the virulence and survival in macrophages have subsequently been found among those regulated by *phoP*.[30] It is of interest that all the regulatory systems identified so far in *Salmonella* have been found to regulate chromosomal genes, while the regulation of the virulence plasmid still remains elusive.

Acknowledgements

The studies in Helsinki have been supported by the Sigrid Jusélius Foundation (P.H.M; M.R.), The Finnish Cultural Foundation (S.S.) and the Academy of Finland (S.T.; M.R.).

REFERENCES

1. International Committee on Systematic Bacteriology Taxonomic Subcommittee on *Enterobacteriaceae*. Minutes of the Meeting, 8 September 1986, Manchester, England. Int. J. System. Bacteriol. 38:223-224 (1988).

2. H. W. Smith and J. F. Tucker. The virulence of salmonella strains for chickens: their excretion by infected chickens. J. Hyg. Camb. 84:479-488 (1980).

3. A. Takeuchi. Electron microscope studies of experimental *Salmonella* infection. I. Penetration into the intestinal epithelium by *Salmonella typhimurium*. Am. J. Pathol. 50:109-136 (1967).

4. A. W. Hohmann, G. Schmidt, and D. Rowley. Intestinal colonization and virulence of *Salmonella* in mice. Infect. Immun. 22:763-770 (1978).

5. G. B. MacKaness, R. V. Blanden, and F. M. Collins. Host-parasite relations in mouse typhoid. J. Exp. Med. 124:573-583 (1966).

6. C. Galanos, M. A. Freudenberg, M. Matsuura, and A. Coumbos. Hypersensitivity to endotoxin and mechanisms of host-response, in: Bacterial Endotoxins: Pathophysiological Effects,

Clinical Significance, and Pharmacological Control, Alan R. Liss, Inc. pp. 295-308 (1988).

7. M. V. Valtonen, M. Plosila, V. V. Valtonen, and P. H. Mäkelä. Effect of the quality of the lipopolysaccharide on mouse virulence of *Salmonella enteritidis*. Infect. Immun. 13:828-832 (1975).

8. M. B. Lyman, B. A. D. Stocker, and R. J. Roantree. comparison of the virulence of O:9,12 and O:4,5,12 *Salmonella typhimurium his+* transductants for mice. Infect. Immun. 15:491-499 (1977).

9. H. Saxén, I. Reima, and P. H. Mäkelä. Alternative complement pathway activation by *Salmonella* O polysaccharide as a virulence determinant in the mouse. Microb. Pathog. 2:15-28 (1987).

10. C.-J. Liang-Takasaki, P. H. Mäkelä, and L. Leive. Phagocytosis of bacteria by macrophages: Changing the carbohydrate of lipopolysaccharide alters interaction with complement and macrophage. J. Immunol. 128:1229-1235 (1982).

11. V. E. Jimenez-Lucho, K. A. Joiner, J. Foulds, M. M. Frank, and L. Leive. C3b generation is affected by the structure of the O-antigen polysaccharide in lipopolysaccharide from *Salmonellae*. J. Immunol. 139:1253-1259 (1987).

12. H. Saxén. Mechanisms of the protective action of anti-*Salmonella* IgM in experimental mouse salmonellosis. J. Gen. Microbiol. 130:2277-2283 (1984).

13. M. V. Valtonen. Role of phagocytosis in mouse virulence of *Salmonella typhimurium* recombinants with O-antigen 6,7 or 4,12. Infect. Immun. 18:574-582 (1977).

14. M. Rhen and S. Sukupolvi. The role of the *traT* gene of the *Salmonella typhimurium* virulence-plasmid for serum resistance and growth within liver macrophages. Microb. Pathog. 5:275-285 (1988).

15. M. Vaara, P. Viljanen, T. Vaara, and P. H. Mäkelä. An outer membrane-disorganizing peptide PMBN sensitzes *E. coli* strains to serum bactericidal action. J. Immunol. 132:2582-2589 (1984).

16. W. H. Benjamin Jr., P. Hall, S. J. Roberts, and D. E. Briles. The primary effect of the *Ity* locus is on the rate of growth of *Salmonella typhimurium* that are relatively protected from killing. J. Immunol. 144:3143-3151 (1990).

17. D. A. Lepay, C. F. Nathan, R. M. Steinmann, H. W. Murray, and Z. A. Cohn. Murine Kupffer cells. Mononuclear phagocytes deficient in the generation of reactive oxygen intermediates. J. Exp. Med. 161:1079 (1985).

18. S. K. Hoiseth and B. A. D. Stocker. Aromatic-dependent *S. typhimurium* are non-virulent and are effective as live vaccines. Nature (London) 219:238-239 (1981).

19. A. Muotiala and P. H. Mäkelä. The role of IFN-γ in murine *Salmonella typhimurium* infection. Microb. pathog. 8:135-141 (1990).

20. P. I. Fields, R. V. Swanson, C. G. Haidaris, and F. Heffron. Mutants of *Salmonella typhimurium*

that cannot survive within the macrophage are avirulent. Proc. Natl. Acad. Sci (USA) 5189-5193 (1986).

21. A. Muotiala, M. Hovi, and P. H. Mäkelä. Protective immunity in mouse salmonellosis: comparison of smooth and rough live and killed vaccines. Microb. Pathog. 6:51-60 (1989).

22. P. H. Mäkelä, M. Hovi, H. Saxén, A. Muotiala, and M. Rhen. Role of LPS in the pathogenesis of salmonellosis, in: Endotoxin Research 1990, Elsevier Science Publishers B.V., Amsterdam (1990 in press).

23. P. A. Gulig. Virulence plasmids of Salmonella typhimurium and other Salmonellae. Microb. Pathog. (1990 in press).

24. S. Taira and M. Rhen. Identification and genetic analysis of mkaA - a gene of the Salmonella typhimurium virulence plasmid necessary for intracellular growth. Microb. Pathog. 7:165-173 (1989).

25. F. Norel, M.-R. Pisano, J. Nicoli, and M. Y. Popoff. Nucleotide sequence of the plasmid-borne virulence gene mkfA encoding a 28 kDa polypeptide from Salmonella typhimurium. Res. Microbiol. 140:263-265 (1989).

26. H. Matsui, K. Kawahara, N. Terakado, and H. Danbara. Nucleotide sequence of a gene encoding a 29 kDa polypeptide in mba region of the virulence plasmid, pKDSC50, of Salmonella choleraesuis. Nucleic Acids Research 18:1055 (1990).

27. R. Curtiss III, R. M. Goldschmidt, N. B. Fletchall, and S. M. Kelly.A Avirulent Salmonella typhimurium del cya del crp oral vaccine strains expressing a streptococcal colonization and virulence antigen. Vaccine 6:155-160 (1988).

28. P. I. Fields, E. A. Groisman, and F. Heffron. A Salmonella locus that controls resistance to microbicidal proteins from phagocytic cells. Science 243:1059-1062 (1989).

29. J. E. Galan and R. Curtiss III. Expression of Salmonella typhimurium genes required for invasion is regulated by changes in DNA supercoiling. Infect. Immun. 58:1879-1885 (1990).

30. S. I. Miller, A. M. Kukral, and J. J. Mekalanos. A two component regulatory system (phoP phoQ) controls Salmonella typhimurium virulence. Proc. Natl. Acad. Sci USA 86:5054-5058 (1989).

EXPERIMENTAL SALMONELLOSIS IN RETROSPECT AND

PROSPECT: 1990

John Stephen, Gillian R Douce, and Iqbal I Amin

School of Biological Sciences, Microbial Molecular Genetics and Cell Biology
Group, University of Birmingham, PO Box 363, Birmingham B15 2TT, UK

Research on *Salmonella spp.* continues unabated for a variety of
different reasons. These organisms cause acute gastroenteritis and systemic
typhoid disease in man, as well as other important infections in domestic
animals. In addition, metabolically crippled *Salmonellae* are likely to be used
increasingly as vectors for delivering extraneous immunogens. The steepling
increase in *Salmonella*-induced gastroenteritis in developed countries has
undoubtedly fuelled the current explosion of research in this area.

The selection of one's model for study is vitally important. While some
non-specific symptoms are common (e.g. nausea and abdominal pain) the
clinical features of salmonella-induced diarrhoea and systemic typhoid
infections are in some respects quite different. For example, gastroenteritis
may follow 8-36 hours after ingestion of contaminated food, whereas typhoid
may follow an incubation period of 10-20 days. Diarrhoea, (which is usually
watery, but may be severe, and sometimes bloody,) is the predominating
feature of gastroenteritis, whereas in the case of adults constipation may occur
in the early clinical stages of typhoid; diarrhoea may occur much later. In
gastroenteritis, fever may occur whereas in typhoid this may be so severe as to
cause delirium. Current perceptions are that gastroenteritis results from the
initial interactions of *Salmonella typhimurium* for example (and/or it's
products) with the gut mucosa, whereas typhoid fever is produced by *S. typhi*
organisms which translocate the mucosa, survive within macrophages,
multiply and release endotoxin which triggers the highly complex endotoxin-
cascade. Gastroenteritis is usually self-limiting, whereas in untreated typhoid
mortality can be as high as 10%.

It follows, therefore, that the choice of organism/host combination for
experimental study and the precise definition of the question(s) to be
addressed, are of crucial importance particularly when, as is the case with
Salmonella spp., virulence is multifactorial. In much contemporary research
on salmonella infections, appreciation of these basic points has become blurred
or ignored. The rest of this article is devoted mainly (but not exclusively) to
work on *S. typhimurium* as an agent of gastroenteritis.

In our laboratory we have made extensive use of the ligated rabbit ileal
loop (RIL) model originally used for *Salmonella* research by Giannella and

coworkers in the early 1970's. The arguments justifying its use have been fully spelt out elsewhere (Stephen *et al.*, 1985) and will not be repeated here. We have continued to use the same strains of *S. typhimurium* as were used in the original American work: strains TML, W118, (virulent, invasive diarrhoeagenic biotype (++)), strains SL1027, LT7 and M206 (avirulent, invasive non-diarrhoeagenic biotype (+-)), and strain Thax-1 (avirulent, non-invasive non-diarrhoeagenic biotype (- -)). One additional virulent strain has been added to the panel of test strains, WAKE (++) which was isolated from a UK outbreak. What follows is a summary overview of recent and current work with *S.typhimurium* (with brief reference to other enteric pathogens) covering studies on attachment to and invasion of the gut, the nature and role of *Salmonella* enterotoxin, the inflammatory response, and the pathophysiological mechanism of diarrhoeal disease.

Initial attempts to study quantitatively the early events of this complex pathogen/host interaction *in vivo,* using the above strains of known different virulence in RILs, were abandoned. The data on attachment to and invasion of intestinal mucosae were wholly irreproducible (unpublished) largely due, it was thought, to the variability of the host response which results in shedding of infected intestinal epithelial cells. Even if this were not the case, sheer logistics would preclude the use of this system for the study of large numbers of strains or mutants derived from them. We therefore developed an organ culture system which allowed short term experiments to be carried out on asymmetrically infected adult rabbit ileum (Worton *et al.*, 1989). It was not possible to discriminate virulent strains from all avirulent strains in terms of their initial attachment to intestinal mucosa *in vitro* (Worton *et al.*, 1989). Viable counts of washed homogenised tissue taken 30 min after challenge showed that virulent strains TML and W118 and avirulent strains LT7 and M206 could not be distinguished from each other. Avirulent strain SL1027 associated less well than the other four strains and Thax-1 associated less well than SL1027; both these strains were non motile. Thus early association with gut mucosa, probably involving interaction with the mucous layer, did not discriminate all avirulent strains from the virulent strains. Qualitative examination of tissues by scanning electron microscopy did not detect strains LT7 and M206 on the mucosal surface whereas strains TML and W118 were readily seen, suggesting that the nature of association of virulent and avirulent strains was different: avirulent strains were readily removed by the processes involved in preparing tissues for electron microscopy. Qualitative examination by transmission electron microscopy of tissues challenged *in vitro* for 120 min showed virulent strains TML and W118 invading epithelial cells in a manner similar to that observed after 120-min challenge *in vivo*. In contrast, invasion by avirulent strains was observed only very rarely (Worton *et al.* 1989). Unfortunately, quantitative studies on tissue invasion could not be carried out. The structural integrity of villi, under the conditions used, was maintained at best only up to 2 h, which was insufficient for quantitative invasion studies. The reason is that freshly excised gut continues to perform its differentiated function of absorbing fluid. In the absence of a functional blood supply and effective lacteal drainage, and, the low hydraulic conductivity of the serosal layer, fluid accumulates under pressure, enterocytes vacuolate, and the epithelium detaches. Attempts to overcome these problems are underway. If these prove successful then a major step forward will have been achieved in the development of a more meaningful system for the quantitative analysis of intestinal epithelial invasion than is currently available in the context of gastroenteritis.

In contrast, Miller, Maskell *et al.,* (1989) have developed another approach using *S. typhimurium* in mice - an excellent model for human typhoid but not for gastroenteritis. They generated 95 Tn*phoA* mutants of virulent *S. typhimurium* strain C5 . Fifteen of 95 mutants were avirulent for mice when administered orally. The importance of the screening system - as exemplified later - cannot be over emphasised. The oral route will identify mutations which affect early events such as gut colonization, attachment to and invasion of the intestinal mucosa and translocation to liver and spleen: these would be missed if parenteral routes or *in vitro* assays were used in the initial screen. It was shown that 9 of the 15 avirulent mutants had defective LPS; the other 6 were smooth. No alterations in the major outer membrane proteins were detected. The mutants differed in their abilities to translocate to the liver and spleen. The further analysis of these mutants is awaited with great interest.

In the interim we have adopted the increasingly used tissue culture cell system in an attempt to develop a quantitative assay for analysis of invasiveness of *S.typhimurium*. Even if the organ culture system were to be successfully developed, a test system based on cell culture would still be highly desirable for handling large numbers of strains in an initial screen for invasiveness; several exist (see below). The basic experimental design is deceptively simple and universally followed. Bacterial suspensions are placed over cell monolayers and organisms gently centrifuged onto the target cells. At various time intervals non-internalised organisms are removed, or killed with gentamicin, intracellular ones are released (usually with TritonX-100) and counted. This is often coupled with analysis of mutants mostly generated by transposon mutagenesis using Tn10 or Tn*phoA*. The object is to identify genes and hence their products, inactivation of which results in the loss of 'invasiveness' or, more strictly, one of its components - attachment, internalisation, intracellular survival or multiplication, and release.

From our experiments (unpublished) using HEp-2 cells and the 7 strains of *S. typhimurium* listed above, several important points have emerged regarding the design and execution of such experiments.

1. We initially used, in line with most workers, an inoculum/tissue culture cell ratio of 100:1 and a 2 h period for organism/cell interaction (Small *et al.,* (1987), *Escherichia coli, S. typhimurium, Yersinia pseudotuberculosis, Y. enterocolitica,* HEp-2 cells; Finlay and Falkow (1988), *S. choleraesuis, Shigella flexneri, Y. enterocolitica,*CHO, HEp-2, and MDCK cells; Elsinghorst *et al.,* (1989), *S. typhi,* Henle cells; Gahring *et al.,* (1990), *S. typhimurium,* Caco2 cells and J774 macrophage cell line; Ernst *et al.,* (1990), *S. typhimurium,* HEp-2 cells). The high multiplicity is used to synchronise events and to increase the number of organisms which invade and hence the meaningfulness of the numbers of intracellular organisms recovered; the numbers tend to be very small. Under these conditions, in agreement with most published data, significantly <1.0 % of the original inoculum was recovered as gentamicin resistant, i.e. internalized.

2. With high inocula the pH of the tissue culture medium became acidic.

3. Microscopic examination of Giemsa-stained monolayers at the point of gentamicin addition revealed rafts of organisms covering the monolayer; in addition from 2 h onwards many HEp-2 cells became rounded. One could not therefore be totally confident about the biological significance of the quantitative data thus obtained.

4. By experiment it was found that an organism cell ratio of 1:1 was the highest which did not lower the pH or cause cells to round. However, the recovery of gentamicin-resistant organisms was still lower than might have

been expected by microscopic examination. After TritonX-100 treatment many organisms remained bound to residual cell nuclei. Use of a rubber policeman after TritonX-100 treatment, and, vigorous pipetting to disperse aggregates of bacteria/cell debris, increased the recoveries of gentamicin-resistant organisms to approximately 10% of the initial inoculum at 2 h, and 20-60% at 4 h depending on strain and experimental conditions. With high inocula (100:1) the recovery was approximately 10% at 2 h with no increase at 4 h.

5. The pattern (but not the absolute values) was highly reproducible on different days and in different hands; Gahring et al., (1990) reported similar findings. However, within any one experiment it has been shown that the mean recovery from 8 or 10 wells of a 24-well tissue culture plate has an associated SEM of ±6.5 % or less. This level of accuracy should be sufficient for comparing strain-invasiveness within any one experiment.

6. From 2 h onwards the distribution of organisms was uneven with many cells having no associated organisms. Microcolonies were seen associated with some HEp-2 cells in a manner reminiscent of that reported by Takeuchi (1967), Takeuchi and Sprinz (1967), Lindquist et al., (1987), and Worton et al., (1989). A possible explanation of this phenomenon was given by Lindquist et al. (1987) who studied the attachment of S. typhimurium to isolated rat enterocytes. They suggested that productive attachment releases a chemotaxin which results in focal aggregation of organisms and the ingestion of a microcolony. This phenomenon either does not happen with high inocula, or it is completely occluded by the blanket mat-like association of organisms.

7. Uptake depends on the batch of foetal calf serum used in the tissue culture medium as reported by Shaw and Falkow (1988) in their studies on uptake of N. gonorrhoea by several cell lines.

8. The phenotype of S. typhimurium is also important. Organisms in early/mid log phase enter more efficiently as has also been reported by Small et al., (1987), Finlay and Falkow (1988), Elsinghorst et al., (1989) and Ernst et al., (1990). We have also clearly shown that organisms grown in Hartley Digest Broth (HDB, Oxoid, UK) are significantly better at invading HEp-2 cells than when grown in Myosate broth (Becton Dickinson, UK, Ltd). Ernst et al., (1990) also demonstrated that S. typhimurium grown anaerobically and added to HEp-2 cells under anaerobic conditions invaded more efficiently.

9. Centrifugation may grossly distort the picture obtained particularly with some avirulent strains.

How far has this approach advanced our knowledge of salmonellosis? To date several studies have been undertaken which by virtue of the choice of the mouse as the animal model used for initial selection of mutants, or, conversely to validate the biological role of putative virulence genes highlighted by tissue culture studies, have more direct bearing on typhoid disease than on acute gastroentroenteritis. Nevertheless, some important points have emerged and four examples are chosen for critical comment. The order reflects our appreciation of the relevance of the work in relation to the progression of organisms from lumen to deeper tissues.

Galan and Curtiss (1989) examined the attachment to and invasion of Henle-407 cells, and mouse intestine, by S. typhimurium. Analysis of TnphoA invasion-deficient mutants revealed an inv locus comprising invA, -B, and -C in one transcriptional unit and invD downstream from this cluster in an independent transcriptional unit. Unlike invasion the ability to attach was not controlled by inv genes. TnphoA mutants which showed reduced invasiveness for Henle-407 cells showed reduced lethality for mice, reduced abilities to colonize Peyer's patches, intestinal wall and spleen when given orally. In contrast, no significant differences between wild type and *inv* mutants were observed as measured by LD_{50}s and ability to colonize the spleen when organisms were administered intraperitoneally. More recently Galan and Curtiss (1990) have shown that the expression of *invA* (and by extension *B* and *C*, since inactivation of *A* abolishes expression of *B* and *C*) and ability to invade MDCK cells (see below) was increased at higher osmolarities. The osmoinducibility was independent of the well characterized *ompR* gene which controls the osmoinducibility of other genes. Several lines of evidence pointed to a direct correlation between the degree of DNA superhelicity and *invA* expression. These data indicate that ability to colonize the gut is a virulence-determining step in disseminated oral infections caused by *S. typhimurium* and that this can, with appropriately selected mutants be analysed meaningfully *in vitro* at a highly sophisticated level.

An elegant model system has been devised by Finlay and co-workers in an attempt to study *in vitro* processes which might be involved in the translocation of *Salmonella* across the gut epithelium. They used MDCK cells (a canine kidney cell line) which retain their differentiated polarized morphology when grown on membrane filters, and, the invasive non-typhoid *S. choleraesuis* (Finlay *et al.*, 1988a). When added to the apical side of these cells wild type organisms attached to brush borders, invaded cells, multiplied, and exocytosed via the basolateral membrane. This process required *de novo* protein synthesis since inhibitors of bacterial RNA and protein synthesis inhibited attachment and invasion. Transcytosis also induced changes in the electrical resistance of the monolayer; the changes were not due to alterations in the tight junctions between enterocytes. Finlay *et al.*, (1988b) found that 42 of 626 Tn*phoA* mutants had lost the ability to transcytose MDCK cells. However, this system did not reflect absolutely the interaction of invasive *S. choleraesuis* with murine intestinal epithelia. The mutants could be grouped into 6 classes of which two classes (3 and 6) were just as lethal for mice as the parent strain when given orally. Again, Finlay *et al.*, (1989) identified 6 induced proteins synthesized when *S. choleraesuis* or *S. typhimurium* interacted with either MDCK or HEp-2 cells; these were regarded as important in the trancytosis phenomenon. But, only class 6 of the 6 classes of non-invasive *S. choleraesuis* mutants did not make these proteins; the other 5 did. It therefore follows that these proteins cannot be the exclusive determinants of transcytosis. Only in the case of 2 Tn10 insertion mutants of *S. typhimurium* did they report a positive correlation between the absence of these proteins and failure to transcytose. However, these two mutants were described as less virulent for mice when injected intraperitoneally. For that reason, the large increase reported in LD_{50}s is more likely to be due to an inability to survive in macrophages than failure to translocate gut epithelium; this leads us to consider the third example.

Fields *et al.*, (1986) generated over 80 Tn10 insertion mutants of *S. typhimurium* which were avirulent for mice by virtue of their inability to resist the lethal effects of phagocytic defensins; resistance to these microbicidal defensins is controlled by the *phoP* gene (Fields *et al.*, 1989). The ability of the Tn10 mutants to invade Caco2 (transformed colonic epithelial) cells was not impaired (Gahring *et al.*, 1990). The sequences of *phoP* (Miller *et al.*, 1989; Groisman *et al.*, 1989) and *phoQ* (Miller *et al.*, 1989) genes have now been obtained. From such analyses and analogy with other systems, an elegant model for intraphagocytic survival has been proposed (Miller *et al.*, 1989; Groisman and Saier, 1990). The perception is of a two component regulon. The *phoQ* protein senses the hostile environment of the phagolysosome and

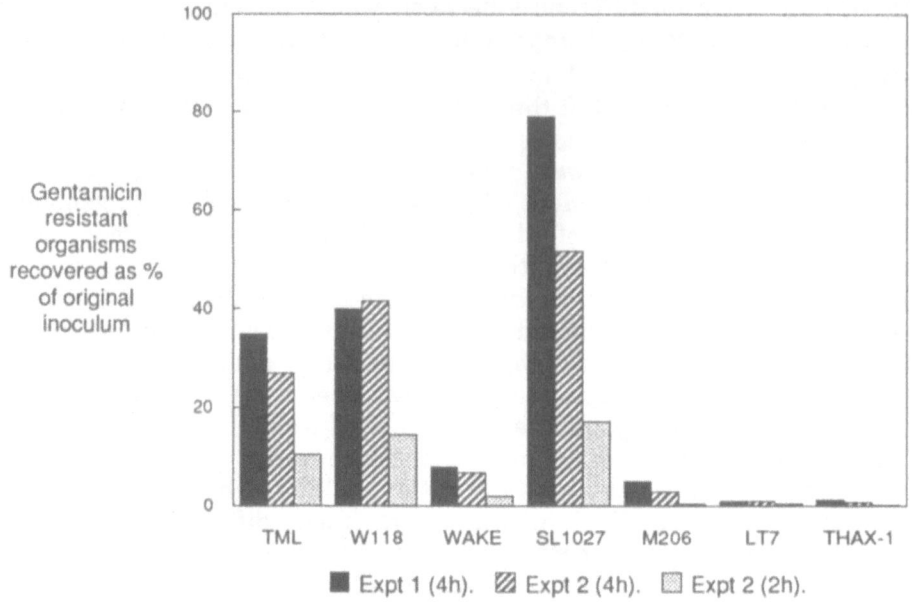

Figure 1. These data are typical of many experiments and have not been pooled and averaged. 24-well tissue culture trays were used which had been seeded with HEp-2 cells such that confluent monolayers (approximately $2-4 \times 10^5$ cells) were formed on the day of usage. Bacterial suspensions were prepared and used as described in the text. No more than 2 strains were inoculated on to one plate; 8-10 wells were used for each strain. No more than 2 strains were handled in any one day. For each strain 2 experiments are described. In experiment 1, randomly selected pairs of strains were examined in a 4 h incubation with HEp-2 cells. In experiment 2, one strain and 2 plates were used. HEp-2 cell monolayers in both plates were prepared from the same cell suspension; bacterial inocula were likewise prepared from the same bacterial suspension. One plate was incubated for 4 h allowing comparison with experiment 1; the other plate was incubated for 2 h.

phosphorylates PhoP which is a transcriptional activator of several genes including *pagC*. Miller *et al.*, (1989) showed that *pagC* was involved in resistance to phagocytic killing and proposed that its product is a membrane protein capable of resisting the toxic effects of defensins.

The fourth example selected for comment is the work of Elsinghorst *et al.*, (1989), which serves as a warning against too simplistic interpretations of data from such genetic experiments. They showed that penetration of Henle-407 cells by *S. typhi* was cytochalasin B and D sensitive and controlled by at least 4 separate chromosomal loci: this genetic element conferred upon non-invasive *E.coli* HB101 the ability to invade (albeit with a lower efficiency) Henle-407 cells. Homologous sequences were identified in and cloned from *S. typhimurium* and shown to contain the same genes. However, these did not confer the invasive phenotype for Henle-407 cells to *E. coli*.

The final example, from our own laboratory, concerns the behaviour in HEp-2 cells of the 7 strains listed above of different virulence in RILs, a model for human gastroenteritis. Fig. 1 summarizes the data obtained from a HEp-2 cell assay under the optimum conditions already described: log phase organisms, low (1:1) multiplicity of infection, low **g** centrifugation, 2 or 4 h incubation, gentamicin treatment, and use of rubber policeman. Several points invite comment. A definite increase was observed in the numbers of intracellular organisms recovered after 4 h as compared to 2 h. It is not yet clear if this represents continued entry or multiplication or both. The 3 virulent strains are not equally invasive. Avirulent strain SL1027 appears to be the most invasive of all; it was also the most variable. On this basis, in the absence of information to the contrary (Wallis *et al.*, 1986*a*; 1989), one would classify strains SL1027 as virulent and WAKE as poorly virulent, maybe avirulent. However, if virulence in this context is multifactorial, then it could well be that strain WAKE is sufficiently invasive to initiate the next event in a biological sequence (induction of the inflammatory response?), whereas SL1027, despite its apparent superior invasiveness, cannot initiate the next event. It is also possible that HEp-2 cells are not a good model for gut epithelia, since it has been shown by immunofluorescence that SL1027 invades the gut very poorly (unpublished). Or, it could be that the test as carried out produces artifacts: in this context we strongly suspect the centrifugation step and this is under active investigation.

Clearly much remains to be learnt about the mechanisms, determinants, routes and significance of invasion in the causation of *Salmonella*-induced disease. Based on the foregoing discussion we feel that most progress has been made to date in developing analytical systems which will enrich our understanding of invasion in the context of systemic infections, particularly in relation to intraphagocytic survival. However, if it is true that "little is known" (with certainty, our comment) "about how *Salmonella* sp. invade animal cells and specifically what genes are required for cell invasion (Gahring *et al.*, 1990)", even less is known with certainty as to why *Salmonella* need to be invasive to cause gastroenteritis with associated diarrhoea. This is particularly so, if as is believed to be the case by many, a cholera-like toxin is involved in the latter. The study of tissue culture cells will never *per se* lead one to the answer to this question. From our own laboratory, some new concepts have been developed which, even though they may not be correct, have the virtue of being experimentally testable.

It is salutary to be reminded that we are dealing with an 'enteritis'. Recently we quantified the magnitude of the leucocyte influx in relation to virulence. This was assessed in rabbit ileal loops with [111]Indium-labelled leucocytes (Wallis *et al.*,1989). Virulent strains (TML, W118, and WAKE) caused a dose-dependent leucocyte influx; avirulent strains failed to induce a significant leucocyte influx. Fluid secretion was never observed in the absence of a leucocyte influx, but leucocyte influx *per se* did not induce fluid secretion. The phenotype of the challenge inoculum influenced fluid secretion since young log-phase organisms induced fluid secretion with a higher frequency than overnight cultures. These findings support earlier evidence implicating leucocytes (see Wallis *et al.*, 1989) whose role is likely to be an interactive one involving bacteria in the genesis of *Salmonella*-induced fluid secretion. The data suggest, though do not prove, that interaction of leucocytes with organisms of appropriate phenotype results in the release of a bacterial or host-derived secretagogue, or both.

The bacterial factor could be the cholera toxin related antigen (CTRA) described by Clarke *et al.*,(1988). They analysed strains TML, W118, SL1027, LT7, M206 and Thax-1, by a fluorescent-antibody-labelling technique involving antibodies to cholera toxin (anti-CT). A rapid increase in the proportion of cells producing CTRA occurred in early log phase (1-4 h growth) followed by a rapid decline during mid-late log phase in each of the six strains when grown in Casamino Yeast Extract (CYE; Oxoid) medium. Culture conditions, composition of medium and size of initial inoculum all affected the expression of CTRA in *S. typhimurium* (Qi *et al.*, 1989). When organisms harvested at 4-5 h were subcultured into fresh CYE medium and incubated for a further 2 h, the numbers of organisms increased, and the decrease in the proportion of organisms expressing CTRA was reduced. Use of HDB in place of CYE medium increased the number of organisms expressing CTRA in all strains tested, in both the uninterrupted and subculture procedures. The higher the initial inoculum, the lower was the proportion of organisms expressing CTRA. The presence of the antigen in cells was detectable for about 18 h after transfer from 37° to 4°C. Such data show that CTRA is expressed in log-phase growth and suggest that it's expression is regulated by a non-growth limiting factor. The nature of CTRA was analysed by immunoblotting using anti-CT. An antigen of mol. wt equivalent to a high mol. wt species of CT B subunit was detected in polymyxin-B extracts of all strains, but greater amounts were observed in the avirulent strains. The relatedness of the salmonella antigen to CT was limited, in line with the genetic analysis of Chopra *et al.*, (1987). The high-mol. wt antigen was not disrupted in the denaturing conditions of SDS-PAGE; nothing was detected by ELISA with either ganglioside or anti-CT as anchor. Earlier, Wallis *et al.*, (1986*a*) had examined the same six strains for enterotoxin production. Enterotoxic activity, assayed in RILs, was detected in polymyxin-B extracts from all strains (with the possible exception of Thax-1) cultured for 6h in CYE medium. The extracts were inactive in tissue-culture assays with CHO, Y-1 adrenal and Vero cells, and in the infant mouse assay for enterotoxin. There was no correlation between enterotoxigenicity *in vitro* and the ability to induce fluid secretion *in vivo*. It is tempting to think of CTRA as an inactive precursor of a CT-like toxin and that the latter is only released and activated under appropriate conditions *in vivo*. However, it is abundantly clear that the virulence of *S. typhimurium* as an agent of gastroenteritis cannot be explained solely, or simply, in terms of enterotoxin.

The foregoing data are compatible with but do not constitute proof that leucocytes play a role in the causation of *S. typhimurium*-induced fluid secretion. In attempting to resolve this controversial matter earlier experiments involving indomethacin and nitrogen mustard (N_2M) were repeated. Pretreatment of rabbits with indomethacin almost completely abolished salmonella-induced secretion and significantly reduced secretion induced by CT as was found by Gots *et al.*, (1974). They suggested that the data implicated prostaglandins of leucocytic origin. However, if *S. typhimurium* and CT share a common or partially common mechanism for inducing fluid secretion, that common part cannot involve neutrophils: there is no evidence for a necessary influx of neutrophils in CT-induced secretion. Moreover, in every case examined by Wallis *et al.*, (1986*b*; 1989; 1990), either quantitatively or qualitatively, virulent strains elicited a leucocyte response but not always a fluid response. Therefore, any common indomethacin-sensitive step must be mediated by cells other than neutrophils (possibly enterocytes) as has been claimed for CT (Duebbert and Peterson, 1985). This leads us to consider pathophysiological aspects of this infection.

Attempts were made to render animals neutropenic since, if neutrophils are involved, such animals should be incapable of mounting a fluid secretion in response to a challenge by fully virulent organisms. In agreement with Giannella (1979), Wallis *et al.*, (1990) showed that rabbits may be rendered leucopenic by N_2M-treatment, and that the ability of such animals to mount a secretory response to challenge by virulent strains of *S. typhimurium* was almost completely abolished. However in two important respects our data differed from those of Giannella (1979). First, as judged both by light and scanning electron microscopy, uninfected villi treated with N$_2$M showed significant morphological damage as compared with untreated villi. Neither uninfected-N$_2$M- or infected-N$_2$M-treated loops showed significant fluid secretion despite N$_2$M-induced damage in both, and, the huge numbers of invading organisms of strains that regularly induced fluid secretion in normal animals. Thus the basic premise that N$_2$M did not damage or upset the gut in a manner prejudicial to the experimental design - claimed by Giannella to have been fulfilled - was not fulfilled in our work. The second major difference was that N$_2$M reduced by 45% the expected fluid secretion induced by CT. While the N$_2$M experiments left the role of leucocytes unresolved they highlight other pathophysiological possibilities summarised as follows.

N$_2$M is used therapeutically in the treatment of certain neoplasias by virtue of its alkylating effect on DNA and hence on dividing cells (Salmon, 1980). It is therefore possible that the reduction of villus height and other morphological changes (*Wallis et al.,* 1990) are a direct consequence of N$_2$M affecting crypt cell mitoses. This would affect normal epithelial cell replacement and hence height of villi, and also crypt functions which, by general consensus (Roggin *et al.,* 1972; Field *et al.,* 1980; Welsh *et al.,* 1982), are involved in normal secretion. Recently, Collins *et al.,*(1988) and Spencer *et al.,* (1990) demonstrated that increased rates of villus base/crypt cell division are involved in the hypersecretory phase of rotavirus-induced diarrhoea in mice. Thus it could be that the inhibition of secretion by N$_2$M seen in this work is due to its ability to inhibit cell division in villus base/crypt assemblies. The significance of this is to be appreciated in the context of a new concept of the pathophysiology of infectious diarrhoea (see Spencer *et al.,* 1990). The intestine responds to insult/injury by shedding affected enterocytes which are rapidly replaced. Dividing cells accumulate transiently high levels of NaCl the secretion of which is perceived to be the basis of fluid loss in diarrhoeal disease.

These pathophysiological principles have been developed most fully for murine rotavirus diarrhoea and extrapolated to bacterial toxin-mediated diarrhoeas (Stephen and Osborne, 1988). The data summarized above strengthen the case for CT (Stephen and Pietrowski, 1986; Stephen and Osborne, 1988) and allow one to extend these ideas to *S. typhimurium*-induced diarrhoea.

In conclusion, the following experimentally testable synthesis can be made to explain *S. typhimurium*-induced gastroenteritis. *S. typhimurium* of the correct virulent genotype invades the intestinal mucosa; the degree of initial invasion is phenotypically determined. An influx of leucocytes, mainly neutrophils, occurs. The latter interact with luminal or invading organisms or both, and if of the correct phenotype these release a processed CT-like toxin. This toxin acts *via* an indomethacin-sensitive step resulting in fluid secretion.

REFERENCES

Chopra, A. K., Houston, C. W., Peterson, J. W., Prasad, R. Mekalanos, J. J., 1987, Cloning and expression of the *Salmonella* enterotoxin gene, *J.Bact.*, 169: 5095.

Clarke, G. J., Qi, G.-M., Wallis, T. S., Starkey, W. G., Collins, J., Spencer, A. J., Haddon, S. J., Osborne, M. P., Worton, K. J., and Stephen, J., 1988, Expression of an antigen in strains of *Salmonella typhimurium* which reacts with antibodies to cholera toxin, *J. Med.Microbiol.*, 25: 139

Collins, J., Starkey, W. G., Wallis, T. S., Clarke, G. J., Worton, K. J., Spencer, A. J., Haddon, S. J., Osborne, M. P., Candy, D.C.A., and Stephen, J., (1988), Intestinal enzyme profiles in normal and rotavirus-infected mice, *J. Pediatr. Gastroenterol. Nutr.*,7: 264.

Duebbert, I. E., and Peterson, J. W., 1985, Enterotoxin-induced fluid accumulation during experimental salmonellosis and cholera: involvement of prostaglandin synthesis by intestinal cells, *Toxicon* 23: 157.

Elsinghorst, E. A., Baron, L. S., Kopecko, D.J., 1989, Penetration of human intestinal epithelial cells by *Salmonella:* molecular cloning and expression of *Salmonella typhi* invasion determinants in *Escherichia coli, Proc. Natl. Acad. Sci. USA*, 86: 5173.

Ernst, R. K., Dombroski, D. M., and Merrick, J. M., 1990, Anaerobiosis, type 1 fimbriae, and growth phase are factors that affect invasion of HEp-2 cells by *Salmonella typhimurium, Infect. Immun.* 58: 2014.

Field, M., Smith, P. L., Bolton, J. E., 1980, Ion transport across the isolated intestinal mucosae of the winter flounder, *Pseudopleuronectes americanus* II; effects of cyclic AMP, *J. Membrane Biol.*, 55: 157.

Fields, P. I., Groisman, E. A., and, Heffron, F., 1989, A *Salmonella* locus that controls resistance to microbicidal proteins from phagocytic cells, *Science*, 243: 1059.

Fields, P. I, Swanson, R. V., Haidaris, C. G., and Heffron, F., 1986, Mutants of *Salmonella typhimurium* that cannot survive within the macrophage are avirulent, *Proc. Natl. Acad. Sci.,USA*, 83: 5189.

Finlay, B. B., and Falkow, S., 1988, Comparison of the invasion strategies used by *Salmonella choleraesuis, Shigella flexneri* and *Yersinia enterocolitica* to enter cultured animal cells: endosome acidification is not required for bacterial invasion or intracellular replication, *Biochimie,* 70: 1089.

Finlay, B. B., Gumbiner, B., and Falkow, S., 1988a, Penetration of *Salmonella* through a polarized Madin-Darby canine kidney epithelial cell monolayer, *J. Cell Biol.*, 107: 221.

Finlay,B. B., Heffron, F., and Falkow, S., 1989, Epithelial cell surfaces induce *Salmonella* proteins required for bacterial adherence and invasion, *Science,* 243: 940.

Finlay, B.B., Starnbach, M. N., Francis, C. L., Stocker, B. A. D., Chatfield, S., Dougan, G., and Falkow, S., 1988b, Identification and characterization of Tn*phoA* mutants of *Salmonella* that are unable to pass through a polarized MDCK epithelial cell layer, *Mol. Microbiol.*, 2: 757.

Gahring, L. C., Heffron, F., Finlay, B. B., and Falkow, S., 1990, Invasion and replication of *Salmonella typhimurium* in animal cells, *Infect. Immun.*, 58: 443.

Galan, J. E., and Curtiss III, R., 1989, Cloning and molecular characterization of genes whose products allow *Salmonella typhimurium* to penetrate tissue culture cells, *Proc. Natl. Acad. Sci. USA*, 86: 6383.

Galan, J. E., and Curtiss III, R., 1990, Expression of *Salmonella typhimurium* genes required for invasion is regulated by changes in DNA supercoiling, *Infect. Immun.*, 58: 1879.

Giannella, R. A., 1979, The importance of the intestinal inflammatory reaction in *Salmonella*-mediated intestinal secretion, *Infect. Immun.*, 23: 140.

Gots, R. E., Formal, S. B., Giannella, R. A., 1974, Indomethacin inhibition of *Salmonella typhimurium, Shigella flexneri,* and cholera-mediated rabbit ileal secretion, *J. Inf. Dis.,* 130: 280.

Groisman, E. A., Chiao, E., Lipps, C. J., and Heffron, F., 1989, *Salmonella typhimurium phoP* virulence gene is a transcriptional regulator, *Proc. Natl .Acad. Sci. USA,* 86: 7077.

Groisman, E. A., and Saier, M. H., 1990, *Salmonella* virulence: new clues to intramacrophage survival, *TIBS,* 15-January 1990, 30.

Lindquist, B. L., Lebenthal, E., Lee, P-C., Stinson, M.W., and Merrick, J. M., 1987, Adherence of *Salmonella typhimurium* to small-intestinal enterocytes of the rat, *Infect. Immun.,* 55: 3044.

Miller, S. I., Kukral, A. M., and Mekalanos,J. J., 1989, A two component regulatory system (*phoP, phoQ*) controls *Salmonella typhimurium* virulence, *Proc. Natl. Acad. Sci. USA,* 86: 5054.

Miller I., Maskell, D., Hormaeche, C., Pickard, D., and Dougan, G., 1989, The isolation of orally attenuated *Salmonella typhimurium* following Tn*phoA* mutagenesis, *Infect. Immun.,* 57: 2758.

Qi, G.-M., Clarke, G. J., Wallis, T. S., and Stephen, J., 1989, The influence of cultural conditions on the expression in *Salmonella typhimurium* of an antigen related to cholera toxin, *J. Med. Microbiol.,* 30: 213.

Roggin, G. M., Banwell, J. G., Yardley J. H., and Hendrix T. R., 1972, Unimpaired response of rabbit jejunum to cholera toxin after selective damage to villus epithelium, *Gastroenterology,* 63: 981.

Salmon, S. E., 1980, Cancer Chemotherapy, *in*: "Review of medical pharmacology," Meyers F.H., *et al.* eds. 7th edn. Lange Medical Publications, Los Altos, CA., p 477.

Shaw, J. H., and Falkow, S., 1988, Model for invasion of human tissue culture cells by *Neisseria gonorrhoea, Infect. Immun.* 56:1625.

Small P. L. C., Isberg, R. R., and Falkow, S., 1987, Comparison of the ability of enteroinvasive *Escherichia coli, Salmonella typhimurium, Yersinia pseudotuberculosis,* and *Yersinia enterocolitica* to enter and replicate within HEp-2 cells, *Infect. Immun.,* 55: 1674.

Spencer, A. J., Osborne, M. P., Haddon, S. J., Collins, J., Starkey, W. G., Candy, D. C. A., and Stephen. J., 1990, X-ray microanalysis of rotavirus-infected mouse intestine: a new concept of diarrhoeal secretion, *J Pediatr Gastroenterol Nutr.* 10: 516

Stephen, J., Osborne, M. P., 1988, Pathophysiological mechanisms in diarrhoeal disease, *in*: "Bacterial infections of respiratory and gastrointestinal mucosae", W Donachie, E Griffiths, J Stephen, eds., special publication of the Society for General Microbiology, vol 24, , UK, IRL Press, Oxford, p 149.

Stephen, J., and Pietrowski, R. A., 1986, "Bacterial Toxins," 2nd. edn. Van Nostrand Reinhold (UK) Co. Ltd., Wokingham, UK.

Stephen, J., Wallis, T. S., Starkey, W. G., Candy, D.C.A., Osborne, M.P., and Haddon, S. J., 1985, Salmonellosis: in retrospect and prospect, *in*: "Microbial toxins and diarrhoeal disease," eds. Evered, D., Whelan, J., Pitman, London (Ciba Foundation Symposium 112) p 175.

Takeuchi, A., 1967, Electron microscope studies of experimental Salmonella infection. I. Penetration into the intestinal epithelium by *Salmonella typhimurium, Am. J.Path.,* 50: 109.

Takeuchi, A., and Sprinz, H., 1967, Electron microscope studies of experimental salmonella infection in the preconditioned guinea pig. II. Response of the intestinal mucosa to invasion by *Salmonella typhimurium, Am. J. Path.,* 51: 137.

Wallis, T. S., Hawker, R. J. H., Candy, D.C.A., Qi, G.-M., Clarke, G. J., Worton, K. J., Osborne M. P. and Stephen, J., 1989, Quantification of the leucocyte influx into rabbit ileal loops induced by strains of *Salmonella typhimurium* of different virulence, *J.Med. Microbiol.,* 30: 149.

Wallis, T. S., Starkey, W. G., Stephen. J., Haddon, S. J., Osborne, M. P., Candy, D. C. A., 1986*a* Enterotoxin production by *Salmonella typhimurium* strains of different virulence, *J. Med. Microbiol.,* 21: 19.

Wallis, T. S., Starkey, W. G., Stephen, J., Haddon, S. J., Osborne, M. P., Candy, D. C. A., 1986*b*, The nature and role of mucosal damage in relation to *Salmonella typhimurium*-induced fluid secretion in the rabbit ileum, *J. Med. Microbiol.,* 22: 39.

Wallis, T. S., Vaughan, A. T. M., Clarke, G. J., Qi, G.-M., Worton, K. J., Candy, D.C.A., Osborne, M. P., and Stephen, J., 1990, The role of leucoytes in the induction of fluid secretion by *Salmonella typhimurium*, *J. Med. Microbiol.,* 31: 27.

Welsh, M. J., Smith, P. L., Fromm M., and Frizell, R. A., 1982, Crypts are the site of intestinal fluid and electrolyte secretion, *Science,* 218: 1219.

Worton, K. J., Candy D. C. A., Wallis T.S., Clarke, G. J., Osborne, M. P., Haddon, S. J., and Stephen, J., 1989, Studies on early association of *Salmonella typhimurium* with intestinal mucosa *in vivo* and *in vitro*: relationship to virulence, *J. Med. Microbiol.,*29: 283.

COLONIZATION OF THE MURINE GASTROINTESTINAL TRACT BY

SALMONELLA TYPHIMURIUM

Roy Curtiss III and Jorge Galán*

Department of Biology
Washington University
St. Louis, MO 63130

*Department of Microbiology
School of Medicine
State University of New York
Stony Brook, NY 11794

INTRODUCTION

Salmonella infection of animal hosts results, in general, from oral consumption of contaminated foods. It is therefore important to understand the series of events necessary for *Salmonella* to exist both outside and inside the host. We have set out to define biochemically and genetically the gene products necessary for *S. typhimurium* to traffic through the animal host and also to determine how *S. typhimurium* regulates the expression of these genes in response to the changing environmental niches occupied during transit through this infection pathway. Ultimately, information from such studies should be useful in the design and construction of attenuated *Salmonella* derivatives to use for immunization to prevent infection.

In this manuscript, we review our studies conducted to define the means by which *S. typhimurium* effectively colonizes the intestinal tract of mice and especially of the gut-associated lymphoid tissue.

MATERIALS AND METHODS

Bacterial Strains

The bacterial strains referred to are listed in Table 1. Strains were grown and maintained by methods previously described (1).

Genetic and Molecular Methods

These have been previously described (1-4).

Table 1. Bacterial strains

Strain	Genotype	Description
Salmonella typhimurium LT2		
DB4673	*galE496 trpB2 flaA66 rpsL120 xyl-404 val metE551 metA22 ΔmalB hsdSA29 hsdL6/F1112*	(see 1)
SR-11		
χ3181	wild type	SR-11 isolated from Peyer's patches of an infected mouse (2)
χ3456	wild type pStSR101 (Tc^r)	χ3181 with Tn mini *tet* inserted into virulence plasmid (2)
χ3642.	*invA ::TnphoA*	(1)
χ3687	*phoP12*	(3)
χ4108	*phoP12 invA ::TnphoA*	P22 HT *int* (c3643) ⇒ χ3687 (4)
χ4109	*phoP12 Y::TnphoA*	Random TnphoA insertion from χ3689 transduced into χ3687 (4)
χ4110	*phoP12 Z::TnphoA*	Random TnphoA insertion from χ3689 transduced into χ3687 (4)
χ4111	*phoP12 proU ::TnphoA*	P22 HT*int* (AD110) ⇒ χ3687
SL1344		
χ3339	*rpsL hisG*	Mouse passage SL1344 (2)
χ3643	*rpsL hisG invA ::TnphoA*	(1)
χ3689	*rpsl hisG phoP12*	(13)

Cell Attachment and Invasion Assays

These have been done by methods previously described (1-4).

Animal Infectivity

Female BALB/c mice were used for all studies. Methods for growth and preparation of bacteria for oral inoculation have been described (1,3,4) as have methods for quantitating the titers of *S. typhimurium* cells in various compartments and tissues (2).

RESULTS

Involvement of Surface Attributes

The presence of LPS O-antigen side chains seems to be important to enable *S. typhimurium* to penetrate the mucin layer covering the intestinal wall such that rough mutants following oral inoculation rapidly pass through the intestinal tract without colonizing. However, neither flagella (5) or Type I pili (Lockman and Curtiss, unpublished) are necessary for intestinal colonization and invasive disease. It is conceivable, though, that *S. typhimurium* makes use of a diversity of adhesive molecules interacting with different receptors such that absence of any one adhesin would have an undetected influence on intestinal colonization and virulence.

Role of Virulence Plasmid in Intestinal Colonization

Jones, et al. (6) initially observed that *S. typhimurium* cured of a large plasmid had reduced virulence when administered by the oral route. Although this observation has been confirmed and extended in numerous studies, it is now clear that plasmid-cured strains as well as their plasmid containing wild-type parents have equal abilities to attach to and invade various cells in culture, to attach to and invade enterocytes of the small bowel and, importantly, of the cells contained in Peyer's patches (2). On the other hand, plasmid-cured strains have a greatly diminished ability to effectively colonize deeper tissues such as the mesenteric lymph node, liver and spleen (2) and are impaired in their ability to survive in some cell types encountered in those tissues which are not present within Peyer's patches or the lining of the intestinal tract.

Identification of *S. typhimurium* Invasion Functions

In the process of screening several *S. typhimurium* strains for their ability to attach to and invade tissue culture cells, we discovered that strain DB4673, a derivative of the commonly used strain LT2, was unable to penetrate a variety of cultured cells although it was fully capable of adhering to the same cells (1).

The introduction of a cosmid library of *S. typhimurium* DNA sequences into DB4673 led to the recovery of several clones which restored the invasion phenotype. Subcloning and transposon mutagenesis coupled with evaluation of mutants for their ability to attach to and invade tissue culture cells permitted us to define a three-gene operon with a closely linked fourth gene that were involved in controlling the ability of *S. typhimurium* to invade tissue culture cells (1). Fig. 1 depicts the arrangement of these genes and the gene products. The *invA*, *invB*, *invC* and *invD* genes encode proteins of 54, 64, 47 and 30 kDa, respectively. The *invA* gene product is most likely an envelope protein since TnphoA insertions in this gene gave rise to productive fusions to alkaline phosphatase (7). TnphoA insertions in *invA* and *invB* completely abolished the invasive phenotype conferred by the recombinant plasmid pYA2219 (Fig. 1) while insertions in *invC* had no effect. Insertions in *invD* diminished the invasive phenotype 5-fold.

invA::TnphoA mutants were generated in the mouse-virulent strains of *S. typhimurium* SR-11 and SL1344. These mutants had a 100-fold reduction in their capacity to invade tissue culture cells although they have wild-type ability to attach to the same cells (1) (Table 2).

Involvement of inv Genes in Intestinal Colonization

Inv^+ and Inv^- strains have identical LD_{50}s following intraperitoneal inoculation but a 50- to 60-fold higher LD_{50} when administered by the oral route (1). When mixed infections of BALB/c mice were carried out with *invA* and wild-type parent *S. typhimurium* strains, larger numbers of the wild-type inv^+ strain were isolated from the intestinal wall and Peyer's patches (Table 3). It is evident from these studies that *S. typhimurium* does not require the *inv* encoded functions when the intestinal barrier is breached in intraperitoneal inoculations. The fact that *inv* mutants of *S. typhimurium* retained considerable virulence suggests the existence of alternative mechanisms of invasion. This should not be surprising considering the variety of hosts that *Salmonella* are capable of infecting.

Regulation of *inv* Gene Expression

We surmised that since *inv* gene functions were specifically needed in the intestinal tract, expression of these genes should presumably be optimal under conditions found in that compartment. We thus investigated the influence of environmental conditions such as osmolarity, temperature, and oxygen concentration on expression of the *inv* genes. We constructed transcriptional and translational fusions to *invA*, the first gene in the *invABC* operon and introduced them into the chromosome of the mouse virulent strains of *S. typhimurium* SR-11 and SL1344 (4). Two random TnphoA fusions that did not alter invasiveness were used as negative controls while a TnphoA fusion to the osmoinducible gene *proU* (8) was used as a positive control. The results depicted in Table 4 reveal that the *invA* gene requires osmotic conditions similar to those of the gut for optimal expression. In addition, it was found that *invA* expression

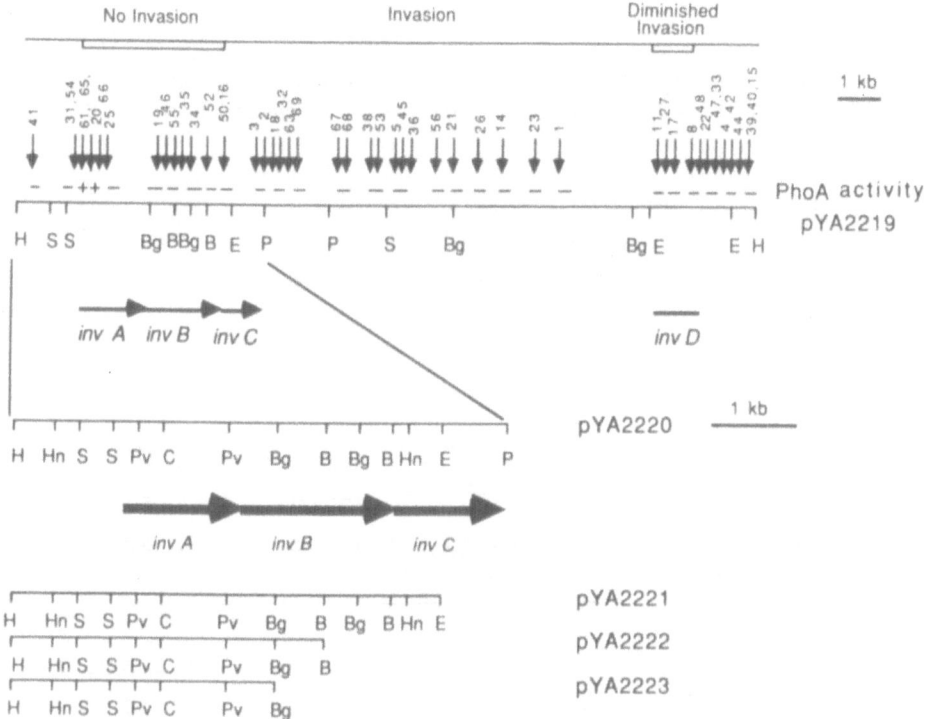

Figure 1. Restriction maps of recombinant plasmids and location of Tn*phoA* insertions. Numbers and vertical arrows above line show positions of Tn*phoA* insertions with their phenotypes indicated above. Letters below horizontal lines indicate restriction enzyme sites: B: *Bam*HI, Bg: *Bgl*II, C: *Cla*I, E: *Eco*RI, H: *Hind*III, Hn: *Hinc*II, P: *Pst*I, Pv: *Pvu*II, S: *Sal*I. Horizontal arrows indicate position of *inv* genes and direction of transcription. From reference 1.

Table 2. Adherence and invasion of *S. typhimurium* strains into Henle-407 cells.

Strain	Relevant phenotype	% Adherence*	%Invasion*
SR-11	inv+	7.33 ± 1.62	35.5 ± 17.5
SL1344	inv+	11.3 ± 0.33	46.1 ± 12.5
χ3642	inv-	3.35 ± 1.09	0.35 ± 0.09
χ3643	inv-	5.70 ± 2.14	0.45± 0.14

*Values are means of triplicate samples ± SD. Similar results were observed in several repetitions of this experiment.

Table 3. Ratios (± SD) of *S. typhimurium* strains χ3456/χ3642 colony forming units in different organs after mixed p.o. and i.p. infections*.

Days after infection	Peyer's patches	Intestinal walls†	p.o. Infection Spleens	i.p. Infection Spleens
1	8.7 ± 5.3	8.0 ± 7.0	–	1.3 ± 0.4
2	2.9 ± 1.2	10.0 ± 8.0	–	1.4 ± 0.3
3	7.8 ± 5.2	4.9 ± 2.4	18.6 ± 10.6	1.0 ± 0.8
4	7.2 ± 5.7	11.0 ± 6.0	8.7 ± 4.6	1.2 ± 0.7
5	4.1 ± 2.5	4.3 ± 2.6	3.4 ± 0.9	1.3 ± 0.9
6	7.8 ± 5.8	3.3 ± 2.7	1.4 ± 0.8	n.d.‡

*   Approximately equal numbers of χ3546 and χ3642 were administered either p.o. (1.8 x 10^9 and 2 x 10^9 CFU respectively) or i. p. (2 x 10^4 and 1.8 x 10^4 CFU, respectively) to eight-week-old BALB/c mice. Values represent means of ratios

± SD for three mice ††

† Small intestine with Peyer's patches removed

‡ not done

Table 4. Effect of osmolarity on the expression of *invA*[a]

Strain	Growth conditions	Relevant genotype	PhoA units
χ3687	K	inv+	12
χ3687	K + NaCl (0.3 M)	inv+	10
χ3687	K + Sucrose (0.6 M)	inv+	14
χ4108	K	*invA* ::Tn*phoA*	31
χ4108	K + NaCl (0.3 M	*invA* ::Tn*phoA*	242
χ4108	K + Sucrose (0.6 M)	*invA* ::Tn*phoA*	256
χ4109	K	Y::Tn*phoA*	140
χ4109	K + NaCl (0.3M)	Y::Tn*phoA*	110
χ4110	K	Z::Tn*phoA*	420
χ4110	K + NaCl (0.3M)	Z::Tn*phoA*	390
χ4111	K	proU::Tn*phoA*	90
χ4111	K + NaCl (0.3M)	proU::Tn*phoA*	2050

[a]Bacterial strains were grown as indicated, cells disrupted by sonication and PhoA activities determined as described. (4). Values representative and are taken from one of three experiments with similar results.

was higher when cells were grown at 37C as opposed to 25C or under anaerobic as opposed to aerobic conditions. However these effects were not as profound as the effect exerted by changes in osmolarity. The *envZ/ompR* two-component regulatory system is responsive to changes in osmolarity and influences the expression of the *ompC* and *ompF* genes among others (9). This regulatory system, however, has no effect on the expression of *invA* under any osmotic condition (4).

Changes in oxygen concentration, temperature and osmolarity are known to result in changes in the degree of DNA negative superhelicity (10-13). To test whether DNA supercoiling was involved, the gyrase inhibitors coumermycin and novobiocin were employed. Both these drugs have the effect of decreasing DNA negative supercoils and diminished *invA* gene expression under conditions of high osmolarity (4). Although all aspects of how the *inv* genes are regulated are not known, it is likely that DNA gyrase, topoisomerase I and the histone-like protein encoded by the *osmZ* (*pilG*, *virR*) gene (14-16) play some role and that the sensing and activation are independent of the *envZ-ompR* two-component regulatory system.

As might be expected, growth of *S. typhimurium* wild-type strains under conditions to maximize expression of *invA* potentiates invasion of tissue culture cells (4).

Role of the *phoP* Gene in Intestinal Colonization

Some years ago, we constructed an *S. typhimurium phoP* mutant to eliminate all phosphatase background for the purpose of using this strain to generate a Tn*phoA* library to screen for mutants with impaired abilities to attach to and/or invade cells in culture or to infect the animal host. Immediately upon constructing derivatives of *phoP S. typhimurium* SR-11 and SL1344, we noted that these strains were totally avirulent yet immunogenic for mice (2). Work by others has demonstrated that the *phoP* gene is part of a two-component regulatory system (17,18) and that *phoP* mutants are inordinately sensitive to killing by macrophages (17). When *S. typhimurium phoP* mutants are used to orally inoculate mice, there is a significantly reduced ability of the *phoP* mutant to successfully colonize the intestinal tract because of impaired ability to survive at least in some cell types present in Peyer's patches (2). It is probable, however, that the *S. typhimurium phoP* mutant is able to attach to and survive in enterocytes lining the small intestinal wall since its ability to attach to and invade Henle 407 cells is identical to the ability of wild-type *S. typhimurium* strains SR-11 and SL1344 (2).

DISCUSSION

We believe that progress is being made in understanding the means by which *S. typhimurium* colonizes the intestinal tract of mice. It is evident that identification of all the adhesins that might be involved in facilitating initial attachment to cells lining the intestinal wall has yet to be accomplished. In addition, one can surmise that there are genes responsive to acidic conditions which, when expressed, might enhance the ability of *S. typhimurium* to survive the gastric acidity barrier. Also, it is becoming increasingly clear that additional systems for regulating these genes have yet to be identified by mutation and/or by gene cloning. A further complication, although with exciting ramifications is our recent discovery that genetic alterations in *S. typhimurium* which either do or do not have an influence on colonization of the murine intestinal tract have a rather different phenotype upon oral infection into one-day-old chicks (Porter, Hassan and Curtiss, unpublished). Furthermore, it is apparent that *S. choleraesuis*, *S. typhi* and the *S. typhimurium-S. enteritidis Salmonella* species have some similar but also some dissimilar mechanisms for colonization, invasion and causation of disease.

ACKNOWLEDGEMENTS

The research was supplied by grant AI24533 from the National Institute of Health. We thank Marion Harris and Jennifer Barry for help in the preparation of this manuscript.

REFERENCES

1) Galan, J. E., and R. Curtiss III. 1989. Cloning and molecular characterization of genes whose products allow *Salmonella typhimurium* to penetrate tissue culture cells. Proc. Nat. Acad. Sci. USA. 86:6383-6387.

2) Gulig, P. A., and Roy Curtiss III. 1987. Plasmid-associated virulence of *Salmonella typhimurium*. Infect. Immun. 55:2891-2901

3) Galan, J. E., and R. Curtiss III. 1989. Virulence and vaccine potential of *phoP* mutants of *Salmonella typhimurium*. Microbial Pathogenesis 6:433-443.

4) Galan, J. E., and Roy Curtiss III. 1990. Expression of *Salmonella* genes required for invasion is regulated by changes in DNA supercoiling. Infect. Immune. 59:1879-1885.

5) Lockman, H. and R. Curtiss III. 1990. *Salmonella typhimurium* mutants lacking flagella or motility remain virulent in Balb/c mice. Infect. Immun. 58:137-143.

6) Jones, G. W., D. K. Rabert, D. M. Svinarich, and H. J. Whitfield. 1982. Association of adhesive, invasive and virulent phenotypes of *Salmonella typhimurium* with autonomous60-megadalton plasmid. Infect. Immun. 38:476-486.

7) Manoil, C., and J. Beckwith. 1985. Tn*phoA*: a transposon probe for protein export signals. Proc. Natl. Acad. Sci. USA 82:8129-8133.

8) Cairney, J., I. R. Booth, and C. F. Higgins. 1986. Osmoregulation of gene expression in *Salmonella typhimurium* : *proU* encodes an osmotically induced betaine transport system. J. Bacteriol. 164: 1224-1232.

9) Hall, M. N., and T. J. Silhavy. 1979. The *ompB* locus and the regulation of the major outer membrane porin protein of *Escherichia coli* K12. J. Mol. Biol. 146:23-43.

10) Goldstein, E., and K. Drlica. 1984. Regulation of bacterial DNA supercoiling:Plasmid linking numbers vary with growth temperature. Proc. Natl. Acad. Sci. USA 81:4046-4050.

11) Kranz, R. G., and R. Haselkorn. 1986. Anaerobic regulation of nitrogenfixation genes in *Rhodopseudomona capsulata*. Proc. Natl. Acad. Sci. USA 83:6805-6809.

12) Pruss, G. J., and K. Drlica. 1989. DNA supercoiling and prokaryotic transcription. Cell 56:521-523.

13) Yamamoto, N., and M. Droffner. 1985. Mechanisms determining aerobic or anaerobic growth in the facultative anaerobe *Salmonella typhimurium*. Proc. Natl. Acad. Sci. USA 82:2077-2081.

14) Maurelli, A. T., and Sansonetti, P. J. 1988. Identification of a chromosomal gene controlling temperature-regulated expression of *Shigella* virulence. Proc. Natl. Acad. Sci. USA 85:2820-2824.

15) Spears, P. A., D. Schauer, and P. E. Orndorff. 1986. Metastable regulation of type I piliation in *Escherichia coli* and isolation and characterization of phenotypically stable mutant. J. Bacteriol. 168:179-185.

16) Dorman, C. J., N. NiBhriain, and C. F. Higgins. 1990. DNA supercoiling and environmental regulation of virulence gene expression in *Shigella flexneri*. Nature 344:789-792.

17) Fields, P. I., E. A. Groisman, and F. Heffron. 1989. A *Salmonella* locus that controls resistance to microbicidal proteins from phagocytic cells. Science 243:1059-1062.
18) Miller, S. I., A. M. Kukral, and J. J. Mekalanos. 1989. A two component regulatory system (*phoP phoQ*) controls *Salmonella typhimurium* virulence. Proc. Nat. Acad. Sci. USA 86:5054-5058.

PLASMID ENCODED VIRULENCE OF YERSINIA

H. Wolf-Watz, Å. Forsberg, R. Rosqvist, I. Bölin, K. Erickson,
L. Norlander, M. Rimpiläinen, T. Bergman and S. Håkansson

Department of Applied Cell and Molecular Biology,
and Department of Cell and Microbiology, FOA4,
University of Umeå, S-901 87 Umeå, Sweden
S-901 82 Umeå, Sweden

Introduction

The three virulent members of the genus _Yersinia_ harbour related virulence plasmids with a molecular weight of about 60-70 kb (11). When exponential phase cultures of these organisms growing in a Ca^{2+} free medium are shifted from 26C to 37C, growth ceases over a period of about 2 generations (10). If however 2.5 mM Ca^{2+} is present in the medium growth continues normally. These bacteria are referred to as being Ca^{2+} dependent (CD). Plasmid cured strains, however, do not show this dependency on Ca^{2+} and are thus, Ca^{2+} independent (CI). Such bacteria are always avirulent. By transposon insertion mutagenesis a 20 kb region of the virulence plasmid has been identified which is involved in this low calcium response (lcr). Such CI-mutants do not require Ca^{2+} for prolonged growth at 37C and they are not virulent. Although the plasmids of _Y. enterocolitica_, _Y. pestis_ and _Y. pseudotuberculosis_ have been subjected to rearrengements, the Ca^{2+} region, however, of the different plasmids is conserved (11).

Expression of virulence plasmid encoded virulence determinants.

When the plasmid containing strain of _Y. pseudotuberculosis_; YPIII(pIB1) is grown in a medium free of Ca^{2+} and shifted from 26C to 37C, the bacteria induce a number of temperature inducible proteins (Yops). These proteins can be recovered in a secreted form as well as in a membrane bound form (5). If Ca^{2+} is added to the culture the rate of synthesis of the Yops is greatly reduced and these proteins can only be detected in minute amounts in the membrane fraction by Western blotting (1,5). Thus, expression of the Yops is regulated by Ca^{2+} and temperature. All three species are able to express the Yops, showing the same regulatory pattern. Upon comparison of the Yops from the three species it is found that they are very similar, showing antigenic relatedness and only minor differences with respect to their respective molecular weights (Table 1) (1). Therefore, it is highly likely that the Yop proteins exhibit similar virulence functions in all three species.

Antibodies directed against the Yops can be detected in convalescent sera obtained from patients recovering from Yersinosis as well as bubonic plaque, demonstrating that the Yops are expressed during infection, which support the idea that at least some of these proteins are essential virulence determinants (2).

Table 1. Molecular weights of the immunologically related YOPs and the V antigen of <u>Yersinia</u> spp.[a]

Protein	Sizes (kDa) of proteins in the following strains:		
	Y.<u>pseudotuberculosis</u> YPIII(pIB1)	Y.<u>pestis</u> EV76	Y.<u>enterocolitica</u> 8081
YopM	45	44	44
YopH	45	45	46
YopB	41-42[b]	41-42[b]	40-41[b]
V antigen	38	38	38
YopD	34	34	36
YopN	34	34	34
YopE	26	26	25.5

[a]Sizes were determined by SDS-PAGE

[b]The range in molecular sizes is due to the presence of two different forms

Mapping and regulation of <u>yop</u> genes

Although the Yop proteins are coordinatley regulated they are encoded by different operons scattered around the virulence plasmid. In most of the cases the genes are contained within monocistronic operons (1,3,6,8). So far only one exception to this has been found. The yopB and yopD genes are members of the same polycistronic operon which in addition contains the lcrGVH genes (9). Analysis of the Ca^{2+} region has revealed that it encodes regulatory elements important for expression of the <u>yop</u>-genes in response to the environmental stimuli temperature and Ca^{2+}. In response to increase in temperature a thermoregulated activator VirF (LcrF) is expressed which activates transcription of the different <u>yop</u>-genes (4). In medium containing high concentrations of Ca^{2+} a repressor is expressed that inhibits <u>yop</u>-gene expression at elevated Ca^{2+} levels. Thus, at 37C in a

Ca^{2+} depleted medium the expression of the yop-genes is derepressed which leads to high expression of the Yops. The regulation of the negative control loop is very complex and involves at least 5 different plasmid encoded regulatory elements (5).

YopH and YopE obstruct the primary host defence

Structural gene mutants unable to express the YopH and YopE proteins are as avirulent as virulence plasmid cured strains (3,5). In contrast to cured strains, virulent strains have the ability to prevent phagocytosis by mouse peritoneal macrophages (12). This ability involves both YopH and YopE and these two proteins act in concert to prevent the pathogen from phagocytosis since only a yopH/yopE double mutant is fully sensitive to phagocytosis in contrast to respective single mutant (Table 2) (11). The mechanisms by which these proteins act are unknown. It was recently reported, however,

Table 2. Inhibition of phagocytosis by different strains of Y. pseudotuberculosis

Strain	Relevant genotype	% extracellular bacteria of total Mo associated bacteria	Cytotoxic for Mo
YPIII	p	6 ± 2 (5)	−
YPIII(pIB29)	$yopH_+$ $yopE^+$	23 ± 3 (5)	+
YPIII(pIB29,pYOP21)	$yopH_+$ $yopE^+$	63 ± 4 (5)	+
YPIII(pIB522)	$yopH_+$ $yopE$	54 ± 7 (3)	+
YPIII(pIB251)	$yopH$ $yopE$	7 ± 3 (3)	−

that YopH exhibits a protein tyrosine phosphatase activity (7). This suggests that YopH interacts with a cell receptor and that dephosphorylation of this receptor mediates a "non-phagocytic" signal. We have obtained results which indicate that YopH in fact interacts with a cell receptor and that this interaction induces a cytotoxic cell-response (11).

Conclusions

- The three virulent species of Yersinia contain homologous virulence plasmids that encode a number of temperature inducible Ca^{2+} regulated proteins (Yops).

- Expression of the Yops is regulated at the level of transcription involving negative as well as positive control elements encoded by the Ca^{2+} region.

- YopH and YopE are essential virulence determinants which obstruct the primary host defence by preventing phagocytosis.

References

1. Bölin, I., Å. Forsberg, L. Norlander, M. Skurnik and H. Wolf-Watz. 1988. Identification and mapping of temperature-inducible, plasmid encoded proteins of Yersinia spp.. Infect.Immun 56: 343-348.

2. Bölin, I., D. Portnoy and H. Wolf-Watz. 1985. Expression of the temperature-inducible outer membrane proteins of Yersinia. Infect.Immun. 48: 234-240.

3. Bölin, I. and H. Wolf-Watz. 1988. The plasmid encoded Yop2b protein of Yersinia pseudotuberculosis is a virulence determinant regulated by calcium and temperature at the level of transcription. Mol.Microbiol. 2: 237-245.

4. Cornelis, G., C. Sluiters, C. Lambert de Rouvroit and T. Michels. 1989. Homology between virF, the transcriptional activator of the Yersinia virulence regulon and araC, the Escherichia coli arabinose operon regulator. J.Bacteriol. 171: 254-262.

5. Forsberg, Å., I. Bölin, L. Norlander and H. Wolf-Watz. 1987. Molecular cloning and expression of calcium-regulated, plasmid coded proteins of Y.pseudotuberculosis. Microb.Pathog. 2: 123-137.

6. Forsberg, Å. and H. Wolf-Watz. 1988. The virulence protein Yop5 of Yersinia pseudotuberculosis is regulated at transcriptional level by plasmid pIB1-encoded transacting elements controlled by temperature and calcium. Mol.Microbiol. 2: 121-133.

7. Guan, K. and J. Dixon. 1990. Protein tyrosine phosphate activity of an essential virulence determinant in Yersinia. Science 249. 553-556.

8. Leung, K. amd S. Straley. 1989. The yopM gene of Yersinia pestis encodes a released protein having homology with the human patlet surface protein GPIba. J.Bacteriol. 171: 4623-4632.

8. Mulder, B., T. Michels, M. Simonet, M-P. Sory and G. Cornelis. 1989. Identification of additional virulence determinants on the pYV plasmid of Yersinia enerocolitica 227. Infect.Immun. 57: 2534-2541.

9. Kupferberg, L. and K. Higuchi. 1958. Role of calcium ions in the stimulation of growth of virulent strains of Pasturella pestis. J. Bacteriol. 76: 120-121.

10. Portnoy, D., H. Wolf-Watz, I. Bölin, A. Beeder and S. Falkow. 1984. Characterization of common virulence plasmids in Yersinia species and their role in the expression of the outer membrane proteins. Infect.Immun. 43:108-114.

11. Rosqvist, R. Å. Forsberg, M. Rimpiläinen, T. Bergman and H. Wolf-Watz. 1990. The cytotoxic protein YopE of Yersinia obstructs the primary host defence. Molec.Microbiol. 4: 657-667.

12. Rosqvist, R., I. Bölin and H. Wolf-Watz. 1988. Inhibition of phagocytosis in Yersinia pseudotuberculosis: a virulence plasmid encoded ability involving the Yop2b protein. Infect.Immun. 56: 2139-2143.

INTERACTIONS OF YERSINIA WITH COLLAGEN

Zsuzsa Kienle, Levente Emödy, Torkel Wadström and Paul O'Toole

Department of Medical Microbiology, University of Lund,
Sölvegatan 23, 223 62 Lund, Sweden; and Department of
Microbiology, University Medical School, Szigeti ut 12,
76 43 Pecs, Hungary

INTRODUCTION

Y. pseudotuberculosis and Y. enterocolitica are intestinal
pathogens which cause yersiniosis. Y. pestis is the causative
organism of bubonic plague. Despite its association with
Y. enterocolitica based on the clinical picture,
Y. pseudotuberculosis is actually almost identical at the DNA
level to Y. pestis. All three Yersinia species pathogenic for
man are able to bind to and invade cultured eukaryotic cells.
This property is conferred by the inv and ail loci (3).
Complications involving connective tissue are relatively frequent
sequelae following infection by a number of micro-organisms
including Salmonella, Shigella, Campylobacter, Chlamydia
trachomatis and Yersinia sp.(2). In the case of yersiniosis
caused by Y. pseudotuberculosis and Y. enterocolitica, ankylosing
spondylitis, erythema nodosum and reactive arthritis are included
in this category. There is an association with the HLA B27
antigen, which is relatively common in Scandinavia. Because of
the known interactions of Yersinia with eukaryotic cells, and the
possible significance of potential tropisms for extracellular
matrix proteins, we initiated the present study of Yersinia
binding to various collagen types. We have evidence for three
separate mechanisms by which Yersinia sp. can bind to this
important and widespread structural protein, with obvious
implications for pathogenesis.

YOP1 (YAD A) MEDIATES BINDING TO COLLAGEN

All three virulent Yersinia species express a panel of proteins
called YOP's under correct environmental conditions (Wolf-Watz et
al, this volume). YOP1 is a thermoinducible protein encoded by
the large virulence-associated plasmid. To it have been ascribed
the functions of haemagglutination, resistance to killing by
human serum, inhibition of protection by interferon and
autoagglutination. We have recently demonstrated that YOP1 is
also responsible for binding to collagen types I, II and IV (1).
The data supporting this conclusion is simplified in Table 1.

DNA sequencing of yopA from Y. pestis has so far shown that this species is incapable of expresssing the gene due to a mutation in the early sequences. It is apparent from the data for Y. enterocolitica and Y. pseudotuberculosis however that loss of the yopA gene results in failure to bind collagen types I and II. This was supported by inhibition of collagen binding by specific anti-YOP1 antiserum. Since the plasmid-free derivatives still bind type IV collagen, we proposed the existence of a chromosomally located gene encoding this function.

Table 1. Binding of radiolabelled collagen by Yersinia and E. coli strains.

	Binding to collagen types		
	I	II	IV
Wild type strains (pYV+)			
Y. enterocolitica (03,08,09)	+	+	+
Y. pseudotuberculosis	+	+	+
Y. pestis	-	-	+
Plasmid cured			
Y. enterocolitica	-	-	-
Y. pseudotuberculosis	-	-	+
Y. pestis	-	-	+
yopA::Tn5	As for cured derivatives		
E. coli K12 / yopA	+	+	+
E. coli K12	-	-	-

GENETIC ANALYSIS OF COLLAGEN BINDING BY Y. PESTIS

A cosmid library of a vaccine strain of Y. pestis, EV76c, was created by ligating unfractionated genomic DNA which had been partially digested with Sau3A to the vector pJB8. Recombinants were screened by colony blotting with radiolabelled type IV collagen. One clone was selected for further study, and the relevant coding sequence was localized by a combination of sub-cloning, transposon mutagenesis and deletion. Analysis of deletions generated from one end of a cloned fragment in conjunction with examination of outer membrane protein profile and collagen binding phenotype allowed the gene to be mapped, and linked the gene product to a 38 kDa protein.

Comparison with the outer membrane protein profile of the parental Y. pestis strain showed that the cloned antigen was in fact a major component of the membrane, and as such, we were surprised that it had not previously attracted attention. Accordingly, we used fine structure restriction mapping to compare the region to known cloned loci, and subsequently we employed limited DNA sequence determination. These experiments showed that the cloned collagen receptor was identical to a previously characterised plasminogen activator (Pla) encoded by a 9 kb plasmid of Y. pestis which also harbours genes for peticin production and immunity (4,6). We confirmed the fidelity of the ability of this protein to mediate collagen binding in a number of quantitative solid phase attachment assays in which the collagen was immobilized in plastic plates.

The affinity of the plasminogen activator for collagen is relatively weak - the E. coli clone containing the gene in a plasmid vector with a copy number exceeding one hundred binds collagen at approximately the same level as the parental Y. pestis strain. If the plasmid carrying the gene in Y. pestis is removed, the level of collagen binding is only slightly diminished, suggesting that the product of the single chromosomal locus is probably the major factor responsible for collagen binding. However, in Pla$^+$ cells, plasminogen is bound in a reaction which drastically inhibits collagen binding, but in Pla$^-$ cells, collagen binding is unaffected by plasminogen. This implies that when Pla is present, the contribution to collagen binding by the product of the chromosomal locus is somehow reduced. A possible explanantion for this observation is the known ability of Pla to degrade outer membrane proteins, including YOP's (5), so it is possible that Y. pestis can use Pla to modulate its own affinity for collagen.

We have recently cloned a chromosomal locus from Y. pseudotuberculosis YPIII which directs expression of collagen binding in E. coli. We are currently sub-cloning and localizing the gene in order to generate mutant strains lacking the collagen binding phenotype for virulence studies. We are purifying the receptor to faciltate immunological studies in reactive arthritis patients and to allow more detailed analysis of the interaction with various collagen types.

References

1. Emödy, L., J. Heesemann, H. Wolf-Watz, M. Skurnik, G. Kapperud, P. O'Toole and T. Wadström. 1989. Binding to collagen by Yersinia enterocolitica and Y. pseudotuberculosis: Evidence for yopA-mediated and chromosomally encoded mechanisms. J. Bacteriol. 171:6674-6679

2. Lamesmaa-Rantala, R., T.H. Ståhlberg, K. Granfors, P. Kuusisto and A. Toivanen. 1990. Serologocal cross-reactions against Y. enterocolitica in patients infected with other arthritis-associated microbes. Clin. Exper. Rheumatol. 8:5-9

3. Miller, L, and S. Falkow. 1988. Evidence for two loci in Yersinia enterocolitica that can promote invasion of epithelial cells. Infect. Immun. 56:1242-1248.

4. Sodeinde, O. and J.D. Goguen. 1988. Genetic analysis of the 9.5 kilobase virulence plasmid of Y. pestis. Infect. Immun. 56:2743-2748

5. Sodeinde, O., A.K. Sample, R.R. Brubaker and J.D. Goguen. 1988. Plasminogen activator/Coagulase gene of Y. pestis is responsible for degradation of plasmid-encoded outer membrane proteins. Infect. Immun. 56:2749-2752

6. Sodeinde, O. and J.D. Goguen. 1989. Nucleotide sequence of the Plasminogen activator gene of Y. pestis: Relationship to opmT of Escherichia coli and Gene E of Salmonella typhimurium. Infect. Immun. 57:1517-1523

THERMOREGULATION OF INVASION GENES IN SHIGELLA FLEXNERI THROUGH THE TRANSCRIPTIONAL ACTIVATION OF THE VIRB GENE ON THE LARGE PLASMID

C. Sasakawa, T. Tobe, S. Nagai, N. Okada,
B. Adler, K. Komatsu, and M. Yoshikawa

Institute of Medical Science, University of Tokyo

Tokyo 108, Japan

Summary

Expression of several invasion (*inv*) genes on the large virulence plasmid of *Shigella flexneri* is subject to thermoregulation. The *ipaBCD* genes are expressed at 37 C but not at 30 C and this expression is regulated by *virF* through the activation of *virB* on the plasmid. To identify the mediator gene for the thermoregulation of the large plasmid, the transcription of *virF*, *virB*, *ipa* and the other two *inv* operons was examined by Northern dot hybridization and S1 nuclease assay methods. The results indicated that transcription of *virF* and *virB* was affected by temperature but the level of *virB* transcription was more strongly controlled by temperature than that of *virF*. On the other hand, transcription of *ipa* and the other two *inv* operons was not dependent on temperature but on *virB* transcription. By overexpressing the *virB* gene with a *tac* promoter, deregulation of invasion phenotype was achieved in the absence of VirF function, suggesting that the transcriptional activation of *virB* mediates the thermoregulation phenomenon.

Introduction

On the large plasmid of *S. flexneri*, seven virulence loci have been identified. Five contiguous segments comprising a 31 kb DNA sequence designated *virB*, *ipaBCD* (*ipa* operon) and Region-3, -4 and -5 are required for primary invasion (Sasakawa *et al.*,1988). The *virG* gene, outside the cluster, encodes a 116 kD immunogenic surface protein and is essential for the ability of invading bacteria to spread into adjacent cells (Makino *et al.*, 1986; Lett *et al.*, 1989). The *virF* gene positively regulates the expression of *virB* and *virG* genes at the transcriptional level. The expression of *ipa*, Region-3, -4 and -5 is activated by *virB* at the transcriptional level. Thus, expression of the invasion phenotype of *S. flexneri* is under dual activation system directed by *virF* and *virB* (Fig. 1) (Sakai *et al.*, 1988; Adler *et al.*, 1989; Nagai *et al.*, manuscript in preparation).

In addition, the invasion phenotype of shigellae is expressed when grown at 37 C but not at 30 C (Maurelli *et al.*, 1984). This temperature-regulated invasion involves a chromosomal gene, *virR* (Maurelli and Sansonetti, 1988), which apparently acts through change in DNA supercoiling (Dorman et al., 1990).

Recently, the transcriptional organization of most of the 31 kb virulence DNA segment on the large plasmid has been determined (Nagai *et al.*, unpublished results). Thus, this made it possible to identify the target gene(s) on the large plasmid which mediate the effect of VirR function on the *inv* genes. In this report, we present evidence that the thermoregulated invasion phenotype is mediated through the transcriptional activation of the *virB* gene on the large plasmid.

Materials and Methods

YSH6000 was a virulent strain of *S. flexneri* 2a carrying the large 230 kb virulence plasmid. YSH6207 and YSH6109 were *virF*::Tn5 and the *virF*-deletion derivatives of YSH6000 respectively (Sakai *et al.*, 1988). The S1 nuclease protection assay and Northern dot hybridization were described previously (Sasakawa et al., 1989). Chl (contact hemolysis) test was used to determine the invasion capacity of *S. flexneri* strains (Sansonetti et al., 1986).

Results

Effect of temperature on the expression of the two positive controllers, virF and virB.

Since the expression of the *ipa* operon is positively regulated at transcriptional level by *virF* through the activation of *virB* transcription (Adler *et al.*, 1989), it seemed likely that one of these two positive regulators mediates the thermoregulation of *inv* gene on the large plasmid. Therefore, the mRNAs of *virF* and *virB* in YSH6000 grown at 37 C or at 30 C were examined by S1 nuclease protection assay. The level of *virF*-mRNA at 30 C was approximately one quarter that at 37 C, but virB at 30 C was greatly reduced to a level one twentieth of that seen at 37 C.

Since *virB* gene expression depends on VirF function, the transcription of *virB* in YSH6000 and in the *virF*::Tn5 mutant, YSH6207, grown at 37 C and at 30 C was compared to examine the effect of temperature on the activation of the *virB* gene by *virF*. *virB*-mRNA in YSH6207 was not expressed at all at either temperatures. YSH6000 produced a low level at 30C, but this was markedly reduced compared with that at 37 C.

The effect of different levels of *virF* transcription and temperature on the activation of *virB* were examined by constructing a series of derivatives of YSH6000 that allowed the production of different amounts of *virF*-mRNA. The levels of *virB*-mRNA expressed in each of the constructs grown at 37 C or at 30 C were then determined. To achieve different mRNA levels of *virF*, the *virF* gene cloned in a pSC101- or a pBR322-derived plasmid or placed under the control of a *tac* promoter, P*tac-virF*, was introduced into the *virF*-deletion mutant YSH6109. The transcriptional analysis of *virF* showed that the levels of *virF*-mRNA increased in accord with the dose of the *virF* gene both at 37 C and at 30 C. The levels of *virB*-mRNA determined by S1 nuclease protection assay also increased in accord with the dose of the *virF* gene at both temperatures, although the levels of *virB*-mRNA at 30 C were significantly lower than those at 37 C. The kinetics of the activation of *virB* transcription by *virF* at 37 C were compared by measuring the radioactivity of the RNA dots and the S1 nuclease protected bands. The results showed that the

activation of *virB* transcription by *virF* was much more efficient at 37 C than at 30 C. This indicated that the full
activation of *virB* transcription could not be achieved solely
by the activation of *virF* at low temperature and that the
virB transcription was strictly dependent on temperature as
well as on VirF function.

Temperature is not directly involved in activation of *inv*
genes on the large plasmid.

A plasmid carrying a P*tac-virB* fusion (pCHR401) was introduced into YSH6109 and the resulting strain was grown at 37 C
or at 30 C in the presence of different concentrations of
isopropyl β-D-thiogalactoside (IPTG) ranging from 0.03 to 1
mM to control the level of *virB*-mRNA. The amounts of mRNAs of
virB, *ipa*, and the other two *inv* operons were quantified by
RNA dot hybridization using four specific DNA probes, and the
amounts of *ipa*-, Region-3 and Region-5 mRNA relative to those
of the wild type at 37 C were plotted against the relative
amounts of *virB*-mRNA. The results showed that the kinetics of
activation of *ipa*, Region-3 and Region-5 transcription were
similar at 30 C and at 37 C, indicating that temperature is
not directly involved in the transcriptional activation of
the three *inv* operons by the *virB* gene.

Deregulation of invasion phenotype of *S. flexneri* by the
transcriptional activation of *virB*.

Increasing *virB* transcription could thus lead to activation
of the three *inv* operons irrespective of growth temperature.
However, there remained the possibility that the expression
of *inv* gene products was temperature-regulated at post-
transcriptional level. To eliminate this possibility and to
confirm the direct role of *virB* activation in the expression
of invasion phenotype, the effect of temperature on the
expression of invasion phenotype with YSH6109 carrying
pCHR401 was examined. Bacteria grown in media containing
different concentration of IPTG at 37 C or at 30 C were
assayed for Chl activity. YSH6109 carrying pCHR401 grown at
37 C at 0.1 mM IPTG expressed hemolytic activity which was 48
% of that of the wild type and at 0.3 mM IPTG the strain expressed about the same level as the wild type. This indicated
that the activation of *virB* transcription induced the bacteria invasive even in the absence of functional VirF. Similarly, the hemolytic activity was also restored by inducing *virB*
transcription at 30 C. At 0.3 mM IPTG the hemolytic activity
was 25 % of that of wild type at 37 C, and when IPTG was
added at 1.0 mM to the bacterial culture the hemolytic activity was about the same level as the wild type at 37 C. The
invasion capacity represented by the hemolytic activity
expressed under different levels of *virB* transcription at
both temperatures showed a good correlation with the level of
transcription of the three *inv* operons.

Discussion

The present study indicates that expression of *inv* genes and
their functions are not directly subjected to thermoregulation, but rather that the transcriptional step of *virB* expression is responsible for mediating the thermoregulation of
the plasmid-coded *inv* genes (Fig.1). It has been proposed
that temperature-regulated expression of *inv* genes on the
large plasmid is under the negative control of *virR* (Maurelli
and Sansonetti, 1988). Recently, Dorman *et al.* (1990) have
examined the role of the VirR function in the thermoregulation of *inv* gene expression and indicated that the *virR* gene

was equivalent to the *osmZ* gene of *Escherichia coli* (Higgins et al., 1988), which has previously been shown to mediate its regulatory effect through changes in DNA supercoiling. Thus, the identity of the mediator gene for the thermoregulated invasion phenotype of shigellae should facilitate the molecular study of how environmental signals such as temperature are transmitted to a set of virulence genes.

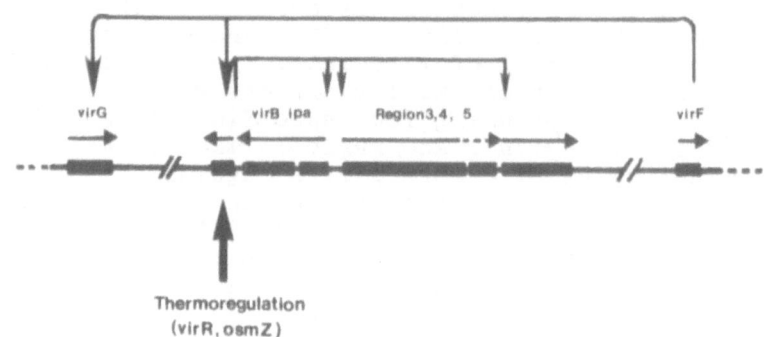

Fig.1 The regulation of plasmid-encoded virulence genes in *S. flexneri* (Sakai et al., 1988; Adler et al., 1989; Tobe et al., unpublished results)

References

Adler,B., Sasakawa,C., Tobe,T., Makino,S., Komatsu,K., and Yoshikawa,M. 1989. A dual transcriptional activation system for the 230 kilobase plasmid genes coding for virulence-associated antigens of *Shigella flexneri*. Mol. Microbiol. 3:627

Dorman,C.J., Bhriain,N.N., and Higgins,C.F. 1990. DNA supercoiling and environmental regulation of virulence gene expression in *Shigella flexneri*. Nature 344:789

Higgins,C.F., Dorman,C.J., Stirling,D.A., Waddell,L., Booth,I.R., May,G., and Bremer,E. 1988. A physiological role for DNA supercoiling in the osmotic regulation of gene expression in *S. typhimurium* and *E. coli* Cell 52:569

Lett,M.-C., Sasakawa,C., Okada,N., Sakai.T., Makino,S., Yamada,M., Komatsu,K., and Yoshikawa,M. 1989. *virG*, a plasmid-coded virulence gene of *Shigella flexneri*: identification of the VirG protein and determination of the complete coding sequence. J. Bacteriol 171:353

Makino,s., Sasakawa,C., Kamata,K., Kurata,T., and Yoshikawa,M. 1986. A genetic determinant required for continuous reinfection of adjacent cells on a large plasmid in *S. flexneri* 2a. Cell 46:551

Maurelli,A.T., Blackmon,B., and Curtiss III,R. 1984. Temperature-dependent expression of virulence genes in *Shigella*. Infect. Immun. 43:195

Maurelli,A.T., and Sansonetti,P.J. 1988. Identification of a chromosomal gene controlling temperature-regulated expression of *Shigella* virulence. Proc. Natl. Acad. Sci. USA 85:2820

Sakai,T., Sasakawa,C., and Yoshikawa,M. 1988. Expression of four virulence antigens of *Shigella flexneri* is positively regulated at the transcriptional level by the 30 kiloDal ton *virF* protein. Mol. Microbiol. 2:589

Sansonetti,P.J. Pyter,a., Clerc,P., Maurelli,A.T., and Mounier,J. 1986. Multiplication of *Shigella flexneri* within HeLa cells: lysis of the phagocytic vacuole and plasmid-mediated contact hemolysis. Infect. Immun. 51:461

Sasakawa,C., Kamata,K., Sakai,T., Makino,S., Yamada,M., Okada,N., and Yoshikawa,M. 1988. virulence-associated genetic regions comprising 31 kilobase of the 230-kilobase plasmid in *Shigella flexneri* 2a. J. Bacteriol. 170:2480

Sasakawa,C., Adler,B., Tobe,T., Nagai,S., Komatsu,K., and Yoshikawa,M. 1989. Functional organization and nucleotide sequence of virulence Region-2 on the large virulence plasmid in *Shigella flexneri* 2a. Mol. Microbiol. 3:1191

ASSOCIATION OF INVASIVE *SHIGELLA* STRAINS WITH EPITHELIAL CELLS

Tibor Pál[a,b], Alf A. Lindberg[b]

[a] Institute of Microbiology, University Medical School, Pécs, Hungary
[b] Department of Clinical Bacteriology, Karolinska Institute, Huddinge Hospital, Huddinge, Sweden

INTRODUCTION

Shigellae are facultative intracellular parasites multiplying in the cytoplasm of the epithelial cells of the large intestine. The capability of *shigellae* to invade epithelial cells grown in tissue cultures showed a close correlation with their virulence (1). This phenotype is regulated by the growth temperature, and is related to the presence of a 140 Md plasmid. The plasmid codes for a set of proteins, some of which (ipaB,C,D) known to be essential for the invasiveness. However, the role(s) of these proteins in the invasion process is not known (for review see 2).

The initial contact between *S. flexneri* and HeLa cells induces the accumulation of polymerized actin and myosin beneath the cell membrane. This is followed by the uptake of the bacterium by directed phagocytosis (3). Recently the participation of clathrin in the internalization of *S. flexneri* was also suggested (4).

It is generally assumed, that there is an intimate attachment between the microbial and the host cell membrane during the engulfment of the bacterium. Using microcinematography, Ogawa showed already in 1968, that there is indeed an adhesion phase in the interaction between the microbe and the HeLa cells (5). This attachment apparently induced a ruffling movement of the host cell membrane leading to the internalization of the bacterium. Later on Hale et al. could demonstrate electronmicroscopically the close attachment between *shigellae* and the HeLa cell membrane during the phagocytosis process (6). They took advantage of the fact that thorium dioxide particles, visible by electronmicroscopy, were excluded from the sites of close contact between the bacterium and the eukaryotic cell.

Molecular Pathogenesis of Gastrointestinal Infections
Edited by T. Wädstrom *et al.*, Plenum Press, New York, 1991

It is still not known, however, whether this attachment, preceeding (or triggering) the engulfment of the bacterium, is mediated by the same bacterial membrane proteins (ipaB,C,D) essential for the invasion of epithelial cells, or other aspecific, or specific factors are also involved. Hydrophobic and ionic forces, as well as interactions mediated by LPS, type 1 pili and non-fimbrial protein adhesins have been implied to contribute to the association between *shigellae* and the epithelial cells. Data concerning these interactions, in the light of their relation to the invasion of the epithelial cells, will be reviewed and discussed below.

MICROBIAL FACTORS INVOLVED IN THE BACTERIUM-HOST CELL INTERACTION

Physicochemical Surface Properties of *Shigellae*

Physicochemical surface properties of *S. typhimurium* have been shown to be important for the adhesion and subsequent invasion of HeLa cells (for review see 7). Phase II *S. sonnei* strains (i.e. rough mutants) were found by Edebo et al. to exhibit hydrophobic and slightly negatively charged surfaces (8), while Kabir et al. reported a more pronounced anionic and weakly hydrophobic character for clinical isolates - presumably S forms - of the 4 *shigella* species (9).

We found, that strains harbouring the invasion plasmid were significantly more hydrophobic compared to their non-invasive derivatives, as measured by their retention on Phenyl-Sepharose CL 4B columns (10). However, the hydrophobicity they exhibited was much less, than that of a piliated *E. coli* RDEC-1 strain used as a positive control. The hydrophobicity of invasive shigellae could be decreased by cultivating them at 30°C, or by pretreating them with trypsin. These data indicated, that temperature regulated proteins coded by the invasion plasmid can contribute to the overall surface properties of invasive *shigella* strains. To conclude, however, that hydrophobic forces are directly involved in the interaction leading to the internalization of *shigellae* by epithelial cells needs further experimental support. We could never observe any invasion of HeLa cells by the very hydrophobic RDEC-1 strain.

That hydrophobicity could contribute in some unspecified way to the overall virulence of *shigellae* was recently indicated by the results of Qadri et al. (11). Testing several clinical isolates, a decreasing order of hydrophobicity was found among strains of *S. dysenteriae* 1. *S. flexneri*, *S. boydii* and *S. sonnei*. This order was in general in a good agreement with the severity of the disease these pathogens cause.

Surface Antigens Mediating Association of Shigellae with Epithelial Cells

LPS and type 1 fimbriae: Bacterial lipopolysaccharide (LPS) has been reported to mediate adhesion of *shigellae* to guinea pig colon epithelial cells via a mucosal adhesin. However, this adhesion was not followed by the invasion of the epithelial cells, and later it was suggested, that this mucosal adhesin represents a defense system in rodents (12,13).

On the other hand, rough *shigellae* or even rough *E. coli* K-12 strains could successfully invade HeLa cells, provided they harboured the invasion plasmid (14,15,16). Recently, using a highly differentiated colon epithelial cell line (CaCo-2) grown on a filter support, we could show, that during a 4 hour long incubation time, these cells were infected not only by a smooth *S. flexneri* 5 strain (M90T), but also by a rough *E. coli* K-12 strain harbouring the invasion plasmid (SP10) (Table 1). The infection could take place from the apical, as well as from the basolateral side. These data indicate, that invasion (and presumably the initial host cell membrane-bacterium contact) can take place without the intact LPS molecule.

Table 1. Apical and basolateral infection of Caco-2 cells

| Strains | Number of intracellular bacteria/monolayer (x1000) | |
	Apical infection	Basolateral infection
M90T	5.4 ± 2.8	15.6 ± 5.0
SP10	11.7 ± 4.0	71.4 ± 30.0

Type 1 fimbriae are effective mediators of adhesion of *shigellae* to colonic, as well as to HEp-2 cells. However, since not all the virulent isolates produce these appendages, and since the *in vitro* invasion of epithelial cells could not be blocked by D-mannose, their direct involvment in the uptake of *shigellae* seems improbable (17,18).

Plasmid-coded, non-fimbrial protein antigens: An interesting model was recently proposed by Daskaleros and Payne (19), concerning the potential involvment of invasion plasmid coded proteins in the interaction of bacteria with the host cell membrane. Wild, invasive strains of *shigellae* can bind the dye Congo Red and this capability is coded by the invasion plasmid (2). They observed, that Congo Red shares the bacterial receptor with the iron containing protein hemin. The molecular weight of the hemin binding protein was later estimated to be 101 kD. Both prebound hemin and Congo Red increased the infectivity of *S. flexneri* for HeLa cells. They proposed, that bacteria covered by the bound hemin adhere to the hem receptor of the epithelial cells, which in turn engulf the bacteria so disguised by the hemin coat (19,20).

Recent data of Sen et al. indicate, that the invasion promoting effect of Congo Red could be host cell specific (21).

We have studied the attachment of *shigellae* to HeLa cells in relation to the expression of the invasion plasmid coded protein antigens (10). HeLa cells were infected with invasive and non-invasive *shigellae*. However, the uptake of bacteria was blocked either by incubating them at 4°C (to decrease membrane fluidity), or by the presence of cytochalasin D, (an active inhibitor of microfilament functions). Both conditions prevents internalization of *shigellae* by HeLa cells (22), so the bacteria still associated with the HeLa cells are extracellular. Both methods provided comparable results in the subsequent experiments.

When the monolayers were infected with a virulent *S. flexneri* 5 strain M90T, about 10 times more bacteria were associated with the HeLa cell monolayers, then in case of its plasmid-less, non invasive derivative, M90T-55. The virulent strain adhered better also to tissue culture plates containing no HeLa cells, as compared to the avirulent clone. However, it clearly showed a preference towards the host cells when tested with graduated number of HeLa cells growing on the plastic surface. With light microscopy it appeared as if bacteria were preferentially associated with the edges of HeLa cells (10).

It should be noted, that with exhaustive washing even virulent strains could be almost completely removed from the cells. This indicates, that under the experimental conditions used (i.e. either at 4°C or in the presence of Cytochalasin D) weak forces were responsible for the observed interaction. Therefore in every experiment, positive and negative control strains were included, and results of the adhesion and invasion assays were expressed as percentage of the positive control.

It was tested, whether this adhesive phenotype showed similar characteristics as the invasiveness, that is, plasmid coded, temperature regulated and protein mediated (Table 2). An isogenic pair of plasmid positive and negative clones representing serotype 2a of *S. flexneri* (M42-43 and M42-43Av) exhibited adhesion comparable to that of the control pair of strains, i.e. the virulent strain was adhesive, while adhesion of the non-invasive one was significantly lower. When the invasion plasmid was mobilized into the negative strain, it regained not only its invasive, but also its adhesive capability (42-43Av(pWR110-R64drd11). Both the invasion and adhesion was reduced when the bacteria were grown at 30°C, as well as after trypsin treatment of the bacteria.

SUMMARY AND CONCLUSIONS

The invasion of the epithelial cells is an important step in the pathomechanism of dysentery. The engulfment of the bacterium takes place after an initial contact has been established between the bacterial and the host cell membrane. This contact appears to be maintained during the process of engulfment (5,6).

Table 2. Invasion and adhesion of *Shigella* strains

Strain	Invasion plasmid	% Invasion[a]	% Adhesion[a]
M90T(control)	+	100	100
M90T-55	-	0	7.7 ± 3.9
M42-43	+	115.0 ± 2.4	132.7 ± 28.2
M42-43Av	-	0	7.6 ± 4.6
M42-43Av(pWR110-R64drd11)	+	92.8 ± 14.7	81.6 ± 41.6
M90T (30°C)	+	30.3 ± 6.7	11.4 ± 2.7
M90T(trypsin)	+	6.7 ± 2.3	16.0 ± 2.6

[a] Results are expressed as percentages of invasion and adhesion of the positive control strain, M90T

Different bacterial surface antigens can mediate the interaction of *shigellae* with epithelial cells. However, not all of these adhesive mechanisms relate to the invasion process. Type 1 pili are not ubiquitous among virulent - invasive - isolates (17). The LPS mediated adhesion is specific to a mucosal adhesin in guinea pigs only (12,13). For human epithelial cells, even for differentiated cell lines, it seems, that the internalization of the bacterium can take place independent of the O specific side chain of the LPS molecule (14,15,16).

After preventing HeLa cells to engulf bacteria, invasive *shigella* strains remain loosly associted with the monolayer, while non-invasive clones can easily be washed off. The preference of *shigellae* to attach to the edges of the cells recalls the observation of Clerc et al. (3), i.e. a virulent, wild type *S. flexneri* 5 strain was preferentially taken up at the edges of the HeLa cells. Furthermore, the accumulation of polymerized actin and myosin beneath the cell membrane - assumed to be a consequence of the contact between the bacterial and host cell membrane - showed a similar pattern.

The adhesive capability of *shigellae* was mediated by surface proteins, it was coded by the invasion plasmid, and seemed to

be under similar temperature control, as the ipaB,C and D
proteins essential for the invasion (2). Based on these
observations it is tempting to speculate, that the initial
contact and subsequent attachment and engulfment could be
mediated by some of the ipaB,C,D proteins or by a combination
of them. Whether the contact itself triggers the uptake of
the bacterium or the contribution of other bacterial factors
are necessary remains to be determined.

We should keep in mind, however, that most of these
observations were made using undifferentiated host cells.
Care should be excersised to adopt these data directly to the
in vivo pathomechanism of shigellosis. Differentiated
epithelial cells, like colon cells may have different
receptors expressed on their surface, which in turn may
require other bacterial surface antigens to interact with
them.

ACKNOWLEDGEMENTS

Part of this work was carried out at the Walter Reed Army
Institute of Research (Washington D.C.) where one os us
(T.P.) was working in the laboratories of Dr T.L. Hale and Dr
S.B. Formal. Their constant support and cooperation is
greatefully acknowledged. Part of this work has been
supported by the Swedish Medical Research Council (grant no
16X-656).

REFERENCES

1. E.H. LaBrec, H. Schneider, T.H. Magnani and S.B. Formal:
    Epithelial cell penetration as an essential step in the
    pathogenesis of bacillary dysentery.*J. Bacteriol.*
    88:1503 (1964).
2. A.T. Maurelli and P.J. Sansonetti: Genetic determinants
    of *shigella* pathogenicity. *Ann. Rev. Microbiol.* **42**:127
    (1988).
3. P. Clerc and P.J. Sansonetti: Entry of *Shigella flexneri*
    into HeLa cells: evidence of directed phagocytosis
    involving actin polymerization and myosin accumulation.
    Infect. Immun. **55**:2681 (1987).
4. P.L. Clerc and P.J. Sansonetti: Evidence for clathrin
    mobilization during directed phagocytosis of *Shigella
    flexneri* by HEp-2 cells. *Microbial Pathogenesis* **7**:329
    (1989).
5. H. Ogawa, A. Nakamura and R. Nakaya: Cinematographic
    study of tissue cell cultures infected with *Shigella
    flexneri*. *Japan J. Med. Sci. Biol.* **21**:259 (1968).
6. T.L. Hale, P.A. Schad and S.B. Formal: The envelope and
    tissue invasion. In: "Medical Microbiology" Vol. 3.,
    E.F. Easmon, Ed., p:87-108, Academic Press, London
    (1983).
7. L. Edebo, E., Kihlström. K.-E. Magnusson and O.
    Stendahl: The hydrophobic effect and charge effects in
    the adhesion of enterobacteria to animal cell surfaces
    and the influences of antibodies of different
    immunoglobulin classes: In: "Cell adhesion and
    motility", A.S.G. Curtis and J.D. Pitts, ed., p. 65-101,
    Cambridge University Press, Cambridge (1980).

8. L. Edebo, K.-E. Magnusson and O. Stendahl: Physico-chemical surface properties of *Shigella sonnei*. *Acta Path. Microbiol. Immunol. Scand. Sect. B.* **91**:101 (1983).

9. S. Kabir, S. Ali and Q. Akhtar: Ionic, hydrophobic and hemagglutinating properties of *Shigella* species. *J. Infect. Dis.* **151**:194 (1985).

10. T. Pal and T.L. Hale: Plasmid-associated adherence of *Shigella flexneri* in a HeLa cell model. *Infect. Immun.* **57**:2580 (1989).

11. F. Qadri, S.A. Hossain, I. Ciznar, K. Haider, A. Ljungh, T. Wadström and D. Sack: Congo red binding and salt aggregation as indicators of virulence in *shigella* species. *Infect. Immun.* **26**:1343 (1988).

12. M. Izhar, Y. Nuchamowitz and D. Mirelman: Adherence of *Shigella flexneri* to guinea pig intestinal cells is mediated by a mucosal adhesin. *Infect. Immun.* **35**:1110 (1982).

13. G. Dinari, T.L. Hale, O. Washington and S.B. Formal: Effect of guinea pig or monkey colonic mucus on *shigella* aggregation and invasion of HeLa cells by *Shigella flexneri* 1b and 2a. *Infect. Immun.* **51**:975 (1986).

14. N. Okamura, T. Nagai, R. Nakaya, S. Kondo, M. Murakami and K. Hisatsune: HeLa cell invasiveness and O antigen of *Shigella flexneri* as separate and prerequisite attributes of virulence to evoke keratoconjunctivitis in guinea pigs. *Infect. Immun.* **39**:505 (1983).

15. M.M. Binns, S. Vaughan and K.N. Timmis: O-antigens are essential virulence factors of *Shigella sonnei* and *Shigella dysenteriae* 1. *Zbl. Bakt. Hyg., I. Abt. Orig. B.* **181**:197 (1985).

16. P.J. Sansonetti, T.L. Hale, G.J. Dammin, C. Kapfer, H.H. Collins, Jr. and S.B. Formal: Alterations in the pathogenicity of *Escherichia coli* K-12 after transfer of plasmid and chromosomal genes from *Shigella flexneri*. *Infect. Immun.* **39**:1392 (1983).

17. J.P. Duguid and R.R. Gillies: Fimbriae and adhesive properties in dysentery bacilli. *J. Path. Bact.* **LXXIV**:397 (1957).

18. V.G. Petrovskaya, V.M. Bondarenko and L.V. Mirolyubova: The problem of interaction of *shigella* with epithelial cells. *Zbl. Bakt. Hyg., I. Abt. Orig. A.* **243**:57 (1979).

19. P.A. Daskaleros and S.M. Payne: Congo red binding phenotype is associated with hemin binding and increased infectivity of *Shigella flexneri* in the HeLa cell model. *Infect. Immun.* **55**:1393 (1987).

20. C.E. Stugard, P.A. Daskaleros and S.M. Payne: A 101-kilodalton heme-binding protein associated with congo red binding and virulence of *Shigella flexneri* and enteroinvasive *Escherichia coli* strains. *Infect. Immun.* **57**:3534 (1989).

21. A. Sen, M.A. Leon and S. Palchaudhuri: Comparative study of attachment to and invasion of epithelial cell lines by *Shigella dysenteriae*. *Infect. Immun.* **58**:2401 (1990).

22. T.L. Hale, R.E. Morris and P.F. Bonventre: *Shigella* infection of Henle intestinal epithelial cells: the role of the host cell. *Infect. Immun.* **24**:887 (1979).

HAEMAGGLUTINATING SHIGELLAE

Ivan Ciznar[1], Firdausi Qadri[2], Shafiqul Haq[2],
Shaik Abu Hossain[2]

[1]Research Institute of Preventive Medicine,
Bratislava, Czech and Slovak Federative Republic,
[2]International Centre for Diarrhoeal Disease
Reserach, Dhaka, Bangladesh

INTRODUCTION

The agglutination reaction between bacteria and erythrocytes has been known for almost a century. However, only recently this reaction was found to be associated with the presence of surface structures called adhesins facilitating interaction of a pathogen with a host receptor[1]. While in noninvasive enteropathogens the association of adhesins with the ability to agglutinate erythrocytes has clearly been demonstrated, in invasive enteropathogens particularly in Shigella species such association deserves to be confirmed by more experimentation and data. We have studied the ability of *Shigella dysenteriae* type 1 isolated from the stool of patients with dysentery to agglutinate mammalian erythrocytes. In the present paper we describe the role of lipopolysaccharide in the haemagglutination reaction.

MATERIALS AND METHODS

Strains were isolated in the Clinical Research Centre of the International Centre for Diarrhoeal Disease Research, Bangladesh (ICDDR,B) in Dhaka. Non-enteropathogenic *E. coli* 08:K25 and a rough mutant of *Shigella dysenteriae* type 1(60R) were obtained from the Center for Disease Control, Georgia, Atlanta. Another rough strain was a laboratory mutant obtained in ICDDR,B. All strains were tested for virulence by Sereny's test[2], by Congo red binding assay and by salt aggregation (SAT)[3].

Bacteria were cultivated in Casamino acid-yeast extract broth in the presence of 1 mM calcium chloride at 37°C for 22 h in a shaker. Trypticase soy agar and broth were also used for the cultivation.

Haemagglutination assays as well as inhibition tests were carried out on slides and in U-bottom microtitration plates (Cooke, Alexandria, Va) using an 0.5% suspension of guinea pig and human erythrocytes.

Molecular Pathogenesis of Gastrointestinal Infections
Edited by T. Wådstrom *et al.*, Plenum Press, New York, 1991

Outer membrane proteins were isolated by a lyzozyme and EDTA extraction procedure[4] or by a simple buffer extraction technique[4]. LPS was extracted by the phenol water procedure[6], O-side chains and core of LPS were separated from Lipid A by acid hydrolysis and by ultracentrifugation. The free LPS released spontaneously from the bacterial cells was separated on a Sepharose 4B column. For characterization of bacterial cell components and products a set of analytical procedures including protein and carbohydrate estimation[7], and chromogenic Limulus Amoebocyte Lysate test (Sigma) were used. Polyacrylamide gel electrophoresis and crossed immuno-electrophoresis were used as well[9].

Henle intestinal 407 cells (Int.407) grown in Basal Eagle Medium supplemented with 15% newborn calf serum and 2 mM glutamine were inoculated with bacterial suspensions and incubated at 37°C in 5% CO_2. Non-adherent bacteria were removed by washing in PBS and cover slips fixed with methanol were stained with Giemsa and examined under a microscope. The effect of various substances, including rabbit antiserum, on the adhesion of bacteria to Int.407 was determined.

Rabbit antisera against LPS were prepared by immunization of rabbits with three doses of a heat-killed suspension of Shigella dysenteriae type 1 containing 10^6 cells/ml. Injections were given i.v. one week apart.

RESULTS AND DISCUSSION

Strains of *Shigella dysenteriae* 1, freshly isolated from the stool of patients revealed haemagglutinating activity (HA) with human and guinea pig erythrocytes only when the bacteria were cultivated in CYA broth enriched with $CaCl_2$ (1 mM). We have tested 26 strains of *Shigella dysenteriae* 1, 26 strains of *Shigella flexneri*, 14 strains of *Shigella boydii* and 15 of *Shigella sonnei*. The pattern of haemagglutination is shown in table 1. Of all media tested CYE with $CaCl_2$ was the best one in promoting production of haemagglutinin in *Shigella dysenteriae* type 1. For *Shigella flexneri* this effect was less pronounced and for *Shigella boydii* and *Shigella sonnei* it was practically absent (table 1).

Table 1. Effect of culture medium on HA of *S. dysenteriae*

Medium	HA
CYE broth	++
CYE agar	−
CYE broth + $CaCl_2$	++++
CYE broth + $FeSO_4$	++
CYE broth + $MgCl_2$	++
Trypticase soy broth	−
AKI broth	−
CFA agar	−

Concentration of salts was 1 mM.
++++ agglutination in 1 min;
++ agglutination in 5 min.

Testing of isolated *Shigella dysenteriae* cell surface components for haemagglutinating activity showed that the isolated outer membrane proteins and cell surface proteins did not possess this activity. Treatment of *Shigella dysenteriae* cells with trypsin or proteinase had no effect on the haemagglutinting activity. It was interesting that a culture supernatant of *Shigella dysenteriae* caused haemagglutination. Since it has been known that a substantial quantity of LPS is released by metabolically active bacteria, we separated the culture filtrate on a column of Sepharose 4B. High molecular fractions containing carbohydrates in concentrations about 1 mg/ml revealed haemagglutinating activity. This fraction contained LPS. Its presence was detected by double diffusion in gel, crossed immunoelectrophoresis, by thiobarbituric acid and by the Limulus test. Heat treatment or treatment with proteolytic enzymes did not affect HA activity of this fraction. HA did not correlate with Congo red binding[10] and both Pcr$^+$ and Pcr$^-$ mutants and their culture filtrates prepared under the conditions described above caused haemagglutination.

Table 2. HA of bacterial components

Cell component	HA	Minim. conc. for HA (ug/ml)+	
Culture sup. *S. dys.*1	+	55.0	($\pm$ 2.8)
Sepharose 4B pool-1	+	34.5	($\pm$ 2.1)
LPS - *S. dys.*	+	19.8	(+ 1.4)
PS	-	4.6	(+ 0.178)
LPS - (*E. coli* 08:K25)	-	-	
LPS - *S. sonnei*	-	-	
LPS - *S. boydii* (1-6)	-	-	
LPS - rough *S. dys.*	-	-	

+   Quantified on the base of carbohydrate content

LPS extracted from *Shigella dysenteriae* 1 by the phenol water procedure and purified by ultracentrifugation revealed agglutinating activity (table 2). Thus, these experiments showed that the main component of the *Shigella dysenteriae* 1 cells involved in agglutinting activity was LPS. However, this activity was not found for LPS isolated from *Shigella flexneri*, *Shigella boydii*, *Shigella sonnei*, *E. coli* and from R-form *Shigella dysenteriae* 1. It has been known that LPS exerts membrane associating activity causing a release of haemoglobin from erythrocytes[11]. This activity can be increased by alkali or heat treatment of LPS and quantitatively it is not equal for all LPS isolated from Gram-negative bacteria. It was interesting to observe that free LPS released spontaneously exerts higher membrane associating activity presented by haemolytic effect on erythrocytes than LPS isolated by phenol-water. This phenomenon has been further studied in our laboratory.

Polysaccharide moiety obtained from LPS by acid hydrolysis showed the HA activity. This activity was found only for the polysaccharides obtained from LPS of haemagglutinating strains of *Shigella dysenteriae* 1. Polyacrylamide gel electrophoresis did not show any difference in banding pattern of polysaccharides from LPS of HA + and HA-strains.

HA positive strains of *Shigella dysenteriae* showed adherence to Int. 407 cells. Incubation at 37°C for 20 min was optimal for this effect. It was interesting that Pcr$^+$ and Pcr$^-$ variants revealed adhesion to the cells at the same level. The only exceptions were rough mutants of *Shigella dysenteriae* which did not adhere to Int. 407 cells (table 3).

Thus it appears that Congo red binding and interaction with Int. 407 cells is mediated by different components. Certainly one component involved in interaction of *Shigella dysenteriae* with Int. 407 cells was LPS. Pretreatment of cells with LPS (0.18 mg/ml) completely inhibited adhesion of *Shigella dysenteriae*.

Table 3. Adhesion to Int. 407 cells of haemagglutinating
(HA) and non-HA variants of strain 14731.

Strains	Adhesion No. of bacteria/cell (± SE)	Adhesion (% of control)	P value
1. Strain 14731, HA	75.8 (7.05)	100	
2. $^+$Strain 14731, HA negative virulent	70.0 (5.69)	92.3	NS^{++}
3. Strain 14731, HA positive virulent	76.0 (5.09)	100.24	NS
4. Rough mutant PSD - 10	12.8 (2.59)	15.5	<0.001
5. Rough mutant 60 - R	15.2 (3.84)	20.0	<0.003

+   Except (2) all strains were grown in CYE medium
++ Not significant (P>0.05)

Similar results regarding the role of *Shigella flexneri* LPS in binding to guinea pig colonic cells were presented by others[12]. Our experiments showed that the agglutination of erythrocytes by LPS from *Shigella dysenteriae* could involve the O-side chains. LPS extracted from rough mutants did not exert agglutinating activity thus confirming the possible role of carbohydrate side chains in this reaction. Apparently this could partially explain why we were not able to find an association between plasmid profile and haemagglutinating activity. An association between plasmid profile of *Shigella flexneri* and adherence to HeLa cells was found by others[13]. HA activity of *Shigella dysenteriae* 1 and isolated LPS was sensitive to rabbit anti-LPS serum and to N-acetyl neuraminic acid or to fetuin obtained from fetal calf serum pointing out that sialic acid could be part of a receptor as it was found for other bacteria [14,15]. Further experiments are needed to clarify the molecular interaction between LPS and erythrocytes.

Our data show that haemagglutinating activity and adhesiveness of *Shigella dysenteriae* to intestinal cells is mediated by components which can be expressed under different culture conditions. LPS is a major component involved in haemagglutination reaction and since *S. dysenteriae* is the most pathogenic among the shigellae, the association of this reaction with LPS and pathogenetic potential raises a question of its relevance for assessment of virulence in Shigella species.

ACKNOWLEDGMENTS

This research was supported by United States Agency for International Development (USAID) and the International Centre for Diarrhoeal Disease Research, Bangladesh (ICDDR,B). ICDDR,B is supported by countries and agencies which share its concern about the impact of diarrhoeal diseases in the developing countries. Current major donors giving assistance to ICDDR,B: The Aga Khan Foundation, Arab Gulf Fund, Australia, Bangladesh, Belgium, Canadian International Development Agency (CIDA), Canadian International Development Research Centre (IDRC), Danish International Development Agency (DANIDA), France, the Ford Foundation, Japan , the Netherlands, Norwegian Agency for International Development (NORAD), SAREC (Sweden), Swiss Development Corporation (SDC), United Kingdom, United Nations Development Programme (UNDP), United Nations Children's Fund (UNICEF), United Nations Capital Development Fund (UNCDF), United States Agency for International Development (USAID), World Health Organization (WHO) and World University Service of Canada (WUSC). We thank Manzurul Haque for secretarial assistance.

REFERENCES

1. E.H. Beachley, Bacterial adherence: adhesin-receptor interaction mediating the attachment of bacteria to mucosal surfaces, *J. Infect. Dis.* 143:325 (1981)
2. B. Sereny, Experimental Shigella conjunctivities, *Acta Microbiol. Acad. Sci. Hung.* 2:293 (1955)
3. F. Qadri, S.A. Hossain, I. Ciznar, K. Haider, Å. Ljungh, T. Wadström, and D. Sack, Congo red binding and salt aggregation as indicators of virulence in Shigella species, *J. Clin. Microbiol.* 26:1343 (1988)
4. K.H. Johnston and E.C. Gotschlich, Isolation and Characterization of the outer membrane of *Neisseria gonorrhoeae*, *J. Bacteriol.* 119:250 (1974)

5. E.V. Oaks, L. Hale, and S.B. Formal, Serum immune response to Shigella antigens in Rhesus monkeys and humans infected with Shigella species, *Infect. Immun.* 53:57 (1986)

6. O. Westphal and K. Jann, Bacterial lipopolysaccharides: Extraction with phenol-water and further applications of procedure, in: "Methods in Carbohydrate Chemistry", 5:83 (1965)

7. M. Dubois, K.A. Giles, J.K. Hamilton, P.A, Rebers, and F. Smith, Colometric method for determination of sugars and related substances, *Anal. Chem.* 28:350 (1956)

8. A. Weissbach and J. Hurwitz, The formation of 2-keto-3-deoxyheptonic acid in extracts of *Escherichia coli*, *J. Biol. Chem.* 234:705 (1959)

9. B. Weeke, Crossed immunoelectrophoresis, in: "A manual of quantitative Immunoelectrophoresis," H.H. Axelsen, J. Kroll and B. Weeke, eds., Universitetsforlaget, Oslo (1973)

10. F. Qadri, S. Haq, and I. Ciznar, Haemagglutination properties of Shigella species, *Infect. Immun.* 57:2909 (1989)

11. I. Ciznar and J.W. Shands, The effect of alkali-treated lipopolysaccharide on erythrocyte membrane stability, *Infect. Immun.*, 4:2580 (1989)

12. M.I. Izhar, V. Nuchamowitz and D. Mirelman, Adherence of Shigella flexneri to guinea pig intestinal cells is mediated by a mucosal adhesin, *Infect. Immun.* 35:1110 (1982).

13. T. Pal and T.L. Hale, Plasmid associated adherence of Shigella flexneri in a HeLa cell model, *Infect. Immun.* 57:2580 (1989)

14. M. Lindahl and T. Wadström, K99 surface haemagglutinin of enterotoxigenic E. coli recognize terminal N-acetyl-galactosamine and sialic acid residues glycophorin and other complex glycoconjugates, *Vetr. Microbiol.*, 9:249 (1984)

15. M. Lindahl, A. Faris, and T. Wadström, Colonization factor antigen on enterotoxigenic Escherichia coli is a sialic acid specific lectin, *Lancet* 11:280 (1982)

MOLECULAR PATHOGENESIS OF *GIARDIA LAMBLIA*:

ADHERENCE AND ENCYSTATION

Gerald T. Keusch, Honorine D. Ward, Eduardo Ortega-Barria, Norma Galindo, Miercio E. A. Pereira

Geographic Medicine and Infectious Diseases
New England Medical Center
Boston, MA U.S.A.

INTRODUCTION

Giardia lamblia, a non invasive protozoan pathogen of the human small bowel, causes a spectrum of infection ranging from asymptomatic carriage, through acute watery diarrhea, to chronic diarrhea and malabsorption. *Giardia* has a simple life cycle, existing in just two forms, the multiplying trophozoite which infects mammalian hosts to cause disease, and the environmentally resistant cyst which serves in the transmission of infection from host to host. Efforts to understand disease pathogenesis and control infection therefore focus on the cell biology of the trophozoite, whereas public health interests target the process of encystation and the potential for control of spread. There are at least three clearly discernible steps in pathogenesis, once the cyst has been ingested and excystation has been triggered by a drop of pH in the stomach. First, the trophozoite has to colonize the small intestinal mucosal surface, second it may disrupt the normal physiology of the gut and produce intestinal symptoms such as diarrhea and malabsorption, and third, it must transform into the cyst in order to reach the external environment and propagate the species. Based on this schema, we can, in general terms, predict the nature of the parasite molecules needed for these events to occur: for example adhesins or lectins for colonization, cytotoxins or other secretogogues for effects on the gut epithelium, or proteins regulating gene expression for stage transformation. Our laboratory has been concerned with adherence and encystation, and these studies will constitute the subject of this paper.

ADHERENCE MOLECULES

Trypsin Activated *Giardia lamblia* Lectin (Taglin)

We began our studies by a deliberate search for a parasite lectin which might be involved in cell-to-cell recognition and adherence events. It was our bias that such a molecule would be surface displayed and able to recognize carbohydrate moieties. We chose the method of mixed agglutination with parasites and erythrocytes to detect this activity and found a mixed hemagglutination (HA) response inhibitable by glucose or mannose,[1]. A significant advance occurred when we discovered that lysates of *Giardia lamblia* contained an inactive HA that could be activated by brief incubation with trypsin or with human small intestinal juice (Figure 1),[2]. This effect was mediated by the catalytic site of trypsin, as activation was prevented by serine protease inhibitors. We named this HA Taglin, for trypsin activated *Giardia lamblia* lectin. Taglin was specific for rabbit erythrocytes, was induced within seconds of trypsin addition, peaked in 10 minutes, remained constant for several hours, and then declined. Activation was dose dependent between 1 and 100 ng of trypsin per 10^6 trophozoites and then diminished with more enzyme,[3]. At least 30,000 fold higher trypsin concentration was needed to completely destroy the HA. This is consistent with a limited proteolysis, an event characteristic of the activation of certain biological processes, for example the complement or coagulation cascades, or conversion of prohormones to hormones or proenzymes to enzymes. The exact mechanism of activation remains uncertain, but it may be a result of cleavage of a small peptide to reveal the erythrocyte binding site or, alternatively, by inducing a conformational change in the protein,[4].

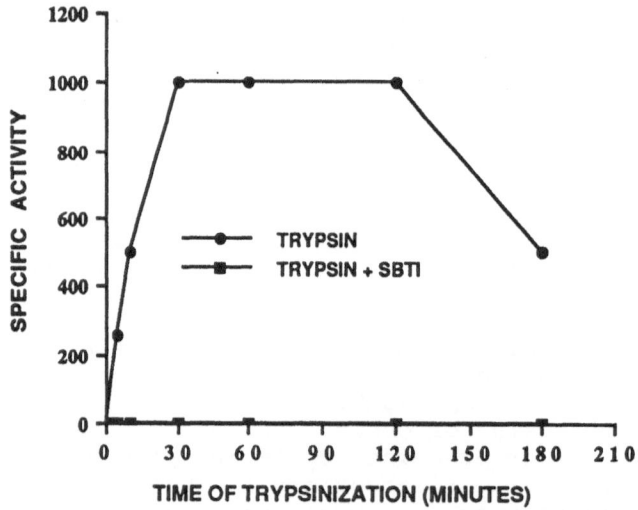

Fig. 1 **Time course of activation of taglin by trypsin.** Aliquots of washed, lysed *Giardia* trophozoites (10^6/ml) were incubated with 25µg/ml of trypsin at 23°C. At various time points, proteolysis was stopped by the addition of 50 µg/ml of soy bean trypsin inhibitor (SBTI) and the lysates were assayed for lectin activity using a hemagglutination assay. Specific activity was expressed as hemagglutinating units (HA) per µg of protein.(from ref. 4)

Table 1. Sugar Specificity of Taglin[a]

Sugar Tested	MIC (mM)	Relative Inhibition[b]
D-mannose	150	1.0
D-mannose 6-phosphate	20	7.5
D-mannosamine	65	2.3
N-acetyl D-galactosamine	100	1.5
N-acetyl D-mannosamine	100	1.5
D-arabinose	150	1.0
D-fucose	150	1.0
D-mannose 1-phosphate	>100	<1.5
D-glucose 1-phosphate	>100	<1.5
D-galactose 6-phosphate	>100	<1.5
D-fructose 6-phosphate	>100	<1.5
D-fructose 1,6-diphosphate	>100	<1.5
D-galactose	>200	<1.3
D-glucose	>200	<1.3
D-glucose 6-phosphate	>200	<1.3

[a]Sugar specificity of taglin was determined by a hemagglutination inhibition assay, .The MIC (minimal inhibitory concentration) is defined as the minimal concentration of a sugar required to inhibit 4 hemagglutination units of lectin activity.
[b]compared to D-mannose

Taglin exhibits well defined sugar specificity, determined by the classical method of hemagglutination inhibition in the presence of simple or complex mono- or oligosaccharides,[2,5]. Mannose-6-phosphate (M-6-P) was the best monosaccharide inhibitor, and was more than 7 times more potent than mannose. Mannose-1-phosphate was inactive at 100 mM, as were a number of other phosphorylated sugars,[5]. Phosphate on the 6 carbon was shown to be important since pretreatment of erythrocytes with alkaline phosphatase to remove the phosphate decreased the HA titer, except in the presence of metavanadate, an alkaline phosphatase inhibitor,[5]. Taglin agglutinated epithelial cells from the mouse small intestine, or isolated rabbit microvillus membranes,[4]. Agglutination was inhibited by addition of M-6-P. For these reasons we have concluded that taglin is a M-6-P specific lectin and have proposed that it is a recognition molecule for *Giardia* in the small bowel.

Further characterization of taglin reveals a rather narrow pH range, unlike most lectins, with an optimum at the precise pH of the intestinal mucosal surface, 6.5,[5]. It requires divalent cations (Ca^{++} or Mn^{++}), and was present in at least 5 axenic human isolates we tested, 1 cat isolate, and the murine parasite *Giardia muris*. We then raised monoclonal antibodies (Mabs) to trypsin activated lysates, screening for HA inhibition. One high titer clone, α-L2, when immobilized on protein A-Sepharose, depleted HA from activated lysates, whereas an isotype matched control had no effect,[5]. α-L2 stained the surface of 5 different *Giardia* isolates by indirect immunofluorescence, which demonstrates its presence on the trophozoite surface,[6]. The antibody recognized a 28/30 kDa doublet (Figure 2) in untreated or trypsin treated lysates by

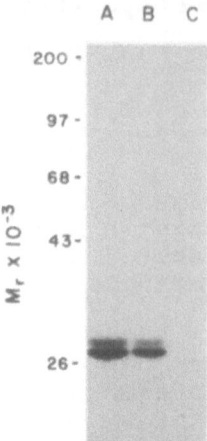

**Fig. 2   Identification of taglin by Western blotting
with monoclonal antibody α-L2**. Aliquots of *Giardia*
lysate, either untreated (lane A) or treated with
trypsin (lane B) or subtilisin (lane C), were
electrophoresed on SDS-polyacrylamide
gels,electroblotted to nitrocellulose, and probed with
monoclonal antibody α-L2. (from ref. 5)

Western blot. When lysates were exposed to subtilisin, which
completely destroys HA activity, this band disappeared. We
were able to confirm that this band is taglin because rabbit
erythrocytes adhered directly to the 28/30 kDa region of SDS-
PAGE separated giardial proteins transfered to, and renatured
on, nitrocellulose (Figure 3). We therefore have proposed that
taglin on the surface of trophozoites in the small bowel is
activated by trypsin present in the small bowel, and that
activated taglin recognizes glycoconjugates on the gut
microvillus membrane and mediates parasite adherence to the gut
epithelial cell.

Giardia lamblia Adherence Molecule-1 (GLAM-1)

    We have extended our studies to an in vitro model for
giardial adherence to IEC-6 cells, as described by McCabe et
al,[7]. We cloned the WB strain of *Giardia* by limiting dilution
in soft agarose, labeled the cloned trophozoites by growth in
[3]H-thymidine, and incubated them with IEC-6 cell monolayers.
The clones differed in attachment ability, assessed by cell
associated radioactivity, from 45% to 10% of input counts
(Table 2).

    We then asked whether or not adherence to IEC-6 cells
correlated with taglin HA activity in parasite lysates; in fact
this was the case. We confirmed this difference by
immunoblotting lysates of clones A and E, and used the Mab α-L2
to quantitate antibody binding after incubating with an
alkaline phosphatase linked second antibody and laser

densitometric analysis of the colored reaction product. The
results indicated that clone A had a significantly higher
amount of taglin than clone E (manuscript in preparation).

We have also generated Mabs by immunizing mice with clone
A and screening first for binding to methanol fixed parasites
by immunofluorescence, and then by inhibition of IEC-6 cell
adherence. One antibody, 2F10 significantly inhibited IEC-6
cell attachment. When reacted with methanol fixed
trophozoites, 2F10 strongly bound to the ventral disc. Western
blots with 2F10 and total parasite lysate, a 0.5 % Triton X-
100 extract, the washed Triton X-100 insoluble pellet, which
contains cytoskeleton elements, and lysed in vitro generated
cysts, demonstrated a 30 kDa band in both the cytoskeleton
preparation and in vitro cysts, consistent with immunofluorescence data. We have designated this band *Giardia
lamblia* adhesion molecule-1 (GLAM-1). Our present working
hypothesis is that initial contact of parasite and small bowel
is mediated by taglin, since it is present over the entire
surface of the parasite, and that the disc specific GLAM-1
(and/or other similar adhesins) present on this organelle
mediates the avid and reversible attachment of the disc to the
target cell surface observed to occur both in vitro and in
vivo.

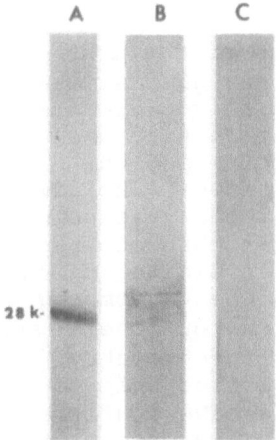

**Fig. 3   Identification of taglin by direct
          erythrocyte binding to nitrocellulose blots.**
          Trypsinized (lanes A and B) as well as untrypsinized
          (lane C) aliquots of a *Giardia* lysate were
          electrophoresed on 10% polyacrylamide gels and
          transferred to nitrocellulose filters. Proteins
          bound to the filter were denatured and renatured.
          Filters (lane B and C) were incubated with 0.25%
          rabbit erythrocytes, and protein bands with lectin
          activity were identified by direct visual observation
          of bound erythrocytes. Lane A was probed with
          monoclonal antibody α-L2. (from ref. 5)

Table 2.   Attachment of Cloned *Giardia* to IEC-6 Cells[a]

Clone	Percent Attachment ± S.D.
A	44.8 ± 11.9
Q	40.8 ± 7.7
S	40.2 ± 3.1
WB	29.6 ± 14.2
R	26.8 ± 2.1
O	20.1 ± 3.7
L	13.5 ± 5.9
E	12.1 ± 6.6
F	10.9 ± 6.9

[a]Clones A, Q, S, R, O, L, E and F.(2.5 x 10^5/well), labelled with ^{3}H thymidine for 48 hours were harvested, washed and incubated with a monolayer of IEC-6 cells for 20 min at 37°C in 96 well plates. Unattached trophozoites were removed by washing with PBS at 37°C.   100µl of 0.5 M potassium hydroxide was added to each well and the amount of radioactivity bound to the cells was quantitated by scintillation counting.   100% attachment represents the radioactivity associated with the initial innoculum.

ENCYSTATION AND CYST SPECIFIC PROTEINS

Our initial studies of the cyst characterized the cyst wall by lectin binding and glycosidase digestion,[8].   The evidence indicated that oligomeric B1->4 linked N-acetyl-D-glucosamine (GlcNAc), or chitin, was the major cyst wall constituent.   We have confirmed the presence of GlcNAc by several additional techniques, including galactosyltransferase labeling, immunochemical analysis using antibodies to chitin oligosaccharides, and gas chromatography/mass spectrometry,[9]. Trophozoites also contain GlcNAc, demonstrated first by the binding of biotinylated wheat germ agglutinin (WGA) and specific inhibition by chitotriose,[10] and later confirmed by gas chromatography/mass spectrometry,[9].   However, there is a subpopulation of trophozoites that do not bind WGA and are deficient in GlcNAc.   We have confirmed this by flow cytometry of live trophozoites stained with FITC-WGA,[10] and by staining with an antibody specific for GlcNAc,[9].   These data suggest the possibility that synthesis of GlcNAc by the trophozoite is a key step in the initiation of encystation.   However, WGA binding in the trophozoite is not sensitive to chitinase, in contrast to the cyst, and is primarily to GlcNAc present in membrane glycoproteins.   Encystation can now be initiated in vitro, as first described by Gillin et al,[11].   When we looked for a change in N-acetyl-D-glucosamine during cyst transformation in vitro, we found a 6 fold increase in the GlcNAc content per µg of membrane protein, but no change in the relative sugar composition (Table 3).   The early cysts also contained N-acetyl-D-galactosamine,[9], absent from the trophozoite, but this is of uncertain importance.

Table 3. Sugar Composition of *Giardia lamblia* Membranes[a]

Sugar	Percent Composition		nmol/100 μg protein	
	A	B	A	B
Man	9	0	1.0	0
Gal	14	26	1.5	25.6
Glc	54	46	6.1	46.2
GalNAc	0	3	0	3.4
GlcNAc	4	3	0.5	3.0
NANA	14	16	1.6	16.0

[a]Sugar composition of *Giardia lamblia* membranes was determined by gas chromatography/mass spectrometry. A = trophozoite solubilized membranes, B = encysting parasite solubilized membranes.

To determine if there was any biological significance of surface GlcNAc in *G. lamblia*, we grew trophozoites in the presence of a variety of lectins including WGA, phytohemagglutinin, *Aaptos papillata* agglutinin, Concanavalin A, and soybean agglutinin. Only WGA had an effect, resulting in around 50% inhibition in trophozoite growth at 100 μg/ml, and a sharp decrease in the number of parasites attached to the glass surface (Table 4),[12]. Growth inhibition was reversible by washing the cells free of WGA, and was dependent on the concentration of lectin. When WGA was administered by daily gavage to *G. muris* infected CF-1 mice, the number of trophozites and cysts in stool was reduced by 30% and 47%, respectively, which is suggestive of in vivo growth inhibition.

Since cell division is controlled at specific points in the cell cycle, we evaluated the effect of WGA on in vitro replication of the parasite by DNA staining and flow cytometry. There were no significant changes in the cell cycle in

Table 4. Effect of Lectins on Growth of *Giardia lamblia* Cultured in vitro[a]

Lectin	Percent Inhibition ± S. D.
WGA	55.1 ± 17.5
WGA + GlcNAc	15.0 ± 4.0
PHA	4.6 ± 0.8
Aap	8.4 ± 4.1
SBA	8.8 ± 0.8
Con A	14.1 ± 2.2

[a]Trophozoites (2.5 x 10^4 in 0.1 ml) were grown in the presence of 100 μg/ml of the following lectins: Concanavalin A (Con A); phytohemagglutinin (PHA); soybean agglutinin (SBA); WGA, *Aaptos papillata* agglutinin (Aap); and WGA plus 2 mg/ml GlcNAc at 37°C for 72 hours. Results are expressed as percentage of growth inhibition relative to control cultures grown in the absence of WGA.

trophozoites grown in the presence of Concanavalin A or lima bean agglutinin, compared to control. However, WGA resulted in a marked arrest in growth at the G2/M phase, which suggests the existence of a regulatory control mechanism which determines the timing of mitosis and that may be blocked by WGA. While the mechanism for this remains uncertain, WGA induces the synthesis of new glycoproteins in *Giardia lamblia* (manuscript in preparation). Compared to control (no WGA), trophozoites grown in the presence of 100 µg/ml WGA expressed two new glyco-proteins, gp215 and gp150, detected by the ability to bind biotinylated WGA, and had significantly lower amounts of gp190. It is our hypothesis that one or more of these glycoproteins may be *Giardia* growth regulators.

To confirm and quantitate these changes in gp215, 190, and 150, we have assessed the binding of iodinated WGA to WGA-induced and uninduced lysates that had been run on SDS-PAGE gels and transfered to nitrocellulose. A dose dependent lectin induction of gp215 and gp150 and the repression of gp190 have been demonstrated.

Encystation Specific Proteins

We have optimized the conditions for large scale in vitro encystation,[13], using as criteria morphology by phase contrast microscopy and resistance to hypotonic lysis in distilled water.

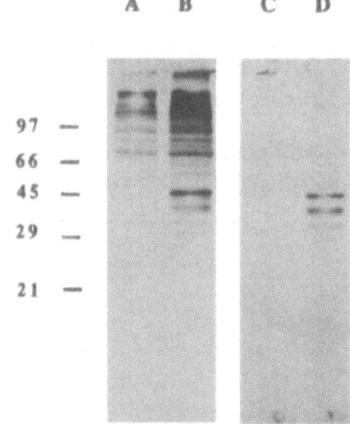

Fig. 4 **Binding of GCSA-1 to cyst specific proteins**. Duplicate aliquots (~20 µg protein) of *Giardia lamblia* trophozoites (A and C) and in vitro cysts (B and D) were resolved by SDS PAGE and electrotransferred to nitrocellulose. Filters (A and B) were probed with polyclonal antiserum from mice immunized with cyst "ghosts", diluted 1:200 (C and D), or with GCSA-1 culture supernatant diluted 1:200. (from ref. 14)

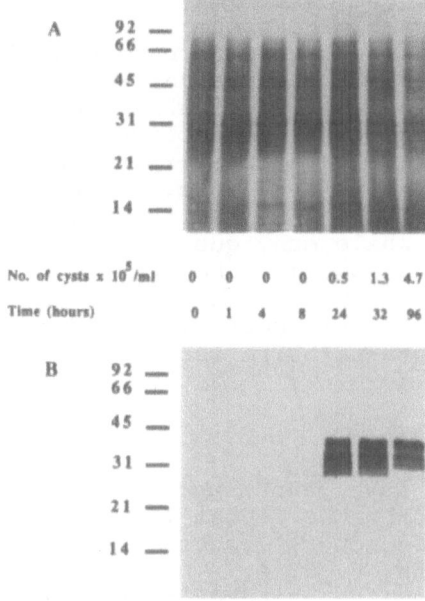

A

| | 92 —
| | 66 —
| | 45 —
| | 31 —
| | 21 —
| | 14 —

No. of cysts x 10^5/ml   0   0   0   0   0.5   1.3   4.7

Time (hours)   0   1   4   8   24   32   96

B

Fig. 5   Time course of appearance of antigens recognized by GCSA-1.   Encysting parasites,were harvested at various time points and ~20μg protein of each were resolved (on duplicate gels) by SDS PAGE and electrotransferred to Immobilon filters. One filter (A) was stained with Coomassie blue and the other (B) probed with GCSA-1 culture supernatant diluted 1:100.   (from ref 14)

We have used in vitro generated cysts to generate a cyst specific monoclonal antibody termed GCSA-1, for *Giardia* cyst specific antibody (Figure 4),[14]. Encysting parasites, harvested at various times after transfer to encystation medium, express several new antigens that are recognized by GCSA-1 (Figure 5). GCSA-1 also immunoprecipitates newly synthesized proteins from encysting parasites that have been metabolically labeled with ^{35}S-methionine. These new labeled antigens are similar in size to those shown by immunoblotting with GCSA-1, and they appear as early as 8 hours after transfer to encystation medium, which demonstrates unequivocally that these genes are being turned on during cyst transformation and that new proteins are being synthesized.

SUMMARY

We have described two putative *Giardia lamblia* adherence molecules. One of these, taglin, is a M-6-P specific lectin, and recognizes glycoconjugates present on gut epithelial cells. The other, GLAM-1, is a disc specific protein that appears to be involved in cell attachment since a monoclonal antibody inhibits attachment to IEC-6 cells in culture. We are now attempting to clone the genes for these proteins. Their availability as recombinant proteins will permit studies necessary to define their role in pathogenesis in vivo. We have also shown that WGA induces and represses certain glycoproteins of trophozoites, inhibits cell replication at the G2/M phase and reduces cyst production in vitro and in vivo. In addition, encysting parasites express other new glycoprotein antigens recognized by a cyst specific antibody, GCSA-1. The nature and role of these antigens in stage transformation are not clear as yet. While many questions remain about these processes in *Giardia lamblia*, the studies reviewed here provide us with some candidate molecules and the molecular tools with which to investigate their possible role in the biology of the organism and the host-parasite interaction.

ACKNOWLEDGEMENTS

These studies have been supported by grants AI 121791, A1 27218, P30 DK 34928 to the Center for Gastroenterology Research on Absorptive and Secretory Processes, from the National Institutes of Health, U. S. A.; Thrasher Research Fund Award 2802-2; a grant from the Whitaker foundation; and a grant from the Rockefeller Foundation under a joint UNDP/World Bank/WHO special program for research and training in tropical diseases.

REFERENCES

1.   M.J.G. Farthing, M.E.A. Pereira, and G.T. Keusch, Description and characterization of a surface lectin from *Giardia lamblia*, Infect. & Immun. 51:661 (1986).

2.   B. Lev, H. Ward, G.T. Keusch, and M.E.A. Pereira, Lectin activation in *Giardia lamblia* by host protease: a novel host-parasite interaction, Science 232:71 (1986).

3.   B. Lev, H. Ward, and M.E.A. Pereira, Prolectin activation in *Giardia*, in: "Microbial Lectins and Agglutinins", D. Mirelman, Ed., John Wiley and Sons, New York (1986).

4.   H.D. Ward, G.T. Keusch, and M.E.A. Pereira, Induction of a phosphomannosyl binding lectin activity in *Giardia*, BioEssays 12:211 (1990).

5.   H.D. Ward, B.I. Lev, A.V. Kane, G.T. Keusch, and M.E.A. Pereira, Identification and characterization of Taglin, a mannose 6-phosphate binding trypsin-activated lectin from *Giardia lamblia*, Biochemistry 26: 8669 (1987).

6.   H.D. Ward, B.I. Lev, G.T. Keusch, M.E.A. Pereira, Induction of lectin activity in Giardia, UCLA Symp. Mol. Cell. Biol. (New Ser.) 42:521 (1987).

7.   R.E. McCabe, G.S.M. Yu, C. Conteas, R.R. Morill, B. McMorrow, Antimicrob. Agents. Chemother., (in press).

8.   H.D. Ward, J. Alroy, B. Lev, G.T. Keusch, and M.E.A. Pereira, Identification of chitin as a structural

component of *Giardia* cysts, <u>Infect. & Immun.</u>, 49:629 (1985).

9.   E. Ortega-Barria, H.D. Ward, J.E. Evans, M.E.A. Pereira, N-acetyl-D-glucosamine is present in cyst and trophozoites of *Giardia lamblia* and serves as receptor for wheatgerm agglutinin, <u>Molec. Biochem. Parasitol.</u>, 1990 (in press).

10   H.D. Ward, J. Alroy, B.I. Lev, G.T. Keusch, M.E.A. Pereira, Biology of *Giardia lamblia*. Detection of N-acetyl-D-glucosamine as the only surface saccharide moiety and identification of two distinct subsets of trophozoites by lectin binding, <u>J. Exp. Med.</u>, 167:73 (1988).

11. F.D. Gillin, D.S. Reiner, M.J. Gault, H. Douglas, S. Das, A. Wunderlich, and J. Sauch, Encystation and expression of cyst antigens by *Giardia lamblia* in vitro, <u>Science</u>, 235: 1040 (1987).

12. E. Ortega-Barria, H.D. Ward, M.E.A. Pereira, Growth inhibition of *Giardia lamblia* by a dietary lectin is associated with arrest of the cell cycle. (submitted).

13. H.D. Ward, A.V. Kane, G.T. Keusch, and M.E.A. Pereira, *Giardia lamblia*: Determination of optimal conditions for in vitro encystation, identification of strain and clone differences in encystation efficiency, and large scale production of in vitro cysts (submitted).

14. H.D. Ward, A.V. Kane, E. Ortega-Barria, G.T. Keusch, and M.E.A. Pereira, Identification of developmentally regulated *Giardia lamblia* cyst antigens using GCSA-1, a cyst specific monoclonal antibody, <u>Molec. Microbiol.</u>, (in press).

THE PATHOGENIC MECHANISMS OF *HELICOBACTER PYLORI* - A SHORT OVERVIEW

Torkel Wadström, Janaki L. Guruge, Shen Wei, Pär Aleljung, and Åsa Ljungh

Department of Medical Microbiology, University of Lund
Sölvegatan 23, S-223 62 Lund, Sweden
Telefax No. +46-46-18 91 17

INTRODUCTION

In recent years *Helicobacter pylori* (previously *Campylobacter pylori)*[1] has become an established pathogen shown to cause the most common form of gastritis in humans (i.e. type B gastritis) and is able to infect human volunteers as well as monkeys and germfree piglets[1-8]. It has also been established that *H. pylori* is a major factor in developing stomach as well as duodenal ulcer disease in humans. Epidemiological studies have shown that *H. pylori* infections occur in early childhood with high incidence of atrophic gastritis predisposing to gastric cancer[3]. Since *H. pylori* cause such diverse entities as histological gastritis and peptic ulcers, it seems most likely that the disease mechanisms are both complex and include several virulence factors. Virulence properties on *H. pylori* known today allowing colonization of the gastric epithelium are summarized in Table 1.

Table 1

Putative virulence properties of *H. pylori* and their relation to pathological conditions

Motility	
Helical shape and flagella	Favours rapid penetration of the gastric mucin layer
Cell surface urease	Neutralizes gastric acid
Haemagglutinins/tissue adhesins	Binding to receptors on epithelial cells (and possibly the mucin layer)
Cytotoxins	Cell lysis (and cytoplasmatic vacuolization)

Penetration through gastric mucin layer

Motile *H. pylori* strains easily infect germfree piglets with 100 percent infection rate[6]. It seems most likely that rapid penetration through the gastric mucin layer is crucial to allow this acid sensitive organism to colonize the gastric epithelium. The ability of *H. pylori* to survive the gastric bactericidal activity in the stomach is primarily attributed to a high surface urease activity leading to the production of ammonium acting as a H+ ion acceptor[9-11].

Interestingly, urease or urease associated proteins have also been proposed as tissue adhesins (see below). Slomiany *et al.*[12] proposed that *H. pylori* rapidly degrades gastric mucins to impair the viscoelastic properties and hydrogen ion retardation capacity of the gastric mucin layer. However, studies in other laboratories have not confirmed high protease mucin degrading capacity in cell surface extracts or extracellular materials of broth or agar grown *H. pylori* cultures (S. Wei and T. Wadström unpublished data). Moreover, our studies have not confirmed that strains of *H. pylori* with various haemagglutinin profiles[13,14] are high producers of lipase or phospholipase A or C activity. Such enzymes could be involved in degradation of the stomach hydrophobic lining[10]. However, Raedsch *et al.*[16] reported on mucosal lipid degradation with *H. pylori* phospholipase A. It cannot be excluded that some strains under certain *in vivo* conditions use such an enzyme and also other enzymes such as endoglycosidases to allow rapid penetration of the mucin layer into the gastric epithelium.

Colonization of the gastric epithelium

H. pylori has a unique tissue tropism for the gastric epithelium and can probably only colonize in ectopic gastric epithelium outside the stomach. Patients with duodenal ulcer disease develop gastric metaplasia necessary for colonization with *H. pylori*. This mechanism is now supposed to be the key factor for development of duodenal ulcer disease in patients after treatment with antacid drugs such as H2 blockers [2-4]. Elegant morphological studies by Bode *et al.*[17] suggest that *H. pylori* can adhere "at distance" to gastric epithelial cells with "fibrillae connected" to target cells (Figure 1).

However, at later stages certain strains are able to cause close cell contacts with motphological changes similar to pedestal or "cup formation" when enteropathogenic *E. coli* (EPEC) adhere to intestinal cells (see Knutton this volume). We know very little today about the tissue colonization at the molecular level. However, from studies published up to now a few alternative models can be proposed for gastric epithelium colonization and tissue penetration.

1) Fibrillar and nonfibrillar surface proteins bind to the different cell surface glycoconjugates and gastric mucin[13,14,18].

2) The complex haemagglutinin/lectin profiles of *H. pylori* strains summarized show strains agglutinating one single erythrocyte species up to all eleven species which indicates different lectin/adhesins surface proteins in various combinations in different strains[13,14,19,20,21].

Recent studies in our laboratory have confirmed that *H. pylori* produce at least two major types of cell surface lectins:

(i)   recognizing sialoglycoconjugates (Class 1) and

(ii)   recognizing certain lactocylceramide derivates.

Interestingly, our studies also showed that strains of different haemagglutination (HA) profiles all bind to Hela cells in tissue culture assay while nonhaemagglutinating strains do not. Fauchere and Blaser[22] have purified a surface associated material (SAM) of *H. pylori* which binds to hela cell, and they have shown that this material, but not N-acetylneuramic acid or fetuin, block adherence better than bovine serum albumin suggesting that SAM-adhesin(s) do not belong to class 1 adhesins. A major antigen of approximately 60kDa was present in this SAM material which sopurified with urease as revealed by immunoblots with an antiurease serum.

From our knowledge today it seems not possible to explain the unique tissue tropism for *H. pylori* from what has been published on *H. pylori* surface proteins up to now. It seems most likely that a combination of adhesins are involved in a multiple step tissue colonization process.

Later stages in gastric tissue colonization

Certain helical shaped organisms not yet properly identified can infect stomach of mice, other rodents and ferrets without causing tissue damage[2] ˙ unlike the human pathogen *H. pylori*. It seems most likely that these organisms thus possess unique properties to penetrate the gastric and duodenal epithelial layers by two alternative strategies;

1) binding to the epithelial cells

2) penetration between these cells and through intercellular junctions.

Both phenomena have been demonstrated by a number of investigators by electron microscopy[17]. At least about 50% of *H. pylori* strains are able to produce a non-lethal heatlabile protein which causes vacuolization in tissue culture cells[23]. Three peptides have recently been identified in such cytotoxin associated material (30 kDa, 95 kDa and 80 kDa) not present in nontoxic strains[24]. However, it cannot be excluded that strains isolated from serious infections cause more tissue damage i.e. as the strains isolated from peptic ulcers are more potent toxin producers and that additional toxins have not yet been discovered besides the cytotoxic phospholipase A enzyme[16].

Since *H. pylori* can penetrate down to the subepithelial basal membrane as shown in histopathological studies[17], we have investigated if this pathogen can interact with connective tissue proteins abundant in subepithelial cell matrices such as fibronectin, collagen type IV (and other collagens) as well as laminin. Interestingly, certain strains of *H. pylori* bind specifically ^{125}I-labelled type IV collagen. This binding is inhibited by unlabelled collagen type IV (T.J. Trust and T. Wadström in preparation). With these findings we recently also screened strains for binding of heparansulphate, a matrix proteoglycan molecule which may have contaminated our collagen preparations. Interestingly, certain strains of *H. pylori* but not strains showing optimal collagen IV binding bind heparansulphate. A cell surface protein of *H. pylori* strain has been purified on heparan Sepharose and is now under further investigation (see also Ascencio *et al.* this volume.

FUTURE PERSPECTIVES

It is only less than a decade since *H. pylori* was established as a human pathogen[1]. Thus, research on virulence mechanisms is still in its infancy and it seems most likely that much more will be learned rapidly about this organism to explain how it can establish a lifelong infection if the patients are not treated with a combination of antibiotics and bismuth salts[25]. It seems likely that bacterial factors not known today are involved to trigger the acute and chronical inflammation process. Regulation of the inflammatory tissue response by release of prostaglandin metabolites by *H. pylori* secretion of phospholipases is only one possibility.

It seems likely that induction of round coccoidal slow-growing dormant forms[26] is triggered by certain nutrient deprivation and is related to synthesis of certain bacterial starvation proteins similar to heatshock and oxidation stress proteins in other organisms such as *E. coli* and various marine vibrios[26,27]. However, preliminary studies in our laboratory have failed to reveal small "sporelike" coccoidal cell forms of *H. pylori* yet with high cell surface hydrophobicity.

Finally, development of genetic systems to study *H. pylori* is still in its infancy including development of cloning vectors for expression of *H. pylori* proteins in other organisms[28].

ACKNOWLEDGEMENTS

This study was supported by a grant from the Swedish Medical Research Council (16x04723).

REFERENCES

1. J.R. Warren and B. Marshall, Unidentified curved bacilli on gastric epithelium in active chronic gastritis. Lancet i:1273-1275 (1983).

2. A. Lee S.L. and Hazell, *Campylobacter pylori* in health and disease: an ecological perspective. Microbial. Ecol. in Health and Dis., 1:1-16 (1988).

3. M.J. Blaser, *Helicobacter pylori* and the pathogenesis of gastroduodenal inflammation. J. Infect. Dis., 161:626-633 (1990).

4. G.E. Buck, *Campylobacter pylori* and gastroduodenal disease. Clin. Microbiol. Rev., 3:1-12 (1990).

5. C.P. Dooley and H. Cohen, The clinical significance of *Campylobacter pylori*. Ann. Intern. Med., 18:70-79 (1988).

6. A. Morris and G. Nicholson, Ingestion of *Campylobacter pylori* causes gastritis and raised fasting gastric pH. Am. J. Gastroenterol., 82:192-199 (1987).

7. B.J. Marshall, J.A. Armstrong, D.B. McGechie and R.J. Glancy, Attempt to fullfill Koch's postulates for pyloric *Campylobacter*. Med. J. Aust., 142:436-439 (1985).

8. S. Krakowska, D.R. Morgan, W.G. Kraft and R.D. Leunk, Establishment of the gastric pathogen *Campylobacter pylori* infection in neonatal piglets. Infect. Immun., 55:2789-2796 (1987).

9. H.L.T. Mobley and R.P. Hausinger, Microbial ureases: Significance, regulation and molecular characterization. Microbiol. Rev., 53:85-18 (1989).

10. B.J. Marshall, Urease protects *Helicobacter pylori* from the bactericidal effect of acid. Gastroenterology, 99:697-702 (1990).

11. B.J. Marshall and I. Surveyor, Carbon-14 urea breath test for the diagnosis of *Campylobacter pylori* associated with gastritis. J. Nucl. med. 29:11-16 (1988).

12. B.L. Slomiany, J. Bilsky, J. Sarosek, V.L.N. Murty, B. Dworkin, K. Van Horn, J. Zielenki and A. Slomiany, *Campylobacter pyloridis* degrades mucin and undermines gastric mucosal integrity. Biochem. Biophys. Res. Commun., 144:307-314 (1987).

13. Å. Carlsson, P. Aleljung, L. Emödy, Å. Ljungh and T. Wadström, Carbohydrate receptor specificity of haemagglutination of *Campylobacter pylori*. In: Gastroduodenal Patholoby and *Campylobacter pylori*. Excerpta Medica, Amsterdam (Eds. F. Mégraud and H. Lamouliatte), pp 375-378 (1989).

14. L. Emödy, Å. Carlsson and T. Wadström, Mannose-resistant haemagglutination by *Campylobacter pylori*. Scand. J. Infect. Dis., 20:353-354 (1988).

15. B.A. Hills, B.D. Butler and L.M. Lichtenberger, Gastric mucosal barrier: hydrophobic lining to the lumen of the stomach. Am. J. Physiol., 244:G561-G568 (1983).

16. R. Raedsch, A. Stiehl, S. Phol and J. Placky, Quantification of Phospholipase A_2-activity of *Campylobacter pylori*. Gastroenterology, 96:A405 (1989).

17. G. Bode, P. Marfertheiner and H. Ditschuneit, Pathogenetic implication of ultrastructural findings in *Campylobacter pylori* related gastroduodenal disease. Scand. J. Gastroenterol., Suppl. 142:25-39 (1988).

18. D.G. Evans, D.J. Evans Jr., J.J. Moulds and D.Y. Graham, N-Acetylneuraminyllactose-binding fibrillar haemagglutinin of *Campylobacter pylori*: a putative colonization factor antigen. Infect. Immun., 56:2896-2906 (1988).

19. J. Huang, C.J. Smyth, N.P. Kennedy, J.P. Arbuthnott and P.W.N. Keeling, Hemagglutinating activity of *Campylobacter pylori*. FEMS letters, 56:109-112 (1988).

20. T. Nakazawa, M. Ishibashi, H. Konishi, T. Takemoto, M. Shigeeda and T. Kochiyama, Hemagglutination activity of *Campylobacter pylori*. Infect. Immun., 57:989-991 (1989).

21. T. Wadström, J.L. Guruge, S. Wei, P. Aleljung and Å. Ljungh, *Helicobacter pylori* haemagglutinins - possible gut mucosa adhesins. In: *Helicobacter pylori* Gastritis and Peptic Ulcer. Springer Verlag (Eds. P. Malfertheiner and H. Ditschuneit), pp 96-103 (1990).

22. J.L. Fauchere and M.J. Blaser, Association of *Helicobacter pylori* with epithelial cells. In: *Helicobacter pylori* Gastritis and Peptic Ulcer. Springer Verlag (Eds. P. Malfertheiner and H. Ditschuneit), pp 110-117 (1990).

23. R.D. Leunk, M.A. Ferguson, D.R. Morgan, D.E. Low and A.E. Simor, Antibody to cytotoxin in infection by *Helicobacter pylori*. J. Clin. Microbiol., 28:1181-1184 (1990).

24. D.T. Smoot, H.L.T. Mobley, G.R. Chippendale, J.F. Lewison and J.H. Resau, *Helicobacter pylori* urease activity is toxic to human gastric epithelial cells. Infect. Immun., 58:1992-1997 (1990).

25. P. Malfertheiner and H. Ditschuneit, *Helicobacter pylori* gastritis and peptic ulcer. Springer Verlag, Berlin, New York (478 pages) (1990).

26. D.M. Jones and A. Curry, The genes of coccal forms of *H. pylori*. In: *Helicobacter pylori* Gastritis and Peptic Ulcer. Springer Verlag (Eds. P. Malfertheiner and H. Ditschuneit), pp 29-37 (1990).

27.   A. Smigielski, B. Wallace and K.C. Marshall, Genes responsible for size reduction of marine vibrios during starvation are located on the chromosome. Appl. Environ. Microbiol., 56:1645-1648 (1990).

28.   A. Labigne, V. Cussac and P. Courcoux, Development of a genetic and molecular approach for the diagnosis and study of the pathogenesis of *Helicobacter pylori*. In: *Helicobacter pylori* Gastritis and Peptic Ulcer. Springer Verlag (Eds. P. Malfertheiner and H. Ditschuneit), pp 19-22 (1990).

SUPERFICIAL COMPONENTS OF *HELICOBACTER PYLORI*

IN RELATION TO ADHERENCE TO EPITHELIAL CELLS

J.L. Fauchère and M. Boulot-Tolle

Faculté de Médecine Necker-Enfants Malades,
156 rue de Vaugirard, 75730 Paris Cédex 15 and
Hôpital Boucicaut, 78 rue de la Convention, 75015 Paris, France

INTRODUCTION

The presence of *Helicobacter pylori* in the human stomach has been highly associated with gastritis and peptic ulcer disease (1-4). Colonization of the mucus layer overlaying gastric tissue is a feature of *H. pylori* infection and is facilitated by the spiral shape and high motility of *H. pylori* (5). Adherence of *H. pylori* to gastric cells of infected patients has been directly observed, and may be important in the maintenance of infection (5-7). The attachment of *H. pylori* to gastric cells may involve bacterial components able to bind to cellular receptors. The cellular receptors may be restricted to the gastric cells which would explain the exclusive association of *H. pylori* with those cells, however the specificity of *H. pylori* for the gastric mucosa may also result from the particular microenvironment of the stomach. In the latter case, the cellular receptors binding *H. pylori* could be present on other human cells (8). It is thus of interest to study the colonization factors of *H. pylori* . Using the microscopic reference method (9,10) or microtiter adherence assays (11), all *H. pylori* strains studied have been found adherent to Hep2, INT407, Y1 or HeLa cells (8, 10). These observations support the hypothesis that the cellular receptor binding *H. pylori* is not restricted to gastric cells and can be studied using epithelial cell lines. Using methods that were developed for the study of colonization factors of *Escherichia coli*, Huang demonstrated hemagglutinating properties in *H. pylori* (12) and Evans described a fibrillar hemagglutinin binding on N-acetyl-neuraminyl-lactose, a receptor not restricted to the gastric cells (13). The N-acetyl-neuraminyl-lactose binding hemaglutinin is a putative adherence factor but other bacterial components may be involved in the fixation of *H. pylori* to gastric cells. Supporting this hypothesis, a glycerolipid unrelated to N-acetyl-neuraminyl-lactose has been found to be another adherence receptor for *H. pylori* (14). Hypothesizing that colonization factors may be surface exposed we extensively studied the composition and the biologic activities of the *H. pylori* superficial material.

MATERIALS AND METHODS

Strains and bacterial procedures

The *H. pylori* strains used in this work were isolated from human gastric biopsies. They were identified and stored as previously described (15). Bacterial cultures were grown on blood agar or Muller Hinton plus horse serum at 37°C under microaerobic conditions for 4 days. For viable counts, appropriate dilutions of the bacterial suspensions in 0.15M NaCl were spread on blood agar, and after incubation colony forming units (cfu) enumerated.

Extraction of bacterial surface exposed material

The surface exposed material was obtain as previously described (11,16). Briefly, bacterial cells of *H. pylori* cultivated for 4 days were suspended in sterile distilled water, bacterial suspensions were checked by viable count and 1 ml was saved and sonicated as total bacterial components (sonicat). The rest of the suspensions were vortex-mixed for 1 min, then centrifuged and the supernatant was designated as the water extract (WE). The glycin extract was obtain by the method of McCoy (17). The protein contained in each extract was determined by the bicinchoninic acid method. The same number of agar plates were incubated without bacteria and treated under the same conditions to obtain a control extract (CE) devoid of bacterial components.

Assessment of adherence of bacterial components on Hela cell membranes by ELISA

This microtiter assay for adherence is based on the immunological detection of bacterial material bound onto HeLa cell membranes as previously described (11). This method produced results well correlated with the microscopic reference method. Briefly, microtiter plates were coated with 10 µg of binding component (HeLa membranes for assay wells and BSA for control wells) and excess binding sites to the polystyrene were blocked with BSA. After washing wells, known quantities of the bacterial component were added, plates were incubated for 1 hr at 37°C and then washed three times. The binding material was then quantitated by an ELISA, as previously described (15), using rabbit anti-*H. pylori* serum. The specific adherence OD (ODs) was the difference between the OD of the assay well and the OD of the control well. For standardization, wells were directly coated with known quantities of the studied bacterial component, and the ELISA assay was performed as described above. The ratio of the ODs of a given well to the ODc of the calibration well directly coated with the same quantity of bacterial component was used to assess the proportion of adhering material and was expressed in micrograms of adhering bacterial component per microgram of original bacterial component.

Determination of urease activity of bacterial components

The urease activity was evaluated by measuring the quantity of ammonia released per min at 37°C in the presence of urea, according to an alcalimetric (18) or an enzymatic method (19).

Electrophoretic analysis of bacterial extracts

The proteic profiles were performed by SDS PAGE, with migration gels of 13% and stacking gels of 4% polyacrilamide. The antigenic profiles were realized by the Western blot procedure. Two dimensional gel electrophoresis (2DGE) was realized according to two different procedures: (1) isoelectro-focusing (IEF) in the first dimension and SDS PAGE in the second dimension and (2) non-denaturating gel electrophoresis (NDGE) in the first dimension and SDS PAGE in the second dimension. IEF was performed in gels containing 4% of ampholytes with a pH gradient between NaOH 20 mM and PO_4H_3 6 mM. NDGE was realized in 4% polyacrylamide gels lacking detergents or denaturating agents. The second dimensions were SDS PAGE in 10-20% polyacrylamide gradients.

Monospecific-rabbit anti-sera

Antigens were obtained by preparative SDS PAGE from 500 µg of crude water extracts. They were given subcutaneously to rabbits with Freund complete adjuvant. After two booster-shots, the rabbits were bled 6 weeks after the first injection. Titers and specificity of the sera were checked by Western blotting against SDS PAGE of water extracts.

Chromatographic separation of bacterial extracts

Bacterial extracts were submitted to gel exclusion chromatography on a Superose 12 column (Pharmacia), as previously described (11). The fractions were eluted in PBS pH 7.4 at a flow rate of 0.5 ml/min and collected every minute.

RESULTS

Proteic and antigenic profiles of the surface exposed material

The water and glycin extracts (WE, GE) from *H. pylori* were found to contain 20 and 6% respectively of the total protein present in the sonicated whole bacterial cells. Proteic and antigenic profiles were similar for the different strains studied and identical for both WE and GE with major components of approximately 60, 50, 40, 30, 25 and 15 kDa M/W.

Two dimensional gel electrophoresis (2DGE) IEF in the first dimension and SDS PAGE in the second dimension, was immuno-revealed with anti-whole bacterial cell or monospecific sera. The antigenic patterns looked more complex with several antigens exhibiting different pHi and similar molecular weights (M/W): (1) M/W = 66 kDa, pHi = 4-4.4 (2) M/W = 60 kDa, pHi = 4.1-4.3; (3) M/W = 50 kDa, pHi = 5.0-5.5; (4) M/W = 48 kDa, pHi = 3.9-4.6; (5) M/W = 42 kDa, pHi = 4.7-5.1, M/W = 31 kDa and pHi = 9.1, 8.8, 8.4 and 7.8.

Adherence and urease properties of the *H. pylori* superficial material

Assuming that the components mediating *H. pylori* adherence are surface-exposed, we studied the adherence of strains before and after treatments designed to extract superficial material as has been demonstrated previously (16). Bacterial cells treated with either water, 0.15M NaCl or glycine buffer retained less than 30% of the adherence material of untreated bacteria. The largest amount of superficial adhering material (SAM) was recovered in the first saline or water extracts (11). Sonicated whole bacterial cells and water extract were assessed for urease activity. Sonicated bacteria exhibited an activity of 0.2×10^{-2} μM NH$_3$/min/μg for the protein versus 0.15×10^{-2} μM NH3/mn/μg for the water extract. Thus the superficial material of *H.pylori* contains approximately 75% of the total urease activity and 70% of the bacterial adhering components.

Since N-acetyl-neuraminyl-lactose (NAN-lactose) has been found to be the cellular receptor of a *H. pylori* hemagglutinin (13), we sought to determine whether the SAM binds to a similar receptor. Using the standard adherence ELISA we compared the ability of various quantities (from 2 to 20 μg) of the water extract from strain 87-263 to bind to HeLa cell membranes, to calf fetal fetuin (rich in NAN-lactose), or to bovine serum albumin (used as a negative control binding component). SAM bound better to HeLa cell membranes (ODs from 0.3 to 1.0) than to BSA (ODs from 0.05 to 0.2) or fetuin (ODs from O.05 to 0.1). Next we attempted to inhibit the adherence of 10 μg of SAM to the HeLa cell membranes by preincubation of the extract with various concentrations (from 1 to 10,000 ng/ml) of either fetuin, BSA, or HeLa membranes. HeLa cell membranes strongly inhibited the binding of SAM at each concentration tested (from 35 to 90% of adherence inhibition) whereas preincubation with BSA or fetuin showed little inhibition (from 10 to 40% and from 0 to 20% inhibition, respectively). Finally, we assessed the effect of neuraminidase pretreatment on the ability of HeLa cell membranes to be a receptor for the *H. pylori* SAM. After the action of neuraminidase HeLa cell membrane preparation was centrifugated, the supernatant was checked to ensure that it contained all the NANA originally present and the pellet, considered to be NANA free, was used as binding component in a standard adherence ELISA. Under these conditions, the treated and untreated HeLa membranes were identical in their ability to be a receptor for binding of the *H. pylori* SAM. In total, these results indicate that SAM bind on a HeLa cell membrane receptor absent from fetuin and BSA and unrelated with NAN-lactose.

Identification of the SAM

Gel exclusion chromatographic separations were performed on the control extract and the water extracts from four *H. pylori* strains and from a *C. fetus* nonadherent control strain. The OD 280 nm recording of the eluates showed identical profiles for each of the *H. pylori* extracts and a different profile for the *C. fetus* extract. The control extracts showed only a small quantity of material in the late fractions. Chromatographic fractions of water extract from strain 87-263 were assayed for protein concentration, proportion of adhering material, and urease activity (Figure 1).

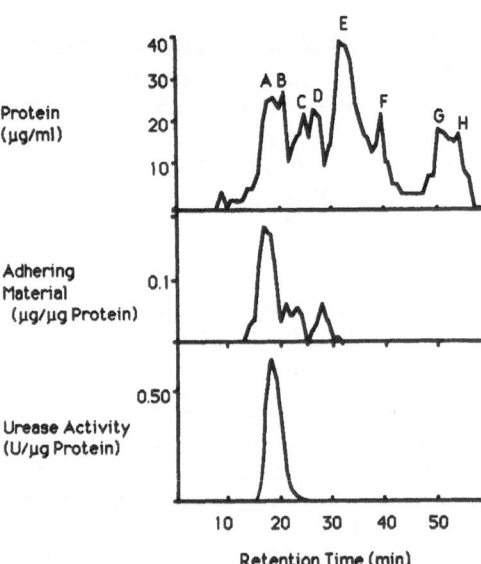

Figure 1. Protein concentration, adhering material and urease activity of the gel exclusion chromatographic fractions from water extract of *H. pylori* 86-263 strain. 200 μl of water extract were injected in a Superose 12 column and were eluted with a 0.15M PBS, pH 7.4 at a flow rate of 0.5 ml/min. Fractions of eluate were collected every minute and assessed for protein concentration (bicinchoninic acid method), proportion of adhering material (ELISA) and urease activity (NADH- dependent coupled enzyme assay). The major peaks of protein concentration were labelled from A to H.

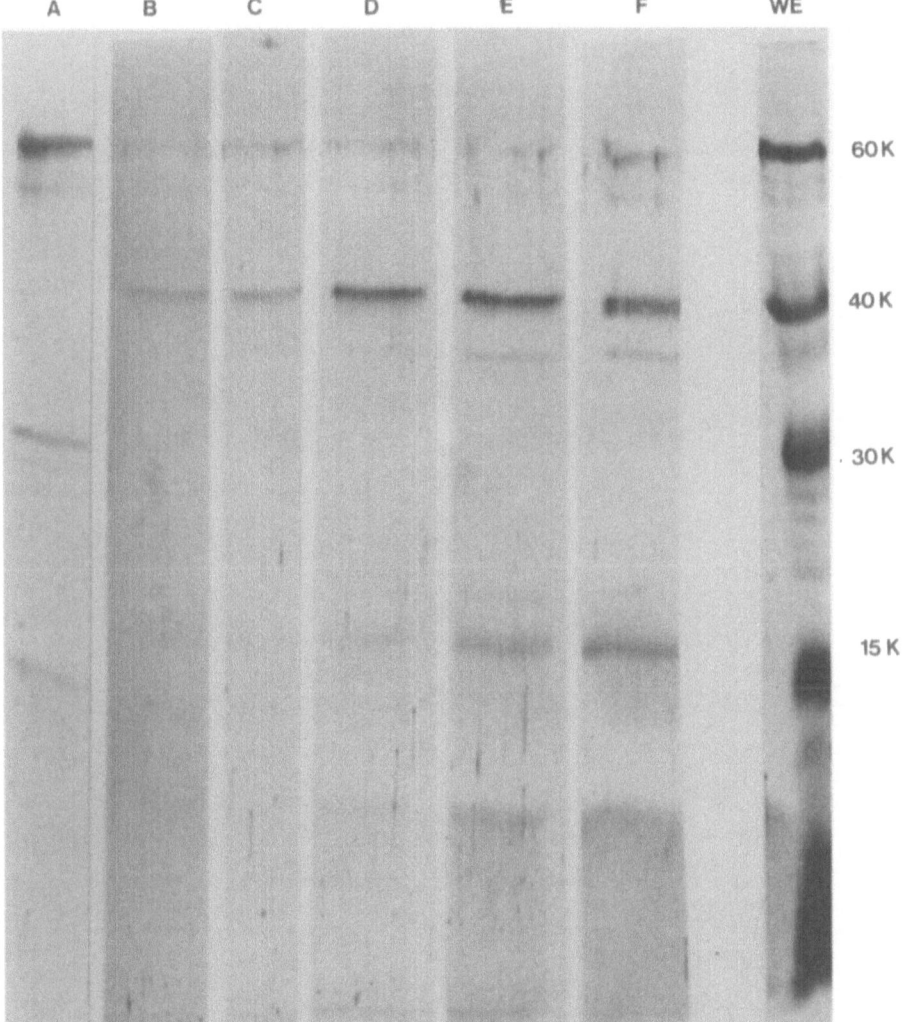

Figure 2 . Western blot of the major peak chromatographic fractions (see Figure 1) of water extract from *H. pylori* 86-263 with antiserum to whole *H. pylori* cells. After chromatographic separation of water extract on a Superose 12 column, the fractions corresponding to the major peaks of protein concentration underwent SDS-PAGE (12% acrylamide) and were immunoblotted with an anti-whole cell serum. Fraction 18 representative of peak A shows a major antigen at 60 kDa and minor bands at 30 and 15 kDa. Fraction 22 representative of peak B and fraction 26 representative of peak C show only one band at 40 kDa. Fraction 28 (peak D) exhibits a major antigen at 40 kDa and minor bands at 15 kDa and < 15 kDa. Fraction 33 (peak E) and 41 (peak F) include antigens of 40, 15 and < 15 kDa. Fractions 50 and 60 representative of the two last peaks (G and H) did not contain any detectable *H. pylori* antigen.

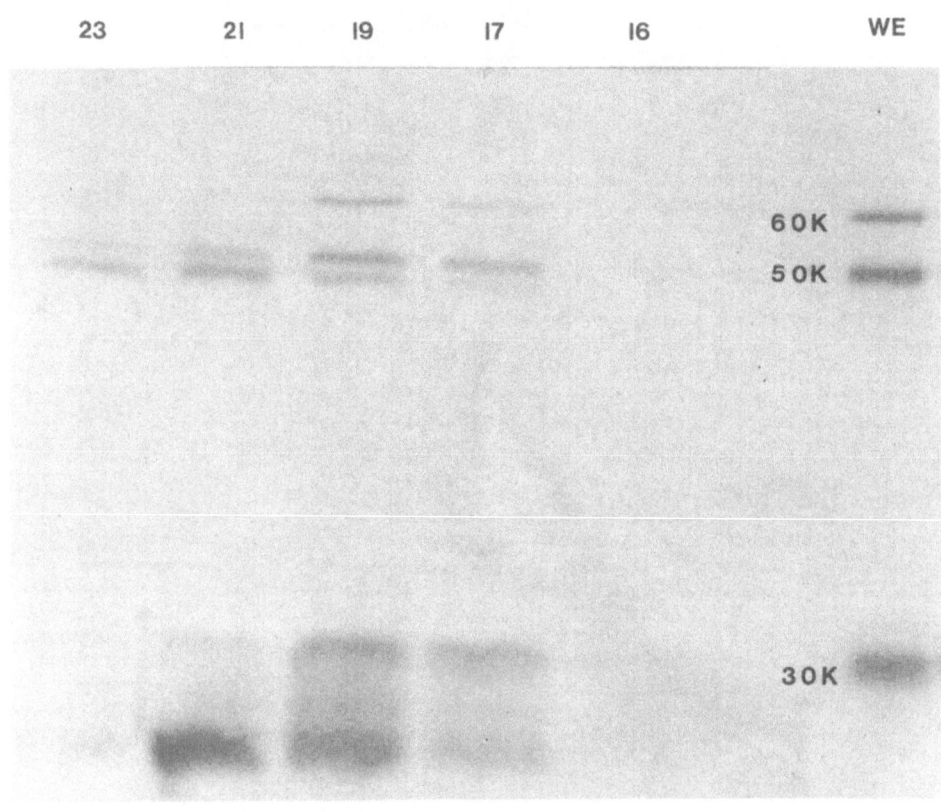

Figure 3 . Western blot of the adhering chromatographic fractions revealed with an anti-urease serum.

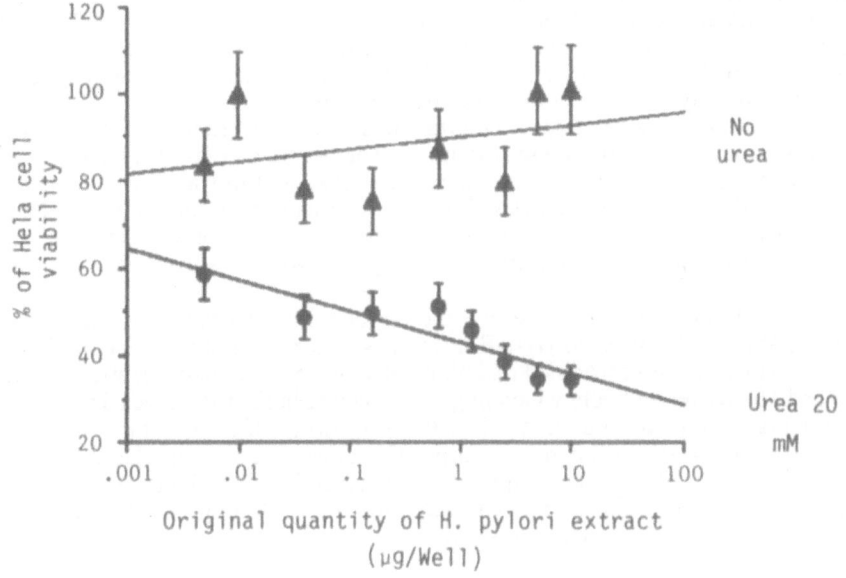

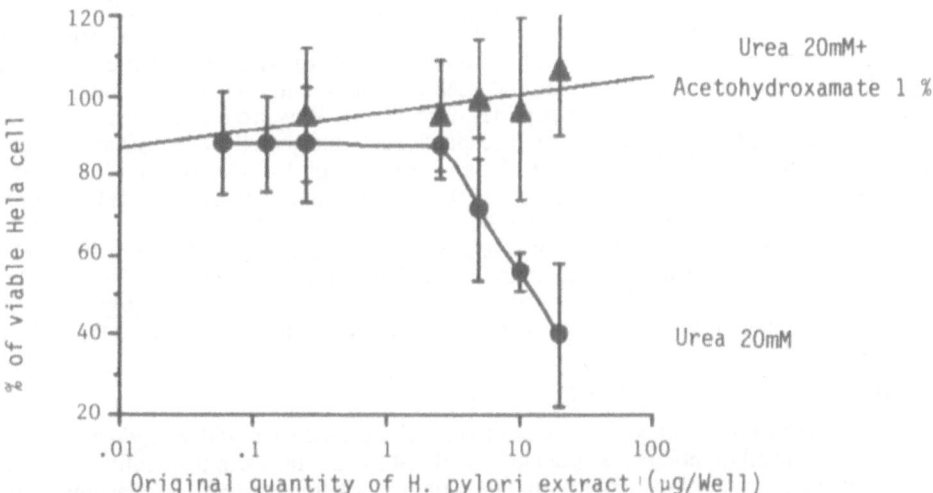

Figure 4 . Cytotoxic activity of superfical material from *H. pylori,* fixed on viable HeLa cells. HeLa cells were coated with SAM, washed and incubated for 12 hrs with urea. The killed cells were eliminated by washing, and the remaining viable cells were assessed colorimetrically. The results showed an urea dependent cytotoxicity related to the quantity of SAM coating the cells. This cytotoxicity was inhibited by hydroxamic acid, a well known urease inhibitor

There were eight major protein peaks (A to H) whereas the control extract (without bacteria) exhibited only one small peak (from fractions 50 to 58). However, adherence activity was mainly found in fractions 17 to 20 and urease activity was concentrated in fractions 17 to 22. Water extract and fractions representative of each peak were then submitted to SDS PAGE and immunobloted with an anti-*H. pylori* serum. Immunoblot of water extract and representative fractions are shown in Figure 2, together with the antigenic profile of water extract. Fraction 18, representative of peak A, shows a major antigen of 60 kDa and minor bands at about 52, 30,and 15 kDa. There bands were not (or only minimally) present in the other fractions from the non-adherent peaks. For further identification, the water extract and fractions 16 to 23 were immunoblotted using specific antiserum to *H.pylori* urease (Figure 3). The 60, 52 and 30 kDa bands, usually associated with urease, were present in the water extract and in fractions 17 to 19 (the most adherent). The 60 kDa antigen was absent in the fractions 21 and 23 whereas the 52 kDa antigen was present in fractions 17 to 23, and no antigen was seen in fraction 16. These results suggest that adhering components co-purify with urease subunits.

In an attempt to better describe the surface exposed bacterial material, we analyzed the WE by 2DGE with NDGE in the first dimension and SDS PAGE in the second dimension. The silver-stained resulting gel showed three major peptidic subunits of 60, 30 and 15 kDa respectively which migrated together in the first dimension and were distinguished only in the denaturating gel (data not shown). This suggest that these three subunits are complexed in their native form. The other fractions of the WE seemed non-complexed. Thus the surface-exposed material of *H.pylori* includes, as a major component, a high-molecular-weight-proteic complex including 60, 30 and 15 kDa subunits in addition to other minor constituents.

In order to identify among the adhering subcomponents the one which is the bacterial ligand binding on the host cell receptor, we semi-purified the three peptidic fractions involved in adherence by preparative electrophoresis in the presence of SDS followed by electroelution and electrodialysis. The obtained peptides exhibited single bands at 60, 30, and 15 kDa by Western blot against an anti-whole bacterial cell serum. Any of these three semi-purified subunits were adherent to HeLa cell membranes. This could be due to the denaturating conditions used in their preparation. We then raised rabbit anti-sera against each of these three peptides. Checked by Western blot against a 2DGE of the water extract, these sera recognized respectively : antigens between 50 and 66 kDa (serum A), antigens of 30 kDa (serum B) and antigens of 15 kDa (serum C). We used these sera in the adherence ELISA to assess the proportion of a given subunit binding on HeLa cell membranes. We assessed both the original quantities and the adhering fractions of each subunit with these monospecific sera or anti-whole cell serum. Of the 50-66 kDa antigens present in the original material 60% ($\pm$ 5 %) were fixed on HeLa cell membranes versus 12% ($\pm$ 2%) of the 30 kDa antigens and < 10% for the 15 kDa antigens. These results suggest that 50-66 kDa antigens include the bacterial ligand binding on the cellular receptor.

Fate of host cells coated with *H.pylori* SAM

We then determined the fate of viable HeLa cells coated with *H. pylori* SAM. A standard HeLa cell monolayer was coated for 1hr at 37°C with various quantities of SAM in a 96-well microtiter plate. After elimination of non-adherent material by washing, HeLa cells were incubated for 12 hr at 37°C in the presence of urea. Cell monolayers were then washed to remove the killed detached cells and the viable cells still adherent on the polystyrene were assessed by Giemsa staining, elution of the stain in 0.1% SDS and measurement of the OD 620 of the eluate. A HeLa cell monolayer treated in the same conditions with MEM replacing SAM and urea represented the 100% cell viability control. In the absence of urea less than 10% of the cells were killed whatever the quantity of SAM coating the cells. In the presence of urea the percentage of killed cells increased with the quantity of SAM coating the cells . This effect was inhibited by aceto-hydroxamic acid (1%), a well known urease inhibitor (Figure 4). Thus SAM seems to exhibit a dose-related cytotoxicity related to urease activity. This toxicity could be due to released ammonia. Supporting this hypothesis we demonstrated in a similar experiment a dose-related cytotoxic effect of ammonium sulfate to HeLa cells (100% of the HeLa cells were killed in the presence of 0.1 nM/cell of ammonium sulfate).

CONCLUSIONS

Our data indicate that *H. pylori* exhibit an abundant surface-exposed material, easily extracted and containing both urease and adherence activities. These two activities should be borne by components of similar molecular weights included in a high-molecular-weight-proteic complex comprising three subunits of 50-66, 30 and 15 kDa, respectively. The bacterial ligand which bind to the cellular receptor is associated with the antigens of 50-66 kDa M/W. The fixation of SAM to the target cell lead to cytotoxic effects dependent on the presence of urea.

It can be hypothesized that the adhering component may be the urease itself acting as an adhesin while it is attached to the bacterial cell, and adhering directly when exported from the cell. Urease fixed to gastric cells may be ideally located to efficiently hydrolyze urea extracted from the bloodstream. The urease activity concentrated at the membrane level of the host cell may participate in cell damage (5). Both hemagglutinins and SAM may be relevant in *H. pylori* colonization of the human gastric mucosa. In favour of this hypothesis is the finding that 75% of colonized patients develop an antibody response against the hemagglutinin and 98% have antibodies against urease (20), proving the close association of these two components with the colonized mucosa.

REFERENCES

1. **Marshall B.J., D.B. McGechie, P.A. Rogers and R.J. Glancy.** Pyloric *Campylobacter* infection and gastroduodenal disease. Med. J. Aust., 1985, **142**: 439-444.

2. **McNulty C.A.M. and D.M. Watson.** Spiral bacteria of the gastric antrum. Lancet, 1984, **1**: 1068-1069.

3. **Rathbone B.J., J.I. Wyatt and R.V. Heatley.** *Campylobacter pylori* -a new factor in peptic ulcer disease. Gut, 1986, **17**: 635-641.

4. **Warren J.R. and B.J. Marshall.** Unidentified curved bacilli on the gastric epithelium in active chronic gastritis. Lancet, 1983, **1**: 1273-1275.

5. **Hazell S.L., A. Lee, L. Brady and W. Hennessy.** *Campylobacter pyloridis* and gastritis: association with intercellular spaces and adaptation to an environment of mucus as important factors in colonization of the gastric epithelium. J. Infect. Dis., 1986, **153**: 658-663.

6. **Chen X.G., P. Correa, J. Offerhaus, E. Rodriguez, F. Janney** *et al.* Ultrastructure of the gastric mucosa harboring *Campylobacter* -like organisms. Am. J. Clin. Pathol., 1986, **86**: 575-582.

7. **Van Spreeuwel J.P., G.C. Duursma, C.J.Meijer, R. Bax, P.C. Rosekrans,** *et al. Campylobacter* colitis: histological, immunohistochemical and ultrastructural findings. Gut, 1985, **26**: 945-951.

8. **Evans D.G., D.J. Evans, Jr., D.Y. Graham.** Receptor-mediated adherence of *Campylobacter pylori* to mouse Y-1 adrenal cell monolayers. Infect. Immun., 1989, **57**: 2272-2278.

9. **Fauchère J.L., A. Rosenau, M. Véron, E.N. Moyen, S. Richard and A. Pfister.** Association with HeLa cells of *Campylobacter jejuni* and *Campylobacter coli* isolated from human feces. Infect. Immun., 1986, **54**: 283-287.

10. **Neman-Simha V and F. Megraud.** *In vitro* model for *Campylobacter pylori* adherence properties. Infect. Immun., 1988, **56**: 3329-3333.

11. **Fauchere J.L. and M.J. Blaser;** Adherence of *Helicobacter pylori* and its surface components to HeLa cell membranes. Microbial Pathogenesis, 1990, **9**: 427-439.

12. **Huang J, C.J. Smyth, N.P. Kennedy, J.P. Arbuthnott and P.W. Napoleon Keeling.** Haemagglutinating activity of *Campylobacter pylori* . FEMS Microbiology Letters, 1988, **56**: 109-112.

13. **Evans D.G., J.E. Doyle, Jr., J.J. Moulds and D.Y. Graham.** N-acetyl-neuraminyl-lactose-binding fibrillar hemagglutinin of *Campylobacter pylori* : a putative colonization factor antigen. Infect. Immun., 1988, **56**: 2896-2906.

14. **Lingwood C.A., A. Pellizzari, H. Law, P. Sherman and B. Drumm.** Gastric glycerolipid as a receptor for *Campylobacter pylori* . Lancet, 1989, **1**: 238-240.

15. **Perez-Perez G.I., B.M. Dworkin, J.A. Chodos and M.J. Blaser.** *Campylobacter pylori* antibodies in human. Ann. Intern. Med., 1988, **109**: 11-12.

16. **Pei, Z.H., P.T. Ellison, R.V. Lewis and M.J. Blaser.** Purification and characterization of a family of high molecular weight surface-array proteins from *Campylobacter fetus* . J. Biol. Chem., 1988, **263**: 6416-6420.

17. **McCoy E.C., D. Doyle, K. Burda, L.B. Corbeil and A.J. Winter.** Superficial antigens of *Campylobacter* (Vibrio) *fetus* : characterization of an antiphagocytic component. Infect. Immun., 1975, **11**: 517-525.

18. **Mobley H.L.T., M.J. Cortesia, L.E. Rosenthal and B.D. Jones.** Characterization of urease from *Campylobacter pylori* . J. Clin. Microbiol., 1988, **26**: 831-836.

19. **Dunn B.E., Campbell G.P., Perez-Perez G.I., Blaser M.J.** Purification and characterization of *Helicobacter pylori* urease. J. Biol. Chem., 1990, **265**: 9464-9469.

20. **Evans D.J., E.G. Evans, K.E. Smith and D.Y. Graham.** Serum antibody responses to the N-acetyl-neuraminyl-lactose-binding hemagglutinin of *Campylobacter pylori*. Infect. Immun., 1989, **57**: 664-667.

DETECTION OF VIRULENCE DETERMINANTS IN ENTERIC *ESCHERICHIA COLI*

USING NUCLEIC ACID PROBES AND POLYMERASE CHAIN REACTION

Ørjan Olsvik, Erik Hornes, Yngvild Wasteson, and Arve Lund

Department of Microbiology and Immunology
Norwegian College of Veterinary Medicine
Post Box 8146 DEP, 0033 Oslo 1, Norway

SUMMARY

As certain strains of *Escherichia coli* are a major cause of diarrhea both in man and animals, diagnosis of the specific etiological agents must be carried out beyond the species level. Several different pathogroups have been identified and for some of them, also the specific virulence determinants. Traditional microbiological assays have to a certain degree been supplemented or replaced with nucleic acid-based methods, like hybridization assays using cloned fragments or oligonucleotides as probes. Recently, the polymerase chain reaction (PCR) has been shown to be a suitable tool for differentiation between *E. coli* strains belonging to the normal enteric flora, and those carrying specific virulence determinants. Probe assays have been established in some diagnostic laboratories and found to be reliable for detection of the different enterotoxins of both the heat-stable-(ST) and the heat-labile-(LT) -families, and the shiga-like cytotoxins (SLTs). Different setups of the PCR principle have also been used to identify and characterize such genes. Genes encoding some of the important adhesion fimbria, characterizing such toxin-producing strains, have been targets for some assays. Several of the genes involved in the invasion process of the enteroinvasive *E. coli* are routinely identified using probe assays, avoiding the use of cell-cultures or laboratory animals. The enteropathogenic *E. coli* strains have for years been defined as a serological cluster, and different probes for both plasmidial and chromosomal genes have now confirmed the presence of common virulence properties. Probes and PCR are becoming very important diagnostic tools in identification and characterization of virulence- or virulence-associated genes in *E. coli*.

BACKGROUND

The different groups of diarrheogenic *Escherichia coli* can be classified into different groups by the presence of certain pathogenicity or virulence factors. The enterotoxigenic, the enteroinvasive, the enteropathogenic and the enterohaemorrhagic strains are among the major causes of *E. coli*-induced diarrhea. Several assays have been developed for identification and characterization of such strains, and nucleic acid hybridization tests have proved to be interesting tools for diagnostic purposes. The first probes were polynucleotide, and table 1 lists some of the most commonly used. However, as several of the genes were sequenced, the use of oligonucleotide probes to demonstrate specific nucleic acid sequences by hybridization has become increasingly important in the characterization and identification of microbial agents or specific pathogenicity factors like toxins. Both entero- and cytotoxin genes in strains of *E. coli* have been identified using oligonucleotides as probes (table 2). The use of radioactive labels for detection and the need for fixation of the nucleic acids on solid membranes before hybridization has, to a certain degree, limited the practical use of this technique. Enzyme-labelled oligonucleotides have been developed for detection of both heat-labile (LT) and heat-stable(ST) enterotoxins from *E. coli* (5), however these probes have relatively low sensitivity and require

Table 1. Polynucleotide probes for diarrhoegenic *Escherichia coli*

Pathogroup	Virulence factor	Polynucleotide fragment
Enterotoxigenic		
	LTI	850-bp *Hinc*II fragment of pEWD 299
	LT II	800-bp *Hind*II-*Pst*I fragment of pCP 2725
	ST Ia	154-bp *Hinf* I fragment of pRIT 10036
		157-bp *Eco*RI-*Bam*HI fragment from pDAS 101
	ST Ib	215-bp *Hpa*II-*Eco*RI fragment from pSLM 004
		215-bp *Bam*HI-*Pst*I fragment from pDAS 100
	STII	460-bp *Hinf*I fragment from pCHL 6
Enteroinvasive		
	Inv	17-kb *Eco*RI fragment of pRM 17
	Inv	1.5-kb *Hind*III fragment of pST 55
	Inv	1750-bp *Eco*RI fragment of pW 22
Enteropathogenic		
	LA (EAF)	1-kb *Bam*HI-*Sal*I fragment from pMAR 22
	DA	450-bp *Pst* I fragment from pW22
Enterohaemorrhagic		
	"EHEC"	3.5-kb *Hind*II fragment of PCVD 419
	SLT I	1145-bp *Bam*HI fragment of pJN 37-19
	SLT II	842-bp *Pst*I-*Sma*I fragment of pNN 110-8
	VT 1	750-bp*Hinc*II fragment of p746
	VT 2	850-pb*Ava*I-*Pst*I fragment of p363

Modified from (1).

a large number (10^4-10^5) of organisms as a source of nucleic acids (2). Amplification of specific nucleic acid sequences as much as 10^6 times using the Polymerase Chain Reaction (PCR) (4, 6, 7, 8) has improved the sensitivity of hybridization methods dramatically. PCR has made it theoretically possible to detect specific nucleic acid sequences from a single microbial organism in a mixture of nucleic acids of different origin without isolating the specific organism . This technique has been utilized to demonstrate the presence of genes encoding both enterotoxins (4, 6), and recently cytotoxins from *E. coli* strains (figure 1).

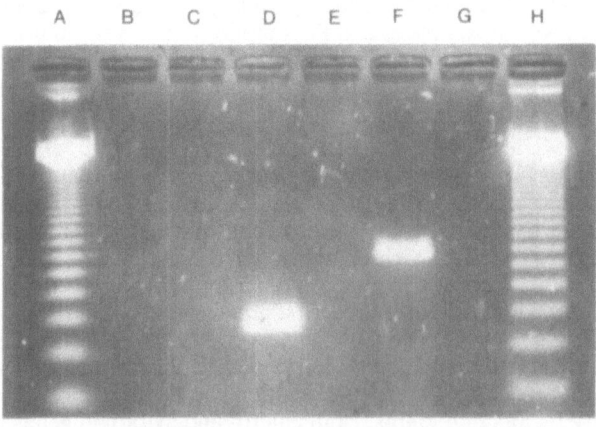

Fig.1.The PCR-generated DNA fragments after using crude boiled bacteria as template DNA, and oligonucleotide primers for SLT I generating a 475 bp fragment or SLT II generating a 863 bp fragment. Lanes A and H contain the 123 base pair ladder. Lane D shows a strain positive for SLT I, and lane F one positive for SLT II. Other lanes were controls. (Olsvik et al, under publ.).

The large number of PCR-produced specific DNA segments are normally visualized by gel electrophoresis followed by ethidium bromide staining. Recently we presented a diagnostic procedure for identification of Shiga-like-toxin (SLT) genes I and II from *E. coli* based on magnetic separation and quantitation of specific double stranded DNA fragments produced by a two-step triple primer PCR (8, Olsvik et al.under publ.). The test principle has been designated DIANA for **D**etection of **I**mmobilized **A**mplified **N**ucleic **A**cid by Wahlberg and co-workers (J. Wahlberg, J. Lundberg, T. Hultman, M. Holmberg and M. Uhlén: Rapid detection and sequencing of specific in vitro DNA sequences using solid phase methods. Mol. Cell. Probes, under publ.). This principle has also been developed to a non-radioactive signal system by introducing parts of the lac operon (*lac*O) in one primer, and using binding of lac inhibitor protein (*lac*I) to the gene. This protein is fused with an enzyme and gives a substrate induced signal.

The principle of DIANA

Primers and labelling of oligonucleotides. A set of 3-4 primers were synthesized using an automatic DNA synthesizer (Applied Biosystems, San Francisco, Calif.) with the automated phosphoramitidite coupling method. Primer 1 was also synthesized with free amino group at the 5´ end to which biotin was attached using a kit from Pharmacia (Uppsala, Sweden).

Table 2.   Oligonucleotides for diarrheogenic *Escherichia coli*

Pathogroup	Virulence factor	Sequence
Enterotoxigenic		
	LTI	5´A CGT TCC GGA GGT CTT ATG CCC AGA GGG CAT ATT
		5´C ACC TCT AAG TAG TTG TTG TTA ATG T
		5´GCG AGA GCA ACA CAA ACC GG
		5´CCC CAG TCT ATT ACA GAA CTA
		5´CGC AAT TGA ATT GGG GGT TTT
		5´CCC GGG CAG TCA ACA TAT AGA
		5´GAG CAC AAT ATA TCT CAG G
	ST IA	5´GAA CTT TGT AAT CCT GCC TGT GCT GGA TGT
		5´CAA CAG TGA AAA AAA ATC AGA AAA T
		5´TTA ATA ACA TCC AGC ACA GGC AGG
		5´GCT GTG AAC TTT GTT GTA ATC C
	ST IB	5´GAA TTG TGT AAT CCT GCT TGT ACC GGG TGC
		5´A AGT AAT AAA AGT GGT CCT GAA AGC
		5´GCT GTG AAT TGT GTT GTA ATC C
		5´GTA GAG TCT TCA AAA GAA AAA ATC AC
Enteroinvasive		
	Inv	5´CCA TCT ATT GAG ATA CCT GTG
Enteropathogenic		
	EAF	5´TAT GGG GAC CAT GTA TTA TCA
Enterohaemorrhagic		
	SLT I	5´CAG TTA ATG TGG TGG CGA AG
		5´CAC AGA CTG CGT CAG TGA GG
		5´CT GCT AAT AGT TCT GCG CAT C
	SLT II	5´CT TCG GTA TCC TAT TCC CGG
		5´GGA TGC ATC TCT GGT CAT TG
		5´CGC TGC AGC TGT ATT ACT TCC

Unpublished results from Department of Microbiology and Immunology, Norwegian College of Veterinary Medicine, and ref. no. 1 .

Hybridization oligonucleotides as probes. Oligonucleotides are normally 5´end labelled with $\gamma^{32}P$ ATP . One μl oligonucleotide (50 ng) was mixed with 38 μl double distilled sterile water, 5 μl of 10x kinase buffer, 1 μl (10 units) T4 polynucleotide kinase (Bethesda Research Laboratories Inc., Gaithersburg, Md.) and 5 μl (100 μg, 50 mCuri) $\gamma^{32}P$-labelled ATP (Amersham, London, England), and incubated at 37ºC for 30 min. Free nucleotide were separated from the oligonucleotide using a 2 ml Sephadex G50 (Pharmacia) spin column. Colony blots of *E. coli* strains were made on Gene Screen membranes (DuPont, NEN Research Products, Boston, Mass.). The blots were placed on filter paper soaked with 100 μg per ml proteinase K (Sigma, St. Louis, Mo) and incubated for 30 min at 37ºC before DNA was fixed by exposure to UV-light for 5 min. The blots were prehybridized in sealed bags at 45ºC in 0.1 ml 2x SSC with 5 x Denhart´s solution per cm^2. One ng labelled probe (approx. 0.7 mCuri) was then added to each ml of prehybridization solution, and hybridization carried out at 45 ºC overnight. The blots were washed at Tm minus 6 ºC , in 2x SSC for 30 min and then in 2x SSC with 0.1 % SDS (same temperature) for 30 min. They were subsequently rinsed for 2 min in distilled water at room temperature, air-dried, and placed in film cassettes with intensifying screens and x-ray films (Kodak, Rochester, N.Y.) at -70ºC for one to two days.

Preparation of template DNA for polymerase chain reaction. Bacteria from blood agar plates were suspended in sterile saline and used to prepare template DNA. Samples containing 10^1 to 10^6 colony forming units (CFU) per ml. were boiled for 20 min, digested at 37 ºC for 20 min with proteinase K (Sigma) at a final concentration of one μg per ml and re-boiled for 20 min before being used.

Amplification of target DNA. The first PCR was carried out in a total volume of 50 μl. The reaction mixture contained 10 mM Tris- HCl pH 8.3; 50 mM KCl; 1.5 mM MgCl$_2$; 0.01% gelatin (w/v); 200 μM of each dNTP; 0.2 μM each of unlabeled primers 1 and 2 or unlabeled primers 1 and 3; variable amounts of template DNA in a volume of 5 μl, and 1.25 U (0.25 μl) thermostable DNA polymerase (Amplitaq, Perkin-Elmer Cetus, Norwalk, CT.). The mixture was overlaid with a drop of mineral oil and incubated for 25 cycles in a DNA Thermal cycler (Perkin-Elmer Cetus). Each cycle consisted of 94 °C for 1 min, 50 °C for 2 min and 72 °C for 1 min. The pre- and post-PCR operations were carried out in separate rooms. In the second PCR, primer 1 was labelled with biotin and primer 2 with $\gamma^{32}P$ ATP or with the lac operon sequences (example from a PCR for ST, the first 23 nucleic acids from 5´ are from the *lac* operon gene: 5´AAT TGT TAT CCG CTC ACA ATT GAT TAC AAC AAA CTT CAC AGC AGT 3´) as , and a five μl aliquot of a 1/100-dilution of the products from the first PCR reaction was used as template DNA, the reaction was carried out for 5 to 20 cycles at conditions identical to those described for the first PCR (8).

Magnetic separation of PCR products. The biotin- and ^{32}P-labelled DNA fragments produced in the last PCR were separated from the solution using streptavidin coated magnetic beads and a magnet (MPC -E, Dynal, Oslo, Norway). Streptavidin coated super-paramagnetic polystyrene beads (Dynabeads M-280 Streptavidin, Dynal) (7,8,9) (see figure 1) were washed once for 2 min in 1M NaCl. One hundred μg (10 μl) of beads were then added to 25 μl of the PCR solution and incubated at room temperature for 20 min with gentle agitation every 5 min. The beads were subsequently washed twice in 500 μl 1x SSC containing 0.1 % SDS and then suspended in scintillation fluid (Opti-Fluor O, Packard Inc, Downers Grove, Ill.) for quantitation of bound radioactivity (cpm) in a scintillation counter (1900 CA, Packard Inc.). All these procedures were carried out at room temperature. In the negative control for the streptavidin-biotin reaction, the biotinylated primer 3 was replaced with a non-biotinylated one. The procedure is illustrated in Fig. 2. Using the *lac* operon in primer 2, the fusion protein *lac*I-glucose oxidase was added to the beads (1ug/ml) in PBS, incubated for 10 min, followed by magnetic separation and two washing steps. The substrate ONPG was finally added and the reaction visually and optically recorded after 10 min at room temperature.

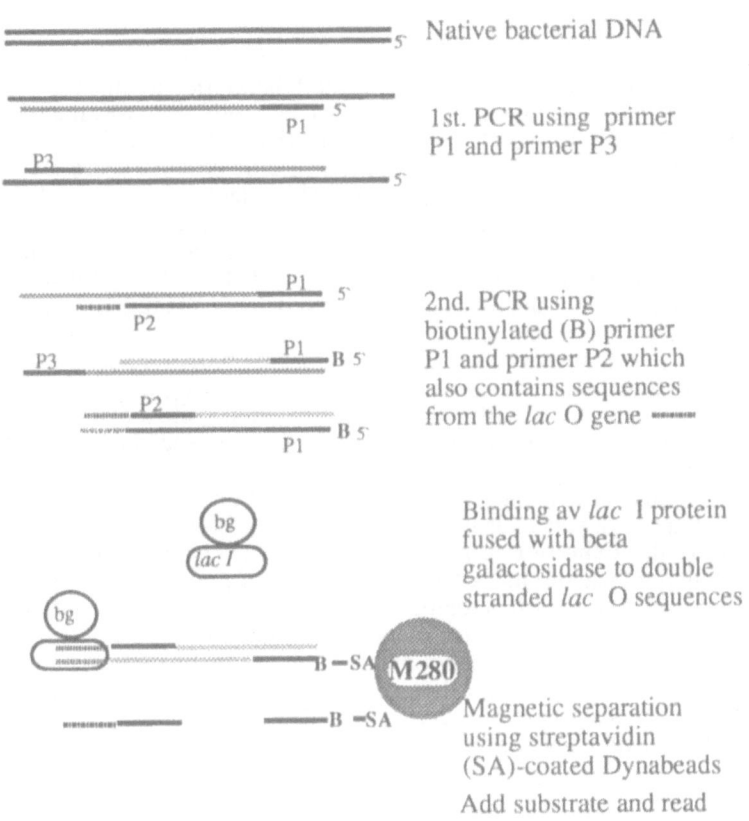

Native bacterial DNA

1st. PCR using primer
P1 and primer P3

2nd. PCR using
biotinylated (B) primer
P1 and primer P2 which
also contains sequences
from the *lac* O gene ▬▬▬

Binding av *lac* I protein
fused with beta
galactosidase to double
stranded *lac* O sequences

Magnetic separation
using streptavidin
(SA)-coated Dynabeads

Add substrate and read

Fig. 2. The principles of the two-step, triple primer PCR. The magnetically isolated, PCR-generated fragments have *lac* operon gene sequenses incorporated. The *lac* inhibitor protein fused with the enzyme bindt to the gene. and the OD after the enzyme reaction reflects the amount of specific double stranded DNA fragments generated in the last PCR step (see also ref. no. 8).

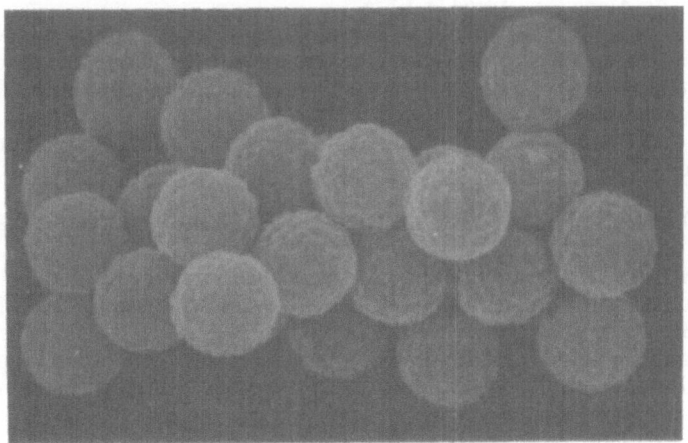

Fig. 3. The super-paramagnetic, streptavidin coated Dynabeads, diameter 2.8 μm, used for binding and separation of double stranded DNA generated fragments.

CONCLUSIONS

Nucleic acid technology entered the diagnostic laboratories by utilization of plasmid profiling for epidemiological purposes (7). For most laboratories this was their first hand on experience with DNA and RNA. However, during the last years, the use of different hybridization techniques for detection of either species specific genes or genes encoding certain virulence factors, have been introduced (1,2,3,4,6). The enterotoxins of *E. coli* were among the first to be exploited, and several of the last years progress in this field has been demonstrated using diarrheogenic *E. coli* strains. The oligonucleotides have several advantages compared to polynucleotide fragments, and through introduction of not so expensive DNA synthesizers, such probes are produced in large quantities. Still, however, the different sequences had to be tested for their suitability in diagnostic work (5). The families of heat-labile-, heat-stable- and shiga-like-toxins are making probes suitable for detection of more than one toxin a difficult task. So has the problem of using radioactive labels. Most laboratories after having tested the different commercial, non-radioactive probes have some questions regarding sensitivity, but recent progress appears to indicate that such might be valuable tools in diagnostic work on enteric *E. coli* strains.

The last years progress utilizing polymerase chain reaction as a diagnostic principle have indicated that there are great potentials in this technology. We have exploited some variants of the PCR for some of the toxins of *E. coli*, previously presented here. The double PCR system reduce the possibility of false results, because the inner set of primers will only function if the outer pair has been used in amplification of the correct sequence, and therefore serves as a control of the first primer set. The number of cycles required in the second PCR is decreased because the number of templates increased highly from the first to the second PCR (7,8,9). However, the assay worked well and the costs were reduced through a combination of microtiter-plate based PCR-apperature, process controlled magnets and application of reagents and washing solutions, a digital readout may largely make this an automatic test procedure.

REFERENCES

1. Escheverria, P., J. Seriwatana, O. Sethabutr, and A. Chatkaeomorakot. Detection of diarrheogenic *Escherichia coli* using nucleotide probes.pp.95-142. In A. J. L. Macario and E. C. deMacario (ed): Gene probes for bacteria Academic Press, New York, 1990.
2. Jablonski, E., E. W. Moomaw, R. H. Tullis, and J. L. Ruth. 1986. Preparation of oligodeoxynucleotide-alkaline phosphatase conjugates and their use as hybridization probes. Nucleic Acids Res. **14**:6115-6128.
3. Jackson, M. P., R. J. Neill, A. D. O'Brien, R. K. Holmes, and J. W. Newland. 1987. Nucleotide sequence analysis and comparison of the structural genes for Shiga-like toxin I and Shiga-like toxin II encoded by bacteriophages from *Escherichia coli* 933. FEMS Microbiol. Lett. **44**:109-114.
4. Karch, H., and T. Meyer. 1989. Single primer pair for amplifying segments of distinct Shiga-like-toxin genes by polymerase chain reaction. J. Clin. Microbiol. **27**:2751-2757.
5. Lo, Y-M. D., W. Z. Mehal and K. A. Fleming. 1988. False-positive results and the polymerase chain reaction. Lancet **ii**:679.
6. Olive, D. M. 1989. Detection of enterotoxigenic *Escherichia coli* after polymerase chain reaction amplification with a thermostable DNA polymerase. J. Clin. Microbiol. **27**:261-265.
7. Olsvik, Ø, and G. Bukholm (ed) Application of molecular biology in diagnosis of infectious diaseases pp.1-134, Norwegian College of Veterinary medicine, Oslo 1990.
8. Rimstad, E., E. Hornes, Ø. Olsvik, and B. Hyllseth. Identification of a double-stranded RMA virus by using polymerase chain reaction and magnetic separation of the synthesized DNA fragments. J. Clin. Microbiol. **28**: 2275-2278, 1990.
9. Uhlen, M. 1989. Magnetic separation of DNA. Nature (Lond.) **340**:733-734.

DNA PROBES AND PCR ANALYSIS IN THE DETECTION OF *CLOSTRIDIUM DIFFICILE* AND *HELICOBACTER PYLORI*

Brendan W. Wren, Christopher L. Clayton and Soad Tabaqchali

Department of Medical Microbiology
St. Bartholomew's Hospital Medical College
West Smithfield, London EC1A 7BE, UK

INTRODUCTION

We have undertaken the molecular characterisation of a variety of species-specific antigens and potential virulence determinants from the enteric pathogens *Clostridium difficile* and *Helicobacter pylori*. The cloning and sequencing of these determinants has allowed the construction of nucleic acid probes and the synthesis of oligonucleotide primers used in PCR analysis for the sensitive and specific detection of these organisms.

Clostridium difficile is well recognised as the aetiological agent of pseudomembranous colitis and antibiotic-associated colitis and diarrhoea. The organism is thought to manifest its pathogenic effect through the production of at least two toxins. Of the toxins characterised, toxin A, an enterotoxin, is thought to play the major role in the clinical manifestation of the disease. The diagnosis of *C. difficile* associated disease depends on the isolation and identification of the organism from stool samples and the detection of toxin by a cytopathic effect on tissue culture cells and its neutralisation by *Clostridium sordellii* antitoxin. However, these procedures are time consuming and it takes up to five days to isolate and identify *C. difficile* and to differentiate between toxigenic and non-toxigenic strains. Clearly, more rapid diagnosis is essential to initiate prompt therapy.

Helicobacter pylori has been implicated in the pathogenisis of gastritis and peptic ulcer disease of humans. The current diagnosis of *H. pylori* associated-disease depends upon endoscopic biopsy of the gastric mucosa and detection of *H. pylori* by a urea hydrolysis test, microscopy and culture. The urease test has been found to give false positive results and along with microscopy has a low sensitivity. Culture of the organism requires at least four days incubation. The use of DNA probes and PCR should provide a rapid means of detecting non-culturable coccoid forms of *H. pylori* and should enable the sources and routes of transmission of the organism to be determined.

METHODS

DNA probes and hybridisation analysis

DNA manipulations, radiolabelling of probes and colony and Southern blot hybridisations were performed as described previously.[1-4,7-10]

Polymerase chain reaction (PCR)

Samples for the PCR were as follows.

i) Bacterial samples:

A single colony or sweep of colonies was added to 0.5 ml sterile water, boiled for 5 min and centrifuged for 3 min.

ii) Samples from faecal specimens:

Approximately 0.1 g of sample was added to 2.5 ml water and vortexed prior to boiling and centrifugation. One µl of sample supernatent was used for PCR analysis.

iii) Gastric biopsies:

A portion (0.1 g) of each biopsy was added to 50 µl of water, vortexed, boiled and 10 µl was used for PCR analyses.

One µl of sample supernatant was sufficient for all PCR experiments. PCR amplifications were carried out using standard conditions with appropriate positive and negative control samples as described previously.[3,5,10]

RESULTS AND DISCUSSION

Detection of toxigenic _Clostridium difficile_ strains using nucleic acid probes

We have previously reported the cloning of the toxin A gene into _E. coli_ K-12[1]. The resultant clone expressed properties consistent with the purified protein. A 4.5 kb _Pst_I fragment encoding an internal region of the toxin A gene was isolated, radiolabelled and used as a probe in colony hybridisation studies against 58 toxigenic and 20 non-toxigenic _C. difficile_ strains.[2] All 58 toxigenic strains showed positive hybridisation, in contrast to the 20 non-toxigenic strains. Southern blot analysis with the toxin A gene probe revealed hybridisation to a single fragment of equal intensity for _Hind_III digested genomic DNA isolated from _C. difficile_ strains of wide-ranging toxin production. These results suggest the presence of a single copy of the toxin A gene on the genome of toxigenic _C. difficile_ strains and the wide variation of toxin expression is not a reflection of gene copy number. The _Hind_III fragment to which the probe hybridised varied in size (9-13 kb) suggesting heterogeniety among toxin genes from different _C. difficile_ strains. The lack of toxin production for non-toxigenic strains can be explained by the absence of at least part of the toxin gene.

The toxin A gene probe failed to hybridise with 76 other strains examined from 12 other enteric species and 11 other clostridial species. The exception was three toxigenic *Clostridium sordellii* strains which were expected to hybridise as the organism produces a toxin which is known to immunologically cross-react with toxin A.

The 4.5 kb *PstI* fragment encoding part of the toxin A gene was sequenced and revealed the presence of several tandemly arranged repeat units of 63, 84 or 90 bp[3]. The most common consensus sequence which was repeated 13 times was

5'-GAAGCAGCTACTGGATGGC<u>AAACTATTGATGGTAAAAAATATTA</u>CTTTAATACTAACACTGCT-3'.

A 25 base oligonucleotide of the underlined sequence was synthesised, end-labelled and used as a probe in hybridisation studies. The colony and Southern hybridisation results were identical to those obtained using the 4.5 kb toxin A gene-specific probe[4]. However, results using the oligonucleotide probe appeared much clearer due to a reduced background which was probably due to non-specific hybridisation from residual vector sequence present on the 4.5 kb *PstI* fragment. Neither of these probes were of sufficient sensitivity for satisfactory testing directly on clinical material.

Detection of toxigenic *Clostridium difficile* strains using PCR

As an alternative approach, PCR was chosen using oligonucleotides primers based on the most frequent 63 bp repeat (the first 19 bases and the complimentary sequence of the last 19 bases, see above). As only two of the 13 repeats were identical and tandemly arranged it was expected that a positive PCR result would yield two bands of 63 and 126 bp[3]. However, a series of amplification products, all of which are multiples of the 63 bp repeat unit, were observed for all 87 toxigenic strains tested. No amplified products were observed for the 24 non-toxigenic strains tested[5]. The series of products resulted in an instantly recognisable ladder of eight or more bands after agarose gel electrophoresis (figure 1). The extra bands are due to the staggered annealing of tandemly repeated units, followed by primer independent extension, generating a series of steadily longer DNA fragments as the number of PCR cycles increases[6]. PCR amplification was successful when as few as 20 toxigenic colony forming bacteria were used and the method could detect this level of toxigenic *C. difficile* in mixed culture and from faecal material.

Detection of *Clostridium difficile* using nucleic acid probes based on a 31kDa species-specific antigens

We have characterised, cloned and sequenced a 31 kDa protein from *C. difficile* which appears to be specific to the species. A 1.8 kb DNA insert from a pUC18 subclone encoding the 31 kDa species-specific antigen (as tested by extensive immunoblotting) was radiolabelled and used in colony and Southern blot studies. The 1.8 kb probe was found to hybridise with a single 2.3 kb *HindIII* fragment from all 20 *C. difficile* strains tested.   The probe did not hybridise with 28 strains from 14 other species tested including *C. sordellii*.

Detection of *Clostridium difficile* using PCR based on sequence data from the 31 kDa species-specific antigen

Nucleotide sequence analysis of the 1.8 kb insert enabled the construction of three primer pairs for PCR which amplified the expected sized fragments (380, 700 and 1200 bp) in all 30 *C. difficile* strains tested   The PCR method was negative for 32 strains from 16 other species tested including *C. sordellii*.

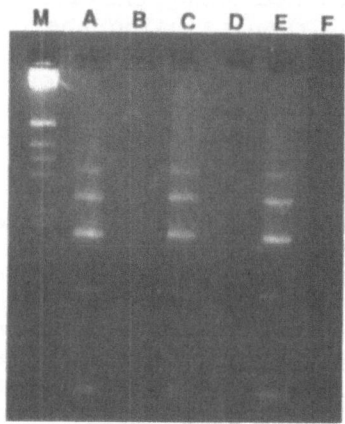

Fig. 1. Identification of toxigenic *Clostridium difficile* by PCR
Lanes: A and C, toxigenic strains; B non-toxigenic strain; D negative faecal sample; E positive faecal sample F negative control and M DNA size marker (1 kb ladder, visible bands 516/506, 394, 344, 298, 220, 154 & 134).

A species-specific probe for the identification of all *C. difficile* strains should prove useful for diagnostic and re-search purposes as it not proven that toxigenic *C. difficile* strains are the sole cause of *C. difficile* related disease.

Identification of *H. pylori* using a urease gene-specific probe

The cloning of *H. pylori* urease genes encoding the 66 and 31 kDa polypeptides gave a 2.6 kb *Taq* I DNA fragment that was found to be specific for *H. pylori* by hybridisation analysis. The probe hybridised with 40 *H. pylori* strains but not with *H.*

mustelae or various species of *Campylobacter*[7]. Southern blot hybridisation against total DNA isolated from *H. pylori* strains showed hybridisation to a 4.4 kb *Hind*III fragment. These hybridisation studies demonstrated that the *H. pylori* urease genes were conserved and species-specific[8].

Identification of *H. pylori* using PCR

The sequencing of the *H. pylori* urease genes[9] enabled the production of specific primers for the sensitive detection of the organism by PCR[10]. The amplified fragment of expected size (400 bp) was obtained for all 40 *H. pylori* strains examined. *H. mustelae* and other urease positive bacteria were PCR negative. *H. pylori* were detected with a level of sensitivity of 100 cells. Further PCR of initial amplified samples was found to increase sensitivity 10 fold. Hybridisation with an oligonucleotide probe derived from a sequence internal to the amplified product was used to confirm the correct products were amplified. Using hybridisation on the amplified PCR products increased sensitivity 10 fold. PCR was found to allow detection of *H. pylori* directly from gastric biopsies.

CONCLUSION

The hybridisation studies effectively identified *C. difficile* strains with the toxin gene or the gene encoding the 31 kDa antigen. *H. pylori* strains were identified with the urease gene-specific probe. Although much genetic information was gained from these studies the procedures were too insensitive to be of practical value in clinical microbiological diagnosis.

The PCR methods, which were rapid and did not require hybridisation technology or a radiolabel, proved more useful for the identification of bacteria from mixed cultures and from clinical specimens. The use of the toxin A and species-specific primers in combination in the PCR should prove particularly useful for research and diagnostic applications for the identification of *C. difficile*. The application of hybridisation analysis in conjunction with PCR for *H. pylori* was highly sensitive for the detection of the organism. The sensitivity was sufficient for the successful detection and identification of *H. pylori* directly from gastric biopsies. This method should be useful for elucidating the epidemiology of this important organism. In general, the complete PCR analysis required just three to five hours from the boiling of the samples to visualisation of the amplified DNA products, which represents a major improvement on existing techniques for the identification of these organisms.

REFERENCES

1.   B. W. Wren, C. L. Clayton, P. Mullany, and S. Tabaqchali, Molecular cloning and expression of the *Clostridium difficile* toxin A gene in E. coli K-12. FEBS. 225: 82-86 (1987).

2.   B. W. Wren, C. L. Clayton, N. Castledine and S. Tabaqchali, Identification of toxigenic *Clostridium difficile* strains using a toxin A gene-specific probe. J Clin Micro 28: 1808-12 (1990).

3.   B. W. Wren, C. L. Clayton, and S. Tabaqchali, Nucleotide sequence of *Clostridium difficile* toxin A gene fragment and detection of toxigenic strains by PCR. FEMS. 70: 1-6 (1990).

4.   B. W. Wren, N. B. Castledine and S. Tabaqchali, A sensitive oligonucleotide probe for the specific identification of toxigenic *Clostridium difficile*. J Med Micro 31: xvii (1990).

5.   B. W. Wren, C. L. Clayton, and S. Tabaqchali, Rapid identification of toxigenic *Clostridium difficile* by PCR. Lancet 423, 335 (1990).
6.   B. W. Wren, M. J. Pallen, and S. Tabaqchali, Beneficial effects of jumping PCR in diagnostic procedures. Lancet 1540, 335 (1990).

7.   C. L. Clayton, B. W. Wren, P. Mullany, A. Topping, and S. Tabaqchali, Molecular cloning and expression of *Campylobacter pylori* species-specific antigens in E. coli K-12. Infection and Immunity, 57: 623-629 (1989).

8.   C. L. Clayton, H. Kleanthous, B. W. Wren, P. Mullany, A. Topping, and S. Tabaqchali, Molecular cloning and characterisation of *Campylobacter pylori* species-specific antigens in E. coli K-12, in Gastroduodenal pathology and *Campylobacter pylori*, F. Megraud, ed., Elsevier, Amsterdam (1990).

9.   C. L. Clayton, M. J. Pallen, H. Kleanthous, B. W. Wren, and S. Tabaqchali, Nucleotide sequence of urease genes from *Helicobacter pylori* encoding for urease subunits, Nucleic acids research, 18: 362 (1990).

10.   C. L. Clayton, M. J. Pallen, H. Kleanthous, B. W. Wren, and S. Tabaqchali, Molecular cloning and sequencing of *Helicobacter pylori* urease genes and detection of *H. pylori* using PCR.encoding for urease subunits, in *Helicobacter pylori*, Gastritis and peptic ulcer, P. Malfertheiner and H. Ditschuneit, eds., Springer-Verlag, Berlin (1990).

THE DEVELOPMENT OF GENETICALLY DEFINED LIVE BACTERIAL VACCINES

Steven Chatfield, Neil Fairweather, John Tite, Ian Charles, Mark Roberts, Mario Posada, Richard Strugnell and Gordon Dougan

Department of Molecular Biology, Wellcome Biotech, Langley Court, Beckenham Kent BR3 3BS, U.K.

INTRODUCTION

It has long been recognised that vaccination with live organisms can induce more effective protection against infectious diseases than the use of dead vaccines presented in conjunction with currently available adjuvants. Live vaccines offer particular advantages in their ability to induce cell-mediated immune responses which are often critical for the establishment of solid protection against certain infectious agents. The route of vaccine delivery is also of importance. Oral vaccines can induce immune responses at mucosal surfaces and such responses are often lacking in individuals who receive parenteral vaccines. The combination of live organisms with oral delivery thus offers potential advantages for creating practical and efficacious vaccines. Many problems were experienced with early live vaccines because of the frequent occurrence of reversion to virulence and batch to batch variation. These problems were difficult to deal with because the attenuating lesions present in avirulent vaccine strains derived from virulent pathogens were uncharacterised at the genetic level. Recent advances in our understanding of the genetics and the mechanisms employed by pathogens to establish infections in the host has allowed the construction of genetically defined attenuated derivatives of many pathogens. The use of modern genetic methods to construct these attenuated vaccine candidates facilitates the elimination of reversion problems and aids the quality control of vaccine lots.

The Concept of Controlled Attenuation

Pathogenic bacteria require many genes in order to induce a disease state in the host. Inactivation of any gene essential for the establishment of the disease should result in an attenuated variant of the virulent pathogen. Since different genes are required at different stages of infection it is theoretically possible to isolate mutants of pathogens blocked at each of these stages. Such mutants have been isolated for a variety of pathogens. A good example are toxin mutants of Vibrio cholerae and Bordetella pertussis (1). Toxin mutants of these organisms show attenuation in animal models and V.cholerae enterotoxin mutants have shown promise in human volunteers as live oral cholera vaccines (16). Pathogenic Salmonella are ideal organisms for studying attenuation since some strains including

Table 1.   Individual genes in which mutations give rise to attenuation in Salmonella sp.

Gene	Enzyme/Function	Reference
aroA, C, D	Aromatic compound biosynthesis	4, 10, 13, 19
purA, B, E, H	Purine biosynthesis	4, 17
galE	UDP galactose 4 epimerase	14
cya/crp	Adenylate cyclase/receptor	7
ompR	Porin regulation	8
phoP	Phosphate regulation	12
htrA	Serine protease	Our unpublished results

Salmonella typhimurium offer excellent systems for genetic analysis in conjunction with a well-characterised mouse infection model. Table 1 lists a variety of genes, mutations in which result in attenuation.

This table illustrates the variety of different genes which can be mutated to develop attenuated variants of a virulent pathogen. Many are not normally classified as classical virulence genes but instead fulfil "household" functions. Some of these genes encode enzymes in essential metabolic pathways which are common to many bacteria. These genes are of particular interest since mutations in them could be used to isolate attenuated variants of a range of different pathogens. A good example of this approach employs strains harbouring mutations in the aromatic biosynthetic pathway known as Aro mutants. Aro mutants are defective in their ability to synthesise chorismic acid from nonaromatic precursors. The chorismate pathway is the only route by which most micro-organisms can synthesise essential aromatic compounds. Since the availability of aromatic compounds including para-aminobenzoic acid is tightly controlled in mammalian tissues, Aro mutant bacteria grow poorly in vivo although they can grow normally on supplemented laboratory medium. Aro mutants of a variety of pathogens including S.typhimurium (10, 13), S.typhi (9), B.pertussis (21) and Yersinia enterocolitica (2) have been constructed and shown to be attenuated. Some but not all of these mutants have shown promise as live vaccines.

Genetic lesions can attenuate a pathogen to different levels. This can be illustrated by introducing mutations into different genes and introducing each gene separately or in combination into the same pathogenic isolate. This point can be well illustrated using a mouse virulent S.typhimurium. Virulence for the mouse can be measured in terms of oral LD_{50}, intravenous LD_{50} or ability to colonise different murine tissues. Results of this type of comparison using S.typhimurium strain SL1344 are shown in Table 2.

Quite clearly different mutations have dramatically different effects on the virulence of this strain. Of further interest is the consequence of combining different attenuating lesions in the same isolate. Combining mutations in aroA with purA produce a strain more attenuated than either parent. Combining aroA with mutations in other genes of the aro pathway such as aroD or aroC creates a strain with a similar level of attenuation.

Table 2.   The effect of different mutations on the
            virulence of S.typhimurium SL1344

Strain	Intravenous Log LD_{50}
SL1344	<1.0
SL1344 aroA	7.1
SL1344 purA	7.7
SL1344 aroA purA	8.7
SL1344 aroA aroC	7.2
SL1344 ompR	5.2
SL1344 ompC ompF	2.2
SL1344 htrA	6.3

This information is important to know since practical vaccine
strains are likely to harbour more than one attenuating lesion for
safety reasons.   It should be remembered that the virulence of
Salmonella strains and mutant derivatives will vary greatly between
different animal species and even between different mouse strains.

Live Attenuated Vaccines Against Salmonellosis

Intensive efforts have been made to develop effective vaccines
against Salmonellae. Live vaccines have proved more effective than
inactivated preparations.   Immune mechanisms involved in protection
against salmonellosis are likely to be very complicated with a clear
involvement of a cellular response. Live vaccine candidates can be
evaluated using the murine model and strains of S.typhimurium,
S.enteritidis or S.dublin. Oral vaccines are required in humans because
of problems of reactogenicity with parenteral vaccines.   The mouse
model can only be used as a guide to the best candidates for use in
other species and proper evaluation of candidate strains will have
to be undertaken separately in each target species.   As described
above there are now a range of attenuated Salmonella strains that
can be considered for use as live oral vaccines. It should be noted
that strain background as well as the nature of the attenuating lesion
can greatly effect the vaccine potential of candidate strains.   Some
attenuated Salmonella strains make poor oral vaccines. Examples are
purA mutants of S.typhimurium and aroA purA mutants of S.typhimurium
(4) and S.typhi (22).   Such vaccines colonise the tissues of mice
poorly after oral feeding and do not stimulate significant cellular
responses.   However, some attenuated Salmonella strains are showing
promise as live oral vaccines. These include aro mutants (10, 19)
and cya, crp mutants (7). Aro mutants of several Salmonella species
have been shown to be effective oral vaccines in different animal
species (10, 15).   Such vaccines can contain single or multiple aro
lesions.   Why are aro vaccines of some Salmonella strains effective
oral vaccines?   Such mutants establish a self-limiting colonisation
of the murine reticuloendothelial system after oral ingestion.   Both
humoral and secretory anti-Salmonella responses can be detected as
well as cellular responses (6). A single dose of an aro vaccine can
protect mice against virulent Salmonella challenge for many months.
It is probable that other attenuated strains will also turn out to
be effective oral vaccines.

Construction of Candidate Human Oral Typhoid Vaccine Strains

Any candidate live vaccine to be used in humans would be subj cted to intense safety and efficacy testing. The lesions responsible for attenuation should be genetically defined at the molecular level. Since aro mutants of Salmonella have shown immense promise as live vaccines we have constructed derivatives of S.typhi strain Ty2 which harbour defined deletion mutations in aroA and aroC. These strains were constructed using cloned S.typhi aroC and aroA genes. Initially both genes were sequenced to fully define their open reading frames (ORF) and then small deletions were introduced within these ORFs. The deletion in aroA extended for 375bp and the deletion in aroC extended for 100bp. The deleted genes were then cloned onto a suicide vector, pGP704 (20), which will replicate in certain E.coli strains but not S.typhi. pGP704 encodes ampicillin resistance and this selectable marker was used to introduce the deleted genes back into the S.typhi chromosome via a single recombinational cross-over which resulted in integration of the suicide vector complete with the deleted aro gene into the S.typhi chromosome at the appropriate homologous site. These cointegrants were used to select for S.typhi Ty2 derivatives carrying deletions in aroA or aroC. Such derivatives were detected amongst ampicillin sensitive variants where the cointegrants had resolved to remove the vector and wild-type aro genes from the chromosome and left behind the deleted aro genes. Derivatives harbouring single aroA or aroC deletions were then subjected to a second cycle of site-directed mutagenesis to create S.typhi Ty2 derivatives harbouring defined mutations in aroA and aroC. The presence of deletions in the chromosomes of these mutants was confirmed using polymerase chain reaction and DNA sequence analysis. These S.typhi double aro mutants are now being evaluated in human volunteers as candidate live oral typhoid vaccines.

Live Bacteria as Vehicles for Delivering Foreign Antigens to the Immune System

Live bacteria harbouring genetically defined attenuating lesions can be developed as effective vaccines against diseases such as salmonellosis. Here individual vaccines are designed to induce protection against specific pathogens. Can we exploit the potent immunogenicity of these live vaccines further? The availability of recombinant DNA techniques means that the potential exists to clone genes encoding protective antigens from heterologous pathogens into live vaccine strains in order to use them as antigen delivery systems. This approach has now been used by many workers. Salmonella strains have proved to be attractive carriers because of their ready availability, potent immunogenicity and use as oral vaccines. Genes from a variety of pathogens including bacteria, viruses and parasites have been used (5). Since Salmonella is genetically closely related to Escherichia coli genes can be readily shuttled between the two species and expressed from extrachromosomal DNA or directly from the Salmonella chromosome. We have developed a simple system for expressing foreign genes at the aroC locus of Salmonellae (23). Delivery using live Salmonella can be used to induce humoral, secretory and cellular responses against heterologous antigens.

Tetanus provides a good example of the use of Salmonella to create a novel oral vaccine. Tetanus disease is an intoxication caused by the tetanus neurotoxin produced by <u>Clostridium tetani</u>. This neurotoxin is a protein with a molecular weight of 150,000 daltons. A polypeptide can be isolated from the carboxy-terminal end of tetanus toxin which is highly immunogenic and non-toxic. This polypeptide, known as C-fragment, is 50,000 daltons in size. Animals vaccinated with C-fragment are well protected against tetanus due to the production of toxin neutralising antibodies. We recently cloned and sequenced the tetanus toxin gene and constructed plasmids which directed the high level expression of C-fragment in <u>E.coli</u> and <u>S.typhimurium</u>. <u>S.typhimurium aro</u> vaccine strains can be constructed which express C-fragment (either from plasmid DNA or from the chromosome) as an intracellular antigen. Oral vaccination with these strains can protect mice both against salmonellosis and challenge with tetanus toxin (11). Vaccinated mice produce anti-toxin antibodies that can neutralise toxin activity. An effective single-dose oral tetanus vaccine would be invaluable for use in developing countries.

Vaccination with oral <u>Salmonella</u> vaccines can also be used to stimulate cell-mediated immune responses to heterologous antigens. Preliminary work using <u>E.coli</u> beta-galactosidase as a model antigen demonstrated a DTH response against beta-galactosidase in mice vaccinated with a <u>Salmonella aro</u> strain expressing this protein (3). This type of analysis was extended to the cellular level using the nucleoprotein of influenzae virus. <u>S.typhimurium aroA</u> mutants were engineered to express high levels of nucleoprotein and these constructs were used to vaccinate B10.5 mice. Strong T-cell responses to the nucleoprotein antigen were detected in proliferation assays using draining lymph node cells. These responses were dependent on $CD4^+$ lymphocytes and recognition of class II MHC gene products. There was no evidence for the priming of class I MHC restricted cytotoxic T-lymphocyte (CTL) responses although in some strains $CD4^+$, class II MHC restricted CTL were detectable (24).

CONCLUSIONS

Modern genetic techniques and a greater understanding of the pathogenic mechanisms of organisms such as <u>Salmonellae</u> has allowed the construction of attenuated strains defined at the molecular level. Some of these genetically defined strains make good live oral vaccines and can be used to deliver foreign antigens to the mammalian immune system. Such bivalent or multivalent oral vaccines may provide a means to vaccinate cheaply and effectively against infectious diseases.

REFERENCES

1. Black, W. J. and S. Falkow. 1987. Construction and characterisation of <u>Bordetella pertussis</u> toxin mutants. Infect. Immun. <u>55</u>: 2465-2470.
2. Bowe, F., P. O'Gaora, D. Maskell, M. Cafferkey, and G. Dougan. 1989. Virulence, persistence, and immunogenicity of <u>Yersinia enterocolitica</u> 0:8 <u>aroA</u> mutants. Infect. Immun. <u>57</u>: 3234-3236.
3. Brown, A., C. E. Hormaeche, R. DeMarco de Hormaeche, M. D. Winther, G. Dougan, D. J. Maskell, and B. A. D. Stocker. 1987. An attenuated <u>aroA S.typhimurium</u> vaccine elicits humoral and cellular immunity to cloned beta-galactosidase in mice. J. Infect. Dis. <u>155</u>: 86-92.
4. O'Callaghan, D., D. Maskell, F. Y. Liew, C. S. F. Easmon, and G. Dougan. 1988. Characterisation of aromatic and

purine-dependent <u>Salmonella typhimurium</u>: attenuation, persistence, and ability to induce protective immunity in BALB/c mice. Infect. Immun. <u>56</u>: 419-423.

5. Chatfield, S. N., R. A. Strugnell, and G. Dougan. 1989. Live Salmonella as vaccines and carriers of foreign antigenic determinants. Vaccine <u>7</u>: 495-498.

6. Collins, F. M. 1974. Vaccines and cell-mediated immunity. Bacteriol. Rev. <u>38</u>: 371-389.

7. Curtiss III, R. and S. M. Kelly. 1987. <u>Salmonella typhimurium</u> deletion mutants lacking adenylate cyclase and cyclic AMP receptor protein are avirulent and immunogenic. Infect. Immun. <u>55</u>: 3035-3043.

8. Dorman, C. J., S. Chatfield, C. F. Higgins, C. Hayward, and G. Dougan. 1989. Characterization of porin and <u>ompR</u> mutants of a virulent strain of <u>Salmonella typhimurium</u>: <u>ompR</u> mutants are attenuated <u>in vivo</u>. Infect. Immun. <u>57</u>: 2136-2140.

9. Dougan, G., D. Maskell, D. Pickard, and C. Hormaeche. 1987. Isolation of stable <u>aroA</u> mutants of <u>Salmonella typhi</u> Ty2: properties and preliminary characterisation in mice. Mol. Gen. Genet. <u>207</u>: 402-405.

10. Dougan, G., S. Chatfield, D. Pickard, J. Bester, D. O'Callaghan, and D. Maskell. 1988. Construction and characterization of <u>Salmonella</u> vaccine strains harbouring mutations in two different <u>aro</u> genes. J. Infect. Dis. <u>158</u>: 1329-1335.

11. Fairweather, N. F., S. N. Chatfield, A. J. Makoff, R. A. Strugnell, J. Bester, D. J. Maskell and G. Dougan. 1990. Oral vaccination of mice against tetanus by use of a live attenuated <u>Salmonella</u> carrier. Infect. Immun. <u>58</u>: 1323-1326.

12. Fields, P. I., E. A. Groisman and F. Heffron. 1989. A Salmonella locus that controls resistance to microbicidal proteins from phagocytic cells. Science <u>243</u>: 1059-1061.

13. Hoiseth, S. K. and B. A. D. Stocker. 1981. Aromatic-dependent <u>Salmonella typhimurium</u> are non-virulent and effective as live vaccines. Nature (London) <u>291</u>: 238-239.

14. Hone, D., R. Morona, S. Attridge and J. Hackett. 1987. Construction of defined <u>galE</u> mutants of Salmonella for use as vaccines. J. Infect. Dis. <u>156</u>: 167-173

15. Jones, P. W., G. Dougan, C. Hayward, N. Mackensie, P. Collins and S. N. Chatfield. 1990. Oral vaccination of calves against experimental Salmonellosis using a double aro mutant of <u>Salmonella typhimurium</u>. Vaccine (in press).

16. Kaper, J. B., H. Lockman, M. M. Baldini, and M. M. Levine. 1984. Recombinant non-toxinogenic <u>Vibrio cholerae</u> strains as attenuated cholera vaccine candidates. Nature (London) <u>308</u>: 655-658.

17. McFarland, W. C. and B. A. D. Stocker. 1987. Effect of different purine auxotrophic mutations on mouse virulence of a Vi-positive strain of <u>Salmonella dublin</u> and of two strains of <u>Salmonella typhimurium</u>. Microbiol. Path. <u>3</u>: 129-141.

18. Mekalanos, J. J., D. J. Swartz, G. D. N. Pearson, N. Harford, F. Groyne, and M. De Wilde. 1983. Cholera toxin genes: nucleotide sequence, deletion analysis and vaccine development. Nature <u>306</u>: 551-556.

19. Miller, I. A., S. Chatfield, G. Dougan, L. De Silva, H. S. Joysey, and C. E. Hormaeche. 1989. Bacteriophage P22 as a vehicle for transducing cosmid gene banks between smooth strains of <u>Salmonella typhimurium</u>: use in identifying a role for <u>aroD</u> in attenuating virulent <u>Salmonella</u> strains. Mol. Gen. Genet. <u>215</u>: 312-316.

20. Miller, V. I., and J. J. Mekalanos. 1988. Novel suicide vector

and its use in construction of insertion mutations: osmoregulation of outer membrane proteins and virulence determinants in <u>Vibrio cholerae</u> requires <u>toxR</u>. J. Bacteriol. <u>170</u>: 2575-2579.

21. Roberts, M., D. Maskell, P. Novotny, and G. Dougan. 1990. Construction and characterization <u>in vivo</u> of <u>Bordetella pertussis aroA</u> Mutants. Infect. Immun. <u>58</u>: 732-739.

22. Stocker, B. A. D. 1988. Auxotrophic <u>Salmonella typhi</u> as live vaccine. Vaccine <u>6</u>: 141-145.

23. Strugnell, R. A., D. Maskell, N. Fairweather, D. Pickard, A. Cockayne, C. Penn, and G. Dougan. 1990. Stable expression of foreign antigens from the chromosome of <u>Salmonella typhimurium</u> vaccine strains. Gene <u>88</u>: 57-63.

24. Tite, J. P., X-M. Gao, C. M. Hughes-Jenkins, M. Lipscombe, D. O'Callaghan, G. Dougan, and F. Y. Liew. 1990. Anti-viral immunity induced by recombinant nucleoprotein of influenza A virus. III. Delivery of recombinant nucleoprotein to the immune system using attenuated <u>Salmonella typhimurium</u> as a live carrier. Immunology, <u>70</u>: (540-546)

DEVELOPMENT OF AN ORAL VACCINE AGAINST ENTEROTOXIGENIC

ESCHERICHIA COLI DIARRHEA

Ann-Mari Svennerholm, Christina Åhrén,
Christine Wennerås and Jan Holmgren

Department of Medical Microbiology and Immunology
University of Göteborg, Göteborg, Sweden

INTRODUCTION

Although enterotoxigenic *Escherichia coli* (ETEC) is the most common cause of diarrhea in developing countries and in travelers to these areas, there is no ETEC vaccine available for use in humans. In developing countries ETEC diarrhea is mainly seen in children below 5 years and it has been estimated that as many as 650 million cases occur annually among these children, resulting in approximately 800.000 deaths [1]. ETEC is also the cause of 30-50% of all diarrheas in travelers to Latin America, Africa and Asia [1,2,3], i.e. between 2 and 3 million cases each year.

The illness caused by ETEC ranges from mild diarrhea without dehydration to severe cholera-like disease [1]. Due to the high incidence of ETEC diarrheas, particularly in developing countries, even a vaccine with a relatively modest protective efficacy may be of a great public health significance.

VIRULENCE FACTORS AND PROTECTIVE ANTIGENS

ETEC produces disease by colonizing the intestine and producing a heat-labile enterotoxin (LT), a heat-stable enterotoxin (ST) or both toxins. Colonization is mediated by colonization factor antigens (CFAs) which usually are fimbriae. In previous studies in animals we have shown that antibodies against LT and CFAs cooperate synergistically in protecting against disease caused by LT-producing *E. coli* carrying the homologous fimbriae [4].

The observation of a decreased illness-to-infection ratio with age in children in endemic areas [1,5] suggests that naturally acquired protective immunity against ETEC diarrhea may develop. Studies in animals and human volunteers have also shown that ETEC infection may give rise to substantial immunity against reinfection with the homologous ETEC strain (6,7,8). Furthermore, clinical ETEC disease evokes intestinal antibody responses against the CFA of the infecting strain as well as against LT (9). These findings suggest that an effective ETEC vaccine should evoke immune responses that interfere with colonization as well as toxin action in the intestine.

CFAs

The three major CFAs in human ETEC isolates are CFA/I, CFA/II and CFA/IV. Whereas CFA/I is a homogeneous protein, CFA/II consists of the coli surface antigen CS3 alone or together with CS1 and CS2 and CFA/IV consists of CS6, usually in combination with CS4 or CS5 [6]. In addition, a number of putative colonization factors (PCFs) have been described, e.g. CFA/III, PCFO159, PCFO166, CS7 and CS17 (M.M. McConnell, this volume). Recent studies in a rabbit nonligated intestine model (RITARD) have suggested that some of these PCFs, e.g. CS7 and CS17, are strong colonization factors (Svennerholm, A-M et.al., to be published). In prospective epidemiological studies in Bangladesh, Mexico and Argentina we have shown the presence of CFA/I, CFA/II and CFA/IV on 50-70% of all clinical ETEC isolates, predominantly on ST only and ST/LT strains. We also observed a marked geographic variation in the prevalence of these different CFAs [5,10, Binztein et.al, to be published]. The relative importance of the different PCFs remains to be evaluated in prospective studies.

Antibacterial immunity against ETEC may to a large extent be ascribed to immunity against the different CFAs, even though antibodies against O-antigens may play a role in protecting against ETEC of homologous O-groups. In both animals and human volunteers intestinal or oral immunization with ETEC strains expressing CFA/I, CFA/II or CFA/IV, has induced protective immunity against subsequent challenge with *E. coli* carrying the homologous CFA/CS-factor [8,11,12]. Furthermore, these studies have suggested that all the individual CS components of CFA/II [11] and CFA/IV [12, unpublished] are colonization factors and protective antigens.

Enterotoxins

A number of epidemiological studies have been performed to assess the enterotoxin profiles of clinical ETEC isolates. Although very different results have been observed, e.g. in different geographic areas, it can be estimated that as a mean, one third of all strains produce LT alone, one third ST alone and one third LT and ST. In a field trial of the oral cholera B subunit (CTB)-whole cell vaccine in Bangladesh [13] it was found that the CTB component provided highly significant, although short-lasting, protective immunity against LT only as well as LT/ST *E. coli* strains. Recent studies in animals have suggested that CTB is equally effective as LTB in inducing protection against LT-producing E. coli (A.-M. Svennerholm, unpublished), supporting that CTB may be used as a toxoid component in an ETEC vaccine.

There has been considerable problems in producing a suitable ST toxoid. The small ST molecule consisting of 18-19 amino acids is not immunogenic unless coupled to a carrier protein, e. g. CTB or LTB. By replacing two of the cysteins in the ST molecule with alanine, synthetic ST peptides that when coupled to B-subunit carriers gave rise to anti-ST as well as anti-LT antibody responses could be achieved; however, the anti-ST antibodies lacked neutralizing ability [14]. Similarly, genetically engineered nontoxic ST-CTB hybrid proteins have induced anti-ST responses but with only weak neutralizing activity [15]. Studies are presently in progress to couple nontoxic ST peptides to various carrier proteins, e.g. CFA/I and CTB, via shorter or longer linker peptides. Due to the problems outlined above it may be questioned whether addition of an ST toxoid to ETEC vaccine candidates is indeed important. Thus, anti-ST antibodies have poor neutralizing capacity, i.e. $\geq$ 33 µg of specific antibody will be needed to neutralize 1 µg of ST. Furthermore, since anti-LT immunity appears to protect against LT only as well as LT/ST strains and a majority of the ST only strains express the most prevalent CFAs a vaccine providing anti-CFA and anti-LT immunity may provide sufficient protective coverage.

POSSIBLE ETEC VACCINES

There are at least three different possibilities of constructing an oral ETEC vaccine providing anti-toxic and anti-colonization immunity in the gut, i.e. 1) purified CFAs + enterotoxoid (CTB or ST • CTB), 2) live bacteria expressing the major CFAs and producing enterotoxoid; or 3) killed CFA-positive E. coli + enterotoxoid.

A vaccine consisting of purified fimbriae may probably be too expensive to prepare and, furthermore, isolated CFAs seem to be very sensitive to proteolytic degradation in the human gastrointestinal tract. Since the different CFAs are normally not expressed on the same strains and it has not yet been possible to clone the genes for different CFAs into the same host organism, a live CFA-vaccine has, at least for the present, to consist of a mixture of different strains. For such vaccines there is a risk of overgrowth of one of the included vaccine strains with supression of the others. Furthermore, live vaccines have the risk of reverting to toxicity by uptake of toxin-encoding plasmids. Therefore, a vaccine consisting of killed ETEC bacteria expressing the most prevalent CFAs and given together with a suitable enterotoxoid is the most promising vaccine candidate within reach. Inactivation of bacteria could either be achieved by colicin E2 treatment [Evans, D.E., personal communication] or by mild formalin treatment [14].

The vaccine we have developed consists of a mixture of killed E. coli bacteria expressing CFA/I and the different CS components of CFA/II and CFA/IV and as enterotoxoid CTB [12]. Strains that belong to common ETEC O-groups and that express the different fimbriae in high concentration have been selected. The concentration of the different CFAs has been estimated using very sensitive monoclonal antibody-based ELISA inhibition tests. The inactivation by mild formalin treatment at different temperatures for several days, causes complete killing of the bacteria without significant losses in the antigenicity of the different CFAs and O-antigens. The CFA antigens of these inactivated organisms have been stable during storage for long times and even after incubation in human gastro-intestinal secretions [14].

EVALUATION OF AN ORAL PROTOTYPE ETEC VACCIN IN HUMAN VOLUNTEERS

A prototype ETEC vaccine, consisting of formalin-killed bacteria expressing CFA/I, CS1, CS2 and CS3, and that is given together with CTB, has been produced by the National Bacteriological Laboratory in Sweden for small scale clinical trials. A study is presently conducted in adult Swedish volunteers to test this vaccine for safety and immunogenicity locally in the gut. Antibody responses in serum as well as production of specific antibodies by peripheral blood lymphocytes will also be assessed. Each of the volunteers have received three oral doses with 10^{11} killed *E. coli* organisms and 1 mg of CTB in buffered bicarbonate solution two weeks apart and intestinal lavages and sera have been collected immediately before and then 9 days after the second and third immunizations.

Hitherto 20 volunteers have received the vaccine. In no instance did immunization with the prototype vaccine result in any adverse local or systemic reactions. Intestinal lavages that were collected from 11 of the vaccinees have been examined for antibodies against CTB/LT, CFA/I, CFA/II and O-antigens of the immunizing strains. Specific ELISA IgA titers divided by the total IgA content of each specimen have been determined. As shown in Table 1 significant IgA antibody responses, were observed against CFA/I, CFA/II, O-antigen as well as CTB in most of the vaccinees. The frequency of responses was comparable to that previously observed [9] in Bangladeshis convalescing from infection with CFA-positive *E.coli* (Table 1.).

Table 1.   Frequency of immune responses[1] in intestine after oral ETEC vaccination and after clinical ETEC disease.

IgA responses to	Vaccinees[2]	Convalescents[3]
CFA/I	10/11	5/8
CFA/II	8/11	
O-antigen[4]	6/11	7/9
CTB/LT	9/11	8/10

[1] Number of volunteers of total number immunized responding with a significant, i.e.> 2-fold, increase in specific IgA titer/total IgA when comparing pre- and post- immunization specimens.

[2] Receiving 3 oral doses of the prototype vaccine

[3] Adult Bangladeshis convalescing from ETEC diarrhea (9)

[4] Specimens from vaccinees were tested against 078 LPS (serogroup of one of the vaccine strains) and from convalescents against O-antigen the infecting strain.

The magnitudes of the antibody responses against CFA/I and CFA/II (Fig 1) were even slightly higher than those previously seen in the convalescents [9] whereas the responses against corresponding O-antigens were somewhat lower after vaccination than after disease (data not shown). The intestinal IgA antibody responses to the CTB component of the vaccine were also higher than the anti-LT responses observed after clinical ETEC disease.

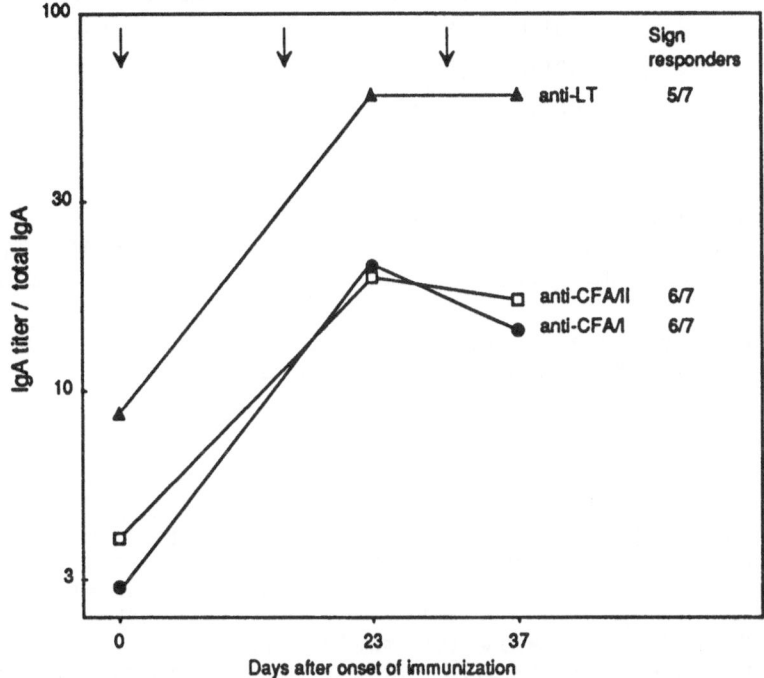

Fig.1.   Swedish adult volunteers recieved three immunizations
          (↓) with the prototype ETEC vaccine on days 0, 14 and
          28 and intestinal lavage specimens were collected
          immediately before and then 9 days after the second and
          third immunization. ELISA IgA titers against purified
          CFA/I, CFA/II (CS1+CS3) and CTB (GM1-ELISA) were
          determined and divided by the total IgA concentration of
          each specimen as described [9].

These results suggest that it is possible to induce substantial CFA antibody responses locally in the intestine by an inactivated vaccine; such responses have previously been difficult to induce by oral immunization with isolated fimbriae. CFA immunity in the gut may enhance the protective effect of anti-LT immunity, which has proven to be effective against LT-producing E. coli in the field [13]. CFA immunity may be particularly important in protecting against disease caused by ST only E. coli strains.

The prototype vaccine also gave rise to significant serum antibody responses against CTB as well as CFA/I in most of the volunteers but the responses against CFA/I and CFA/II were relatively modest (Table 2). Preliminary results also suggest production of antibodies against CFA/I as well as CFA/II by peripheral blood lymphocytes from most of the vaccinees.

Table 2. Antibody responses in serum after oral
immunization with the prototype ETEC vaccine

Day of immunization[1]	Geometric mean ELISA titers, U/ml					
	Anti-CTB/LT		Anti-CFA/I		Anti-CFA/II	
	IgA	IgG	IgA	IgG	IgA	IgG
0	450	850	200	800	450	3200
9	3250	4500	400	1100	450	3700
23	5000	9300	450	1200	700	4000
37	4200	1600	400	1200	700	4100
No of responders per no. immunized:	11/12	12/12	8/12	4/12	6/12	4/12

[1] On the day of the first immunization and 9 days after
the first, second and third vaccination

Provided that the continued studies in Swedish volunteers
support that the formalin-killed CFA/CTB ETEC vaccine is safe
and affords substantial local antibody production in intestine
against the major CFAs, studies to evaluate the protective
efficacy of the vaccine will be undertaken within short. Due
to the high incidence of ETEC infections in travelers to ETEC
endemic areas, protective efficacy trials could probably be
performed using relatively small numbers of volunteers.
Hopefully, an effective ETEC vaccine will soon be available.

ACKNOWLEDGEMENTS

These studies were supported by grants from the Swedish
Medical Research Council, grant 16X-09084, SAREC and WHO. The
skillful technical assistance by Ms Gudrun Wiklund and Kerstin
Andersson is gratefully acknowledged.

REFERENCES

1. R.E. Black, The epidemiology of cholera and
    enterotoxigenic *E. coli* diarrheal disease,
    in: "Development of vaccines and drugs against diarrhea,
    "11th Nobel Conference, Stockholm, J. Holmgren, A.
    Lindberg and R. Möllby, eds., Studentlitteratur, Lund,
    p23 (1986).

2. R.L. Guerrant, Microbial toxins and diarrhoeal disease:
    introduction and overview, in: "Microbial toxins and
    diarrhoeal disease," Ciba Foundation Symposium 112,
    Pitman, London, p1 (1985).

3. R. Steffen, Epidemiologic studies of travelers' diarrhea,
    severe gastrointestinal infections, and cholera, *Rev. of
    Infect. Dis.* 8(suppl 2):122 (1986).

4.   C. M. Åhrén and A.-M. Svennerholm, Synergestic protective effect of antibodies against *Escherichia coli* enterotoxin and colonization factor antigens, *Infect. Immun.* 38:74 (1982).

5.   Y. Lopez-Vidal, J.J. Calva, A. Trujillo, A.P. de Léon, A. Ramos, A.-M. Svennerholm, and G.M. Ruiz-Palacios, Enterotoxins and adhesins of enterotoxigenic *Escherichia coli*: Are they risk factors for acute diarrhea in the community? *J. Infect. Dis.* 162:442 (1990).

6.   M.M. Levine, Escherichia coli that cause diarrhea: enterotoxigenic, enteropathogenic, enteroinvasive, enterohemoarrhagic and enteroadherent, *J. Infect. Dis.* 155:377 (1987).

7.   M.M. Levine, R.E. Black, M.L. Clements, D.R. Nalin, L. Cisneros, and R.A. Finkelstein, Volunteer studies in development of vaccines against cholera and enterotoxigenic *Escherichia coli*: a review. in: "Acute Enteric Infections in Children. New prospects for treatment and prevention", T Holme, J. Holmgren, M.H. Merson, R. Möllby, eds., Elsevier/North-Holland Biomedical Press, Amsterdam, p 443 (1981).

8.   C.M. Åhrén and A.-M. Svennerholm, Experimental enterotoxin-induced *Escherichia coli* diarrhea and protection induced by previous infection with bacteria of the same adhesin or enterotoxin type, *Infect. Immun.* 50:255 (1985).

9.   B.J. Stoll, A.-M. Svennerholm, L. Gothefors, D. Barua, S. Huda, and J. Holmgren, Local and systemic antibody reponses to naturally acquired enterotoxigenic *Escherichia coli* diarrhea in an endemic area, *J. Infect. Dis.* 153:527 (1986).

10. L. Gothefors, C. Åhrén, B. Stoll, D.K. Barua, F. Ørskov, M. Salek, and A.-M. Svennerholm, Presence of colonization factor antigens on fresh isolates of fecal *Escherichia coli*. A prospective study, *J. Infect. Dis.* 152:1128 (1985).

11. A.-M. Svennerholm, C. Wennerås, J. Holmgren, M.M. McConnell, and B. Rowe, Roles of different coli surface antigens of colonization factor antigen II in colonization by and protective immunogenicity of enterotoxigenic *Escherichia coli*, *Infect. Immun.* 58:341 (1990).

12. A.-M. Svennerholm, Y. Lopez-Vidal, J. Holmgren, M.M. McConnell, and B. Rowe, Role of PCF8775 antigen and its coli surface subcomponents for colonization, disease, and protective immunogenicity of enterotoxigenic *Escherichia coli* in rabbits, *Infect. Immun.* 56:523 (1988).

13. J. Clemens, D. Sack, J.R. Harris, J. Chakraborty, P.K. Neogy, B. Stanton, et al., Cross-protection by B subunit-whole cell cholera vaccine against diarrhea associated with heat-labile toxin-producing enterotoxigenic

Escherichia coli: results of a large-scale field trial, *Infect. Dis.* 158:372 (1988).

14. A.-M. Svennerholm, J. Holmgren, and D.A. Sack, Development of oral vaccines against enterotoxinogenic *Escherichia coli* diarrhoea, *Vaccine* 7:196 (1989).

15. J. Sanchez, A.-M. Svennerholm, and J. Holmgren, Genetic fusion of a non-toxic heat-stable enterotoxin-related decapeptide antigen to cholera toxin B-subunit, *FEBS Lett.* 241:110 (1988).

COMPARISON OF VIRULENCE FACTORS IN DIFFERENT FRESHLY ISOLATED STRAINS OF ENTMAOEBA HISTOLYTICA

G.D. Burchard[1], D. Mirelman[2]

[1]Bernhard-Nocht-Institute for Tropical Medicine
2000 Hamburg 36, F.R.G.
[2]Department of Biophysics, The Weizmann
Institute of Science
Rehovot 76100, Israel

Introduction

To produce liver abscesses, virulent Entamoeba histolytica have to cope with several problems: colonization of the gut, adherence to colonic epithelium, destruction of epithelial layer, adherence to and lysis of host inflammatory cells, and resistence to immune defence mechanisms. E. histolytica trophozoites obtained from symptomatic patients (designated as pathogenic) have been found to differ from those isolated from asymptomatic carriers (nonpathogenic) by a number of biochemical criteria such as isoenzyme electromigration, interaction with monoclonal antibodies or hybridization with DNA probes (Bracha et al. 1990). The reason for the difference in the pathogenic behaviour of the two classes of ameba is, however, not well understood, partly because only few studies were performed with nonpathogenic strains which cannot be grown in axenic cultures.

In the present study we have compared a number of virulence properties in ten, freshly isolated pathogenic, and forty nonpathogenic strains of E. histolytica. The assays included determination of adherence to target cells, rates of destruction of target cells, and phagocytosis.

Methods and results

Adherence to PMNs was determined by the rosetting technique described by Ravdin and Guerrant (1981). In short, 2×10^4 trophozoites and 1×10^6 neutrophils were centrifuged at 150 g and incubated at 4 °C for 30 min. The percentage of trophozoites with three or more attached PMNs (rosettes) was counted.

In all strains adherence mechanisms probably mediated by amebic lectins (Petri et al. 1989) were inhibited by specific carbohydrates (galactose, N-acetyl-galactosamine, lactose, and asialofetuin). Mannose and glucose did not inhibit adherence. Pretreatment of the PMNs with galactosidase (50 mU/ml) in order to cleave off galactose residues resulted in an inhibiton of adherence. Preincubation of the PMNs with neuraminidase (0.1 U/ml) in order to remove sialic acid and to unmask membrane galactosyl residues led to an increase of adherence.

It was also tried to inhibit other possible adherence mechanisms like those mediated by neutrophil receptors. Pretreatment of washed trophozoites with fibronectin resulted in an increase of adherence to PMNs (preincubation 30 min with 200 ug/ml: 148% + 35%, with 1000 mg/ml: 155% + 31%, control = 100%). Adding monoclonal antibodies to trophozoites and PMNs in the adhesion assay resulted in some degree of unspecific inhibition. Monoclonal antibodies against the leucocyte adhesion molecules CD11a, CD11b, CD11c and CD18 did not lead to a further increase of this unspecific inhibition of adherence. In the nonpathogenic, freshly isolated strains adherence to PMNs was identical to that in pathogenic strains and no difference in the adherence mechanisms could be found.

Experiments were also performed with different E. histolytica strains to determine if functional actin microfilaments and microtubules are a requirement for adhesion. Trophozoites were preincubated with cytochalasin B in concentrations between 1 and 100 ug/ml. Concentrations up to 40 ug/ml did not affect the viability but did interfere with adhesion. Preincubation of the trophozoites with nocodazole, a synthetic microtubule inhibitor was without effect.

Cytopathogenicity was tested by measuring the destruction of tissue culture monolayers grown in microtiter plates (Burchard et al. 1988). Rat endothelium lung cells, baby hamster kidney cells and U937 cells were tested, but most of the studies were done with Chinese hamster ovary cells. The attached cells were exposed to the ameba for 1 to 2 hours. The cells remaining on the plate were stained with methylene blue. The incorporated dye was extracted by incubation with HCl, and the resulting supernatants were read by spectrophotometry at 660 nm. Compared to virulent strains from patients with invasive amebiasis strains from asymptomatic carriers were less cytopathogenic (fig. 1).

The phagocytic capacity was measured according to Trissl et al. (1978). In brief, trophozoites were fed with erythrocytes and phagocytized RBCs were counted. The rate of erythrophagocytosis in nonpathogenic strains was far less than in pathogenic strains (fig. 2).

Discussion

Laboratory grown adherence-deficient E. histolytica mutants have been shown to be less virulent (Orozco et al. 1982, 1987). Our results indicate that defect adherence mechanisms are not a common feature of nonpathogenic strains, because our freshly isolated, nonpathogenic strains did not exhibit deficient adherence to PMNs. Our results also indicate that adherence between E. histolytica and PMNs is not exclusively mediated by amebic lectins and Gal- and GalNac-oligomers on the PMNs. Our results indicate that fibronectin - for which E. histolytica has been shown to possess a

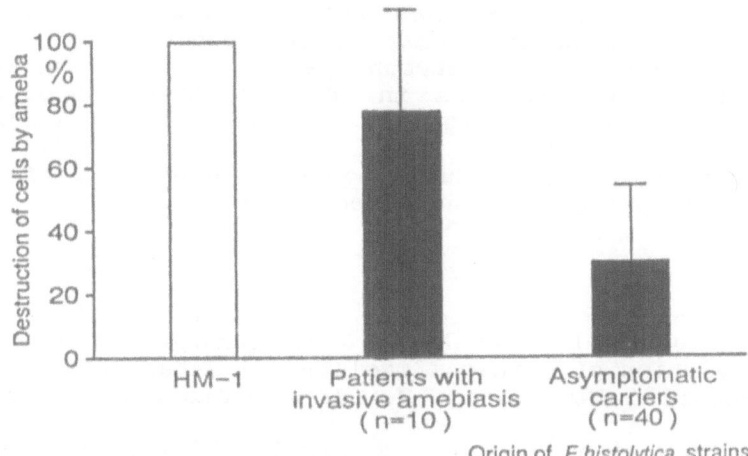

Figure 1. Destruction of CHO cells by different
E. histolytica isolates, compared to
destruction by HM-1

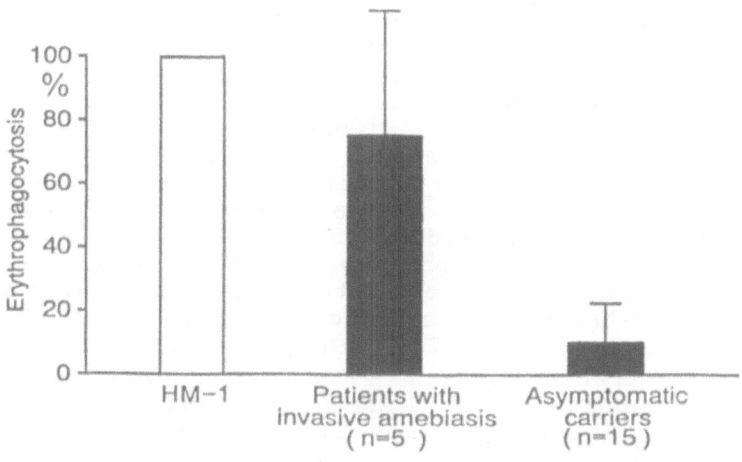

Figure 2. Erythrophagocytosis of different
E. histolytica isolates, compared
to erythrophagocytosis of HM-1

receptor (Talamas-Rohana et a. 1988) - can mediate adherence between trophozoites and PMNs. Similar findings were reported concerning adherence mechanisms of Trypanosoma cruzi (Ouassi et al. 1986). On the other hand, we could not find an effect of monoclonal antibodies against CD11/CD18 on the adherence between E. histolytica and PMNs as has been described for Histoplasma capsulatum and macrophages (Newman et al. 1990).
The important difference between pathogenic and nonpathogenic strains of E. histolytica was in the defective phagocytosis and lower destructive property against target cells of nonpathogenic strains. The reason for the lower phagocytic capacity of these strains is not known, the reduced aggressiveness could be due to a lower content of proteinases and amoebapore.

References

Bracha, R., Diamond, L.S., Ackers, J.P., Burchard, G.D., Mirelman, D.: Differentiation of clinical isolates of Entamoeba histolyt ica using specific DNA probes. J. Clin. Microbiol. 28 (1990) 680-684
Burchard, G.D., Mirelman, D.: Entamoeba histolytica: virulence potential and sensitivity to Metronidazole and Emetine of four isolates possessing nonpathogenic zymodemes. Exp. Parasitol. 66 (1988) 231-242
Newman, S.l., Bucher, C., Rhodes, J., Bullock, W.E.: Phagocytosis of Histoplasma capsulatum yeasts and microconidia by human cultured macrophages and alveolar macrophages. Cellular cytoskeleton requirement for attachment and ingestion. J. Clin. Invest. 85 (1990) 223-230
Ouaissi, M.A., Cornette, J., Afchain, D., Capron, A., Gras-Masse, H., Tartar, A.: Trypanosoma cruzi infection inhibited by peptides modeled from a fibronectin cell attachment domain. Science 234 (1986) 603-607
Orozco, E., Rodriquez, M.A., Murphy, C.F., Salata, R. A., Petri, W.A., Smith, R.D., Ravdin, J.I.: Entamoeba histolytica: Cytopathogenicity and lectin activity of avirulent mutants. Exp. Parasitol. 63 (1987) 157-165
Orozco, E., Martinez Palomo, A., Robles, A.G., Guarneros, G., Galindo, J.M.: Las interacciones entre lectina y receptor median la adherencia de e. histolytica a células epiteliales. Relación de la adhesión con la virulencia de las cepas. Arch. Invest. Méd. (Mèx.) 13 (Supl.3) (1982) 159-167
Petri, W.A., Chapman, M.D., Snodgrass, T., Mann, B.J., Broman, J., Ravdin, J.I.; Subunit structure of the galactose and N-acetyl-D-galactosamine-inhibitable adherence lectin of Entmoeba histolytica. J. Biochem. Chem. 264 (1989) 3007-3012
Ravdin J.I., Guerrant, R.L.: The role of adherence in cytopathogenic mechanisms of E. histolytica. Study with mammmalian tissue culture cells and human red blood cells. J. Clin. Invest. 68 (1981) 1305-1313
Talamas-Rohana, P., Meza, I.: Interaction between pathogenic amebas and fibronectin: substrate degradation and changes in cytoskeleton organization. J. Cell Biol. 106 (1988) 1787-1794
Trissl, D., Martinez Palomo, A., de la Torre, M., de la Hoz, R., Perez de Suarez, E.: Surface properties of Entamoeba: increased rates of human erythrocyte phagocytosis in pathogenic strains. J. Exp. Med. 148 (1978) 1137-1145

A SYSTEM FOR PRODUCTION AND RAPID PURIFICATION OF LARGE AMOUNTS OF THE SHIGA TOXIN/SHIGA-LIKE TOXIN B SUBUNIT

Arthur Donohue-Rolfe[1], David W.K. Acheson[1], Gerald T. Keusch[1], Marcia B. Goldberg[2], Stephanie A. Boyko[2], and Stephen B. Calderwood[2]

Division of Geographic Medicine and Infectious Diseases, New England Medical Center Hospitals, Boston, MA, USA[1], Infectious Disease Unit, Massachusetts General Hospital, Boston, MA, USA[2]

INTRODUCTION

Shiga toxin produced by *Shigella dysenteriae* type 1 is a potent protein toxin which has cytotoxic, neurotoxic and enterotoxic properties. The toxin inhibits eukaryotic protein synthesis by acting as an *N*-glycosidase, cleaving an adenine residue at nucleotide position 4324 of the 28S rRNA of the 60S ribosomal subunit. Some strains of *Escherichia coli* produce toxins which are biologically similar to Shiga toxin and are termed Shiga-like toxins (SLTs).

Shiga toxin and SLTs consist of an A chain and multiple copies of a B chain. The B subunits of Shiga toxin and SLT-I are identical, and a key functional role of the B subunits is to mediate binding of the toxin to surface receptors on susceptible cells. The cell surface receptors for the B subunit of Shiga and Shiga-like toxins are glycolipids containing the disaccharide Galα1-4Gal. To date Shiga toxin/SLT-I B subunit has been obtained by treating isolated holotoxin with strong denaturing substances such as urea or formic acid, separating A and B chains and then renaturing activity. In this study we report the construction of an *E. coli* strain that hyperproduces SLT I/Shiga toxin B subunit and a simple purification scheme for the isolation of the B subunit. The immunologic and biologic activities of the purified B subunit have also been characterized.

METHODS

Plasmid expression vectors: Plasmid pSBC31 and pSBC32 were constructed by subcloning a portion of the *slt*-I-coding sequence from plasmid pSC4 (1) into the plasmid expression vector pKK233-2 (Pharmacia LKB Biotechnology, Inc.) which contains the inducible *trc* promoter. The 1150 bp fragment contained a 3' portion of the coding sequence of *slt*-I-A (18 amino acids), the Shine-Delgarno sequence and the entire coding sequence of slt-I-B and additional downstream DNA (pSBC32). pSBC31 was constructed in the inverse orientation of the insert to act as a negative control. Both plasmids were expressed in *E. coli* JM105.

Preparation and purification of the B subunits: Whole cell and periplasmic extracts were prepared from exponentially growing cells of JM105 containing either of the two plasmids. Isopropyl-β-D-thiogalactopyranoside (IPTG) (final concentration of 1 mM) was added to growing cells at an OD_{600} of 0.4-0.6. For large scale production of B subunit JM105(pSBC32) was grown in an 18 liter fermentor culture with IPTG. The B subunit released by polymyxin treatment was purified by using a receptor analog affinity chromatography procedure, utilizing P_1 glycoprotein from hydatid cyst fluid, used previously for the purification of Shiga toxin and Shiga-like toxins (2).

Immunologic and biologic properties of the B subunit: The B subunit was quantitated by ELISA using a mouse monoclonal antibody specific to the B subunit of Shiga toxin as the capture molecule, and rabbit anti-Shiga toxin polyclonal antiserum to detect bound B subunit (3). The B subunit was crosslinked with dimethypimelimi-date and the crosslinked material subjected to SDS-PAGE (15% acrylamide). Purified B subunit was also chromatographed on a Bio-Gel P-30 (BioRad) column equilibrated with phosphate buffered saline and compared with ovalbumin (45kd) and chymotrypsinogen A (28kd).

The ability of purified B subunit to inhibit the binding of ^{125}I-Shiga toxin to HeLa cells was examined by adding various amounts of the B subunit to fixed quantities of ^{125}I-Shiga toxin and adding this mixture to the cells. Bound counts were quantitated following a 2 h incubation. HeLa cell cytotoxicity was determined by measuring ^{3}H-leucine incorporation into protein (2).

RESULTS

Whole cell and periplasmic extracts of JM105(pSBC32) showed IPTG inducible expression of the B subunit of Shiga toxin/SLT I as determined by immunoblotting using polyclonal anti-Shiga toxin antiserum. JM105(pSBC31) showed no immunoreactive protein as expected. A periplasmic extract from the cell pellet of the 18 liter culture was applied to the P_1 affinity matrix, after which SDS-PAGE of the 4.5 M $MgCl_2$ eluate, revealed a major band with a molecular weight less than 10 kd, corresponding to Shiga toxin/SLT I B subunit. A minor band was also seen at 14 kd corresponding to B subunit dimers.

The ELISA to quantitate B subunit was able to detect as little as 30 pg, and gave a linear response up to 250 pg. IPTG-induced cultures of JM105(pSBC32) yielded 1.75 µg B subunit per ml culture compared with uninduced cultures which yielded 0.41 µg/ml. The periplasmic extract of the 18 liter culture yielded 36 mg of pure B subunit.

Gel filtration of the B subunit on a Bio-Gel P-30 column gave a homogeneous peak with an apparent molecular weight of 30,000 indicating that the purified B subunit consts of a multimer. Following crosslinking with the homobifunctional agent dimethyl-pimelimidate the B subunit, when run on SDS-

300

PAGE, demonstrated a series of bands corresponding to 7 kd multiples equivalent to mono-mers, dimers, trimers, tetramers and pentamers (Figure 1). Immuno-blotting, with the B subunit specific monoclonal antibody, of the crosslinked material following SDS-PAGE, revealed bands up to 8B.

Purified B subunit, at concentrations as high as 10 μg/ml showed no cytotoxic activity on HeLa cells. The purified B subunit did however inhibit the binding of ^{125}I-Shiga toxin to HeLa cells in a concentration dependent manner.

CONCLUSION

We have constructed a plasmid-based expression system which allows the production of large amounts of Shiga toxin/SLT-I B subunit under the control of the inducible *trc* promoter. The B subunit interacts normally with both monoclonal antibody to Shiga toxin B subunit and polyclonal antiserum to Shiga toxin. The B subunit appears to be assembled into multimers despite the absence of an intact A subunit. The purified B subunit is biologically active in terms of blocking the binding of ^{125}I-Shiga toxin to HeLa cells and as expected is not cytotoxic. Using this inducible expression system we have been able to purify 38 mg of Shiga toxin/SLT-I B subunit. This large quantity has enabled us to make sizeable crystals of the B subunit material which are currently being analyzed by x-ray crystallography.

REFERENCES

1. Calderwood, S.B., Auclair, F., Donohue-Rolfe, A., Keusch, G.T., Mekalanos, J.J., 1987, Nucleotide sequence of the Shiga-like toxin genes of *Escherichia coli*, Proc. Natl. Acad. Sci. USA, 84:4364.

2. Donohue-Rolfe, A., Acheson, D.W.K., Kane, A.V., Keusch, G.T., Purification of Shiga toxin and Shiga-like toxins I and II by receptor analog affinity chromatography with immobilized P_1 glycoprotein and production of cross-reactive monoclonal antibodies. Infect. Immun., 57:3888.

3. Donohue-Rolfe, A., Keusch, G.T., Edson, C., Thorley-Lawson, D., Jacewicz, M., Pathogenesis of *Shigella* diarrhea. IX. Simplified high yield purification of *Shigella* toxin and characterization of subunit composition and function by the use of subunit-specific monoclonal-polyclonal antibodies. J. Exp. Med., 160:1767.

EXPRESSION AND POSSIBLE BIOLOGICAL FUNCTIONS OF CURLI ON
INFANTILE DIARRHOEA ESCHERICHIA COLI ISOLATES

Levente Emödy[1], Åsa Ljungh[2], Tibor Pal[1], Géza Sarlós[1],
and Torkel Wadström[2]

[1]Department of Medical Microbiology, University of Pecs
Pecs, Hungary, and [2]Department of Medical Microbiology
University of Lund, Lund, Sweden

INTRODUCTION

Fröman et al.(1) reported on fibronectin (Fn) binding to
enterotoxigenic Escherichia coli and demonstrated two different
binding mechanisms. Studies by Ljungh et al.(2) showed that also
human enteropathogenic E.coli (EPEC) of different O:H serotypes
as well as uropathogenic E.coli and strains isolated from
septicaemia and other human infections bind Fn.

A new surface appendage called curli were described on bovine
mastitis E.coli strains as conglomerates of filamentous surface
appendages (3). An infantile diarrhoea E.coli strain NG7C which
we had earlier shown to bind Fn, vitronectin (Vn) and various
types of collagens (Cn) (2), was also found to express curli
(S.Normark, personal communication). This observation prompted us
to investigate the conditions of curli expression, the possible
biological functions of curli and the geographical distribution
of curliated E.coli strains.

MATERIALS AND METHODS

E.coli strain NG7C and other E.coli strains from patients with
infantile diarrhoea in the Gambia, Papua, New Guinea and Vietnam
were used (2). Bacteria were grown overnight at 37°C or 20°C
either in broth or agar phase of Nutrient, Tryptic Soy, CFA, BHI
and Luria media (4). One medium with 1% Bacto Tryptone, pH 7.2,
with or without 1.5% Bacto Agar was also employed.
The presence of curli was visualized by electronmicroscopy
(5). Cell surface hydrophobicity and binding to iodinated
connective tissue proteins was assayed as earlier described
(6,2). Binding of Congo red was studied in CFA agar plates with
100 µg/ml Congo red (Sigma Chemical Co., St Louis,Mo). Attachment
to HEp-2 cells was studied as earlier described (7).

RESULTS AND DISCUSSION

Curli were well expressed on bacterial cells grown on Tryptic Soy Agar, Colonization Factor AGAR (CFA) and Luria Agar plates overnight at 37°C as shown by electron microscopy. Nutrient Agar and Brain Heart Infusion Agar did not permit the expression of curli. Growth in liquid form of these media or incubation at 25°C were not favourable for curli production. Optimal expression was achieved on a simple medium containing 1% Bacto Tryptone and 1.5% Bacto Agar, pH 7.2. Also in the liquid form of this medium curli were expressed. Two percent sodium chloride in the growth medium prevented the expression of curli.

Curli were coexpressed with (i) high cell surface hydrophobicity (i.e. autoaggregation in 0.02 M KPB), (ii) binding to connective tissue proteins (i.e. various types of Cn, Fn, Vn, and laminin), (iii) binding of Congo red, and (iv) attachment to epithelial cells (Hep-2) (Table, Fig.1A).

One spontaneous curli negative variant of E.coli NG7C was isolated, NG7C/1, and shown to be negative for Congo red binding and epithelial cell attachment (Table, Fig.1B). The strains expressed low cell surface hydrophobicity and binding to connective tissue proteins was markedly reduced.

Table   Characteristics of NG7C and NG7C/1 E.coli strains

Character	NG7C	NG7C/1
Presence of curli	+	−
Congo red binding	+	−
Autoaggregation	+	−
Hep-2 cell attachment	3.48*	0.02*
Binding of Cn type I	65%**	23%**
Fn	81%**	17%**
Laminin	78%**	29%**

Bacterial cells were grown on CFA at 37°C over night.
* Attached bacteria in percent of total added bacteria.
**Bound protein in percent of total added ^{125}I-labelled protein.

A number of intestinal pathogens have been shown to express high cell surface hydrophobicity upon growth on various media under conditions which usually also favour expression of different colonization factors (8). It seems most likely that curly are surface filamentous appendages which may well be involved in intestinal colonization by interacting with molecules in the mucin layer and on intestinal epithelial cells. Studies to explore this possibility are now under way in several laboratories including exploring the role of curli in Salmonella colonization of the gastrointestinal tract.

CONCLUSIONS

1. Interaction of curli with the mucosal epithelial lining or with connective tissue components might facilitate the colonization of the host by curli expressing E.coli .
2. Strains expressing curli production are widely distributed in various geographic areas.
3. Curli expression by E.coli might remain undetected by routine faecal cultures.

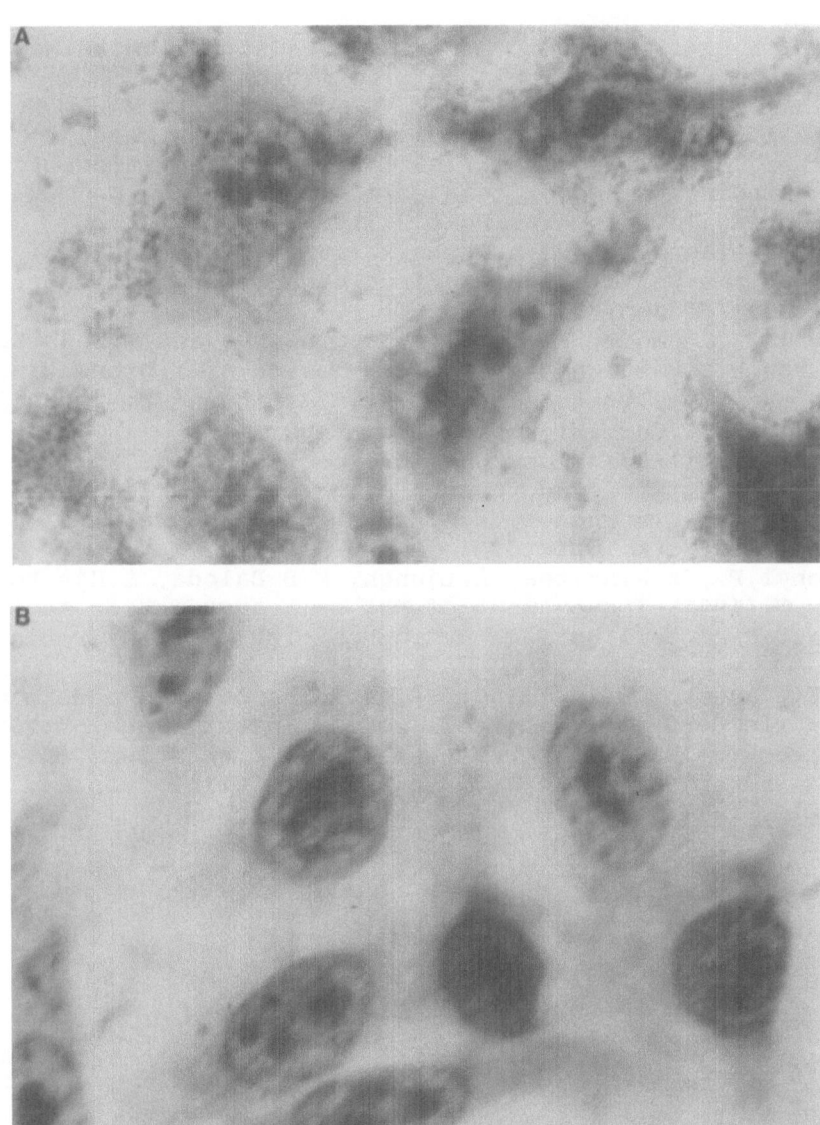

Fig.1 HEp-2 cell attachment by overnight <u>E.coli</u> cultures at
37°C. Magn. 1200. A: <u>E.coli</u> NG7C   B: <u>E.coli</u> NG7C/1.

ACKNOWLEDGEMENTS.

This study was supported by a grant from the Swedish Medical Research Council (16x-04723) and from the Nanna Svartz' foundation.

REFERENCES

1. Fröman,G.,L.M.Switalski, A.Faris, T.Wadström, and M.Höök (1984) Binding of Escherichia coli to fibronectin. J.Biol.Chem.259:14899.
2. Ljungh,Å., L.Emödy,H.Steinruck, P.Sullivan, B.West, E.Zetterberg, and T.Wadström (1990) Fibronectin,vitronectin and collagen binding to Escherichia coli of intestinal and extraintestinal origin. Zbl.Bakt.273:000.
3. Olsén,A.,A.Jonsson and S.Normark (1989) Fibronectin binding mediated by a novel class of surface organelles on Escherichia coli. Nature (London) 338:652.
4. Ljungh,Å., L.Emödy, P.Aleljung, O.Olusanya and T.Wadström (1991) Growth conditions for the expression of fibronectin, collagen type I, vitronectin and laminin binding to Escherichia coli strain NG7C.Curr.Microbiol.,in press.
5. Lounatmaa,K.(1985) Electron microscopic methods for the study of bacterial surface structures. In: Enterobacterial surface antigens (Eds.T.K.Korhonen, E.A.Dawes and P.H.Mäkelä),Elsevier, Amsterdam, New York, OXford.
6. Rozgonyi,F., K.R.Szitha, Å.Ljungh, S.B.Baloda, S.Hjertén and Wadström,T.(1985) Improvement of the salt aggregation test to study bacterial cell surface hydrophobicity. FEMS Microbiol.Lett. 30:131.
7. Pàl,T., and L.H.Hale.(1989). Plasmid-associated adherence of Shigella flexneri in a HeLa cell model. Infect.Immun.57:2580.
8. Wadström,T.(1989). Adherence traits and mechanisms of microbial adhesion in the gut. Baillière's Clin. Trop.Med.& Commun.Dis.3:417.

STUDIES ON HEAT-STABLE ENTEROTOXIN TYPE II (STII OR STB)

FROM *ESCHERICHIA COLI*

Carina Handl and Jan-Ingmar Flock

Center for Biotechnology
Karolinska Institute, NOVUM
S-141 52 HUDDINGE
SWEDEN

Background

Acute diarrhoeal disease caused by Enterotoxigenic *Escherichia coli* (ETEC) is a serious health problem amongst young children and animals. Great economical losses are suffered when a significant number of piglets die or loose weight due to diarrhoeal disease.

ETEC produce one or several of three types of enterotoxins; one heat-labile toxin (LT) and two heat-stable toxins (STI also called STa and STII also called STb)(1, 3). In contrast to LT and STI, STII has not been thoroughly studied, mainly due to the lack of a good, reliable and convenient assay. The only available assay for STII during recent years was an *in vivo* pig intestinal loop assay (8). In 1988 our group developed an ELISA for STII (5) and recently Whipp (11) published a rat model using trypsin inhibitor to avoid degradation of the toxin by intraluminal protease. It is important to note that the only features STI and STII have in common are their heat resistance and their low molecular weight (2 and 5 kD respectively). Their nucleotide sequences, biochemical characteristics and mechanism of action seem to differ totally (4, 6, 7, 10).

Epidemiological studies on STII have been limited, however. It has been reported that ETEC producing STII are mainly found amongst piglets, especially during the weaning period. STII does not seem to be a major problem in human diarrhoea (2, 9, own preliminary results).

The gene encoding STII has been cloned and the sequence determined (6). The deduced protein sequence is 48 amino acids long.

A new sensitive assay: STII-ELISA

Due to limitations of the pig intestinal loop assay system, we developed an Enzyme Linked ImmunoSorbent Assay (ELISA) for STII, using the cloned STII gene (5).

STII is a molecule of only 5kD, which suggests that it is unlikely to be immunogenic. We therefore fused the STII gene to genes for carrier proteins (Staphylococcal protein A and ß-galactosidase from *E.coli*) in two different expression systems. ProteinA-STII fusion protein was purified on IgG-Separose directly from the culture supernatant and ß-gal-STII fusion protein was enriched as inclusion bodies in the cells. The cells were lysed, washed and ß-gal-STII was finally solubilised in 8 M urea. Slow removal of urea allowed ß-gal-STII to remain in solution.

Rabbit antisera raised against proteinA-STII fusion protein crossreacted with ß-gal-STII and native STII in a Western blot test.

Attempts to set up a direct ELISA for STII yielded inconsistent results. We therefore developed an inhibition ELISA as shown in figure 1. In this test, solubilised ß-gal-STII fusion protein is immobilised in microtiter wells and the sample is added to the coated well together with antisera raised against proteinA-STII. STII in the sample will inhibit binding of antibodies to the coated surface. The binding of the antibodies is monitored with alkaline phosphatase conjugated anti-rabbit antibodies.

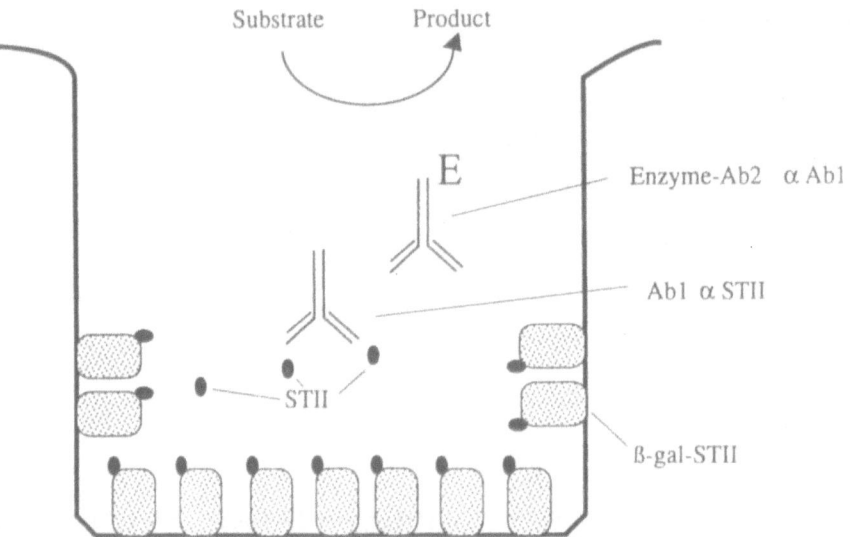

Figure 1.   Inhibition ELISA for STII

Good correlation between three STII assays

A radiolabelled DNA probe for STII was made from the entire STII gene.

The three tests now available; pig intestinal loop for STII biological activity, ELISA for determination of immunogenicity of STII and DNA probe for determination of the gene, were used in a screen of 201 ETEC isolates from young pigs (of which about half were STII producers) to assess their comparability. The results correlated very well in the three different test methods.

Ninety six % of the strains gave the same result in the STII-ELISA and DNA probe test, and 88% correlated in all three tests.The pig loop test was the most unreliable test due to varying sensitivities to the toxin amongst piglets used. A sensitivity comparison between the loop test and ELISA, showed that the ELISA is about two orders of magnitude more sensitive.

Since no "silent genes" were found we recommend the DNA probe test for large scale epidemiological studies and the STII-ELISA for small scale ETEC screening and for protein work. The pig loop test is still needed for biological studies of the toxin, since the ELISA may theoretically be positive also with a biologically non-functional STII.

STII producing *E.coli* are common in Swedish pigs

We are presently investigating the frequency of STII producing *E.coli* strains isolated from pigs with enteric disorders in Sweden during 1989. Since the correlation between the three test methods was found to be satisfactory, the DNA probe test is being used.

Twenty six % of all (446) *E.coli* strains isolated were STII positive as shown in figure 2. About 90 % of ETEC isolated from pigs of weaning age produced STII .

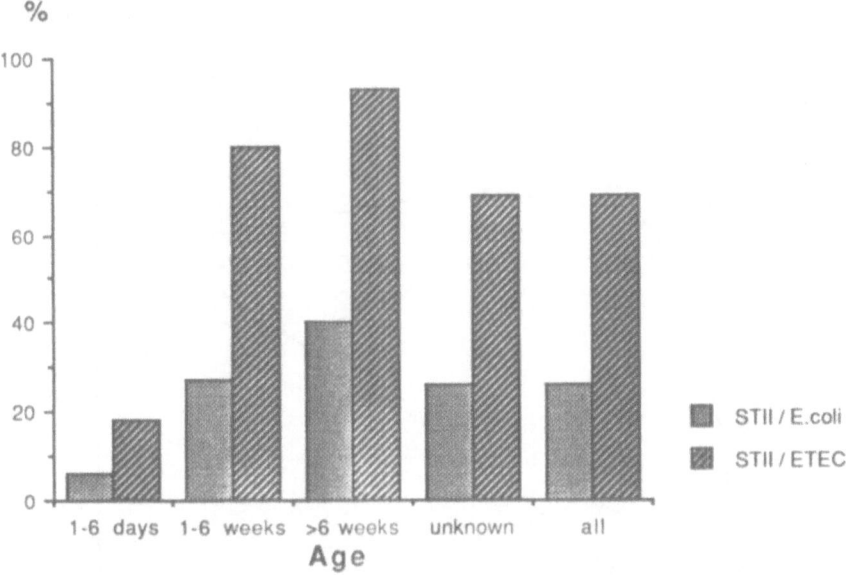

Figure 2.   Frequency of STII producing *E.coli* connected with diarrhoeal disorders.

E.coli isolated from children in tropical countries rarely produce STII

E.coli strains isolated from children in Nicaragua, Bangladesh, Gambia and New Guinea were tested for STII production using the STII-ELISA.
Preliminary data indicated that there was only 1 positive strain of 218 *E.coli* isolates from infants with enteric disorders from these countries.

STII is constitutively and extracellularly produced and not catabolite repressed

A clinical isolate of an ETEC producing only STII was cultivated in L-broth.
The amount of STII produced was measured in the ELISA and values given are dilution factors giving 50% inhibition. Figure 3 shows that production seems to be constitutive and that STII is accumulated during log phase and stationary growth phase.
Cultivation in M9-minimal media supplemented with 0,4 % glucose or 0,4 % glycerol did not give a significant difference in production level of STII. STII production therefore does not seem to be catabolite repressed.

Characteristics of STII

In studies performed prior to purification attempts we found the following:
STII is heat-stable at 80° for more than 15 minutes and very pH stable. It retained its immunological activity after precipitation with $(NH_4)_2SO_4$ and after lyophilisation.
However, significant loss of activity was found upon concentration with dialysis membrane (cut off 1000) using high pressure. Also filtration through 0.22 µm Millipore filter to clarify and sterilize, resulted in nearly 90% loss of activity in the filtrate.
Attempts were made to purify STII on FPLC Mono Q (Pharmacia) column using different pH values with and without urea. STII eluted in numerous fractions together with other proteins and the method did not lead to any obvious purification. We thus conclude that STII seems to form aggregates leading to difficulties in purification.

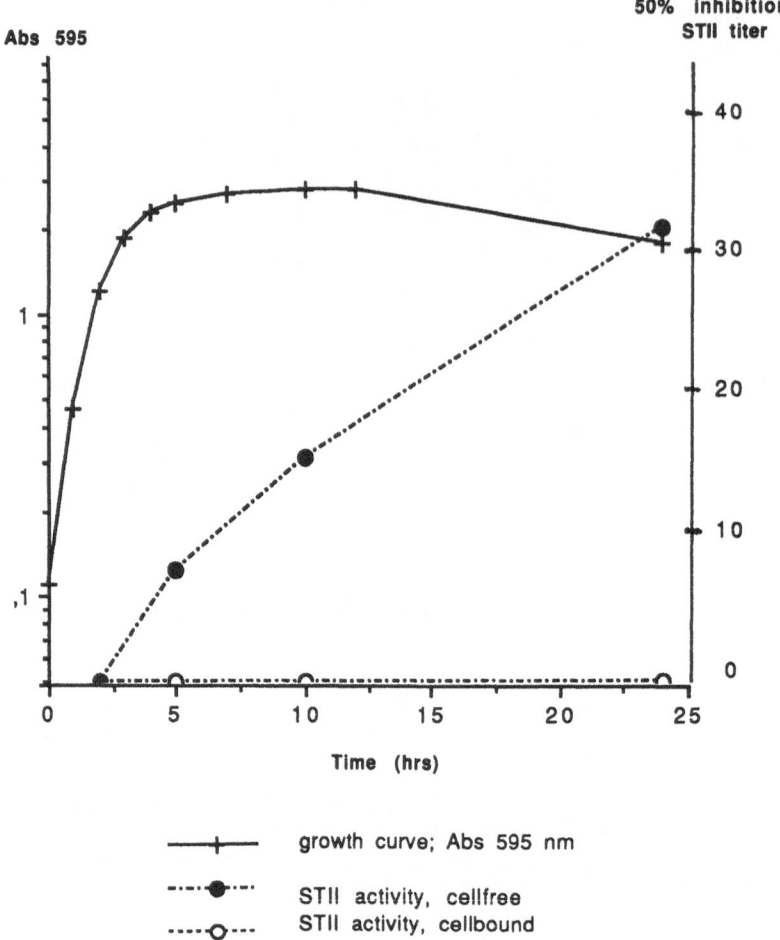

Figure 3.   STII production during growth of an ETEC strain.

Immunological response to STII in pigs

Immunisation of young pigs (3 months of age) with proteinA-STII fusion protein resulted in specific STII antibody production. This was demonstrated as crossreactive bands to ß-gal-STII and native STII by Western blot technique.
Antisera from these pigs were mixed and incubated with STII containing culture supernatant and thereafter injected into pig intestinal loops. No neutralising activity to the STII toxic effect was found.
Unlike human placenta, swine placenta are not permeable to immunoglobulins. In order to investigate whether piglets can gain immunity via colostrum and milk against STII from sows, we immunised pregnant sows with STII fusion proteins. Only very low STII specific titers were found in serum and colostrum.

References

1) M. N. Burgess, R. J. Bywater, C. M. Cowley, N. A. Mullan and P. M. Newsome. 21: 526 (1978)
2) P. Echeverria, J. Seriwatana, D. N. Taylor, C. Tirapat, W. Chaicumpa and B. Rowe. 151: 124 (1985)

3) D. G. Evans, D. J. Evans and N. F. Pierce. 7: 873 (1973)
4) R. L. Guerrant and C. S. Weikel. 4: 181 (1988)
5) C. Handl, B. Rönnberg, B. Nilsson, E. Olsson, H. Johansson and J.-I. Flock. 26: 1555 (1988)
6) C. H. Lee, S. L. Moseley, H. W. Moon, S. C. Whipp, C. L. Gyles and M. So. 42: 264 (1983)
7) S. L. Moseley, M. Samadpour-Motalebi and S. Falkow. 156: 441 (1983)
8) E. Olsson and O. Söderlind. 11: 6 (1980)
9) O. Söderlind, B. Thafvelin, R. Möllby. 26: 879 (1988)
10) C. S. Weikel, H. N. Nellans and R. L. Guerrant. 153: 893 (1986)
11) S. C. Whipp. 58: 930 (1990)

Tel.no. ++46 8 779 52 00
Tel.fax.++46 8 774 55 38

IMMUNE RESPONSE TO *VIBRIO CHOLERAE* INFECTION IN RABBITS WITH

SPECIAL REFERENCE TO ANTIBODIES AGAINST *IN VIVO* SPECIFIC

ANTIGENS

Gunhild Jonson, Ann-Mari Svennerholm,
and Jan Holmgren

Dept. of Medical Microbiology and Immunology
University of Göteborg, Göteborg, Sweden

INTRODUCTION

In *Vibrio cholerae* 01 bacteria, the causative agent of cholera, changes in the *in vitro* environment have been found to have marked influences both on the production of cholera toxin and of cell-surface associated structures, e.g. haemagglutinins, outer membrane proteins and a toxin-coregulated pilus (TCP)[1,2]. We have previously shown that the intestinal milieu provide an environment in which *V. cholerae* 01 of both El Tor and classical biotype expresses new surface antigens, that are not found after *in vitro*-growth[3]. We have also recently shown, by studying *in vivo*-grown bacteria, that both TCP and a putative adhesin on El Tor vibrios, i.e. the mannose-sensitive haemagglutinin (MSHA), are expressed by *V. cholerae* during infection[4].

In this study we have examined whether the *in vivo*-specific antigens of *V. cholerae* induce an immune response during experimental infection and whether TCP and MSHA are immunogenic during growth *in vivo*.

METHODS

Rabbits were intestinally infected twice with El Tor (strain T19479) or classical (strain Cairo 48) *V. cholerae* 01 bacteria using the rabbit non-ligated intestinal model (RITARD)[5]; an immunizing dose was followed 2 weeks later by reinfection with a higher inoculum (> ID_{100} in previously infected rabbits). Rabbits immunized with rough or non-01 *V. cholerae* were reinfected with strain T19479. The antibody responses in serum after 1 and 2 "immunizations" were tested by ELISA and/or immunoblotting using LPS, cholera toxin (CT), *V. cholerae* 01 bacteria grown *in vitro* (Trypticase Soy Broth, $37^{\circ}C$) or *in vivo* (rabbit intestine) as antigens.

Antibody responses to the TcpA subunit of the toxin-coregulataed pilus TCP and to MSHA were studied by immunoblotting using whole bacteria of strain 569B (AKI-medium, $30^{\circ}C$) expressing TCP, and a crude preparation of MSHA (from El Tor strain Phil 6973), respectively as antigens.

RESULTS AND DISCUSSION

All rabbits infected once with O1 or non-O1 vibrios were protected against rechallenge with *V. cholerae* O1 bacteria given in a dose that caused severe diarrhea or death in non-immunized animals. None of the animals were excreting cholera vibrios for > 2 days after challenge suggesting protection against colonization.

Rabbits infected with El Tor vibrios responded with serum antibodies against CT (not shown), LPS and several antigens present in both *vitro-* and *vivo*-grown bacteria (Fig.1A). Antibody responses were consistently observed against 3 to 5 antigens with a molecular mass of < 20 kDa, some of which were *in vivo*-specific. Post-infection sera showed increased reactivity also with high molecular weight *in vivo*-specific, 62 to 200 kDa antigens[3]. Rabbits infected with classical vibrios responded weakly to LPS and CT (not shown) in sera; in several rabbits no LPS response could be detected. However, a strong antibody response against a group of < 20 kDa antigens was seen also in these rabbits (Fig.1B); 3 of the antigens in this region were *in vivo*-specific (*). Antibody responses against < 25 kDa antigens have also been observed in intestinal fluids from cholera convalescents[6].

Antisera from rabbits infected once with rough or non-O1 vibrios, and lacking detectable LPS antibody responses, reacted with several low molecular weight antigens (< 20 kDa) from classical vibrios; some of these antigens were not present in *in vitro*-grown bacteria (not shown).

Infection with classical vibrios also induced weak antibody responses against TcpA in some rabbits (not shown). These results are in agreement with the poor antibody responses observed in sera and jejunal fluids from humans after oral infection with TCP-positive *V. cholerae* O1 bacteria[7].

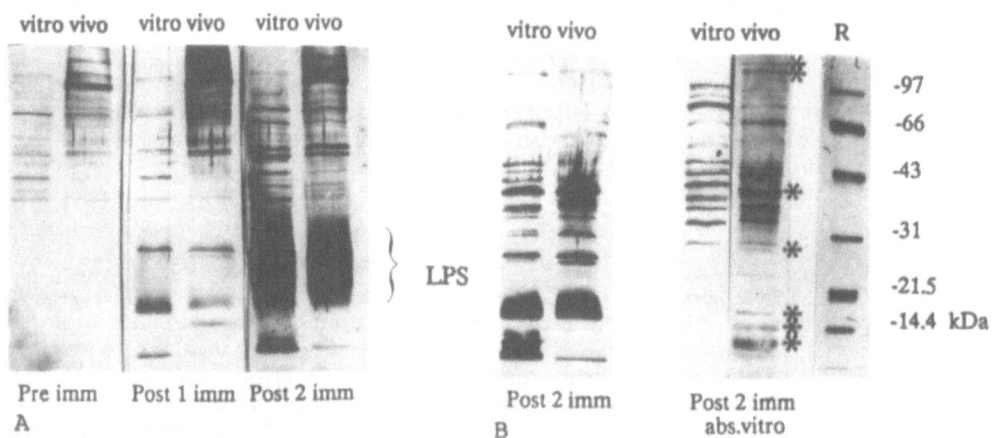

Fig.1. Immunoblot showing the antibody response, after infection with the homologous strain, against *in vitro*-grown (vitro) and *in vivo*-grown (vivo) El Tor T19479 (A) and classical Cairo 48 (B) bacteria. Post immunization sera (Post 1 imm) were taken 2 weeks after the initial and (Post 2 imm) 1 week after the second infection. *In vivo*-specific immunogens (*) are indicated in Fig. 1B and were detected after absorption with *vitro*-grown bacteria.

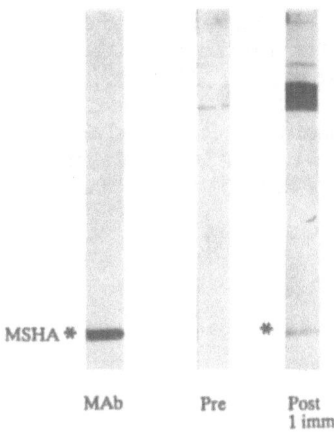

MSHA *

MAb   Pre   Post
                      1 imm

Fig.2. Serum antibodies against MSHA in a rabbit infected with
V.cholerae. A crude MSHA preparation from El Tor strain Phil
6973 was used as antigen; the position of MSHA is indicated by
a monoclonal antibody specific for this antigen.

All El Tor strains and many non-O1 *V. cholerae* express
MSHA on the cell-surface both *in vitro* and *in vivo* [1,4]
Parenteral immunizations of rabbits and mice have yielded both
polyclonal and monoclonal antibodies against MSHA that only
agglutinate MSHA expressing vibrios. Antibodies against MSHA
were induced both after infection with O1 and non O1 cholera
vibrios as shown in Fig. 2.

ACKNOWLEDGEMENTS

We thank Christina Olbing for her skilled technical
assistance. The studies were financially supported by the
Swedish Medical Research Council (project 16 X-3382).

REFERENCES

1.   G. Jonson, J. Sanchez, and A-M. Svennerholm, Expression
      and detection of different biotype-associated cell-bound
      hemagglutinins of *Vibrio cholerae* O1. *J. Gen Microbiol*.
      135:111 (1989).
2.   R.K. Taylor, V.L. Miller, D.B. Furlong, and J.J.
      Mekalanos, Use of *phoA* gene fusions to identify a pilus
      colonization factor coordinately regulated with cholera
      toxin. *Proc.Natl.Acad.Sci.U.S.A.* 84:2833 (1987).
3.   G. Jonson, A-M. Svennerholm, and J. Holmgren, *Vibrio
      cholerae* expresses cell surface antigens during intestinal
      infection which are not expressed during *in vitro* culture.
      Infect. Immun. 57:1809 (1989).
4.   G. Jonson, A-M. Svennerholm, and J. Holmgren, Expression
      of virulence factors by classical and El Tor *Vibrio
      cholerae in vivo* and *in vitro*. *FEMS Ecology* (1990).
5.   N. Lycke, A-M. Svennerholm, and J. Holmgren, Strong
      biotype and serotype cross-protective antibacterial and
      antitoxic immunity in rabbits after cholera infection.
      Microb. Pathog. 1:361 (1986).

6. K. Richardson, J.B. Kaper, and M.M. Levine, Human immune response to *Vibrio cholerae* O1 whole cells and isolated outer membrane antigens. *Infect. Immun.* 57:495 (1989).

7. M.M. Levine, J.B. Kaper, D. Herrington, G. Losonsky, J.G. Morris, M.L. Clements, R.E. Black, B. Tall, and R. Hall, Volonteer studies of deletion mutants of *Vibrio cholerae* O1 prepared by recombinant techniques. *Infect. Immun.* 56:161 (1988).

EXPRESSION OF TYPE 1 FIMBRIAE BY E. COLI F18 IN THE STREPTOMYCIN-TREATED MOUSE LARGE INTESTINE

K.A. Krogfelt[1], B.A. McCormick[2], R.L. Burgoff[2],
D.C. Laux[2] and P.S. Cohen[2]

[1]Department of Microbiology, Technical University of
Denmark, 2800 Lyngby,
[2]Department of Microbiology, University of
Rhode Island, Kingston, RI 02881, USA

SUMMARY

E. coli F18, isolated from the feces of a healthy human, is an excellent
colonizer of the streptomycin-treated mouse large intestine. Adhesion of E. coli F18
to specific receptors present in cecal and colonic mucus was inhibited by D-
mannose, suggesting that the E. coli F18 Type 1 fimbriae may mediate adhesion to
intestinal mucosal components. Recently, we reported that E. coli F18 and E. coli
F18 FimA⁻, an isogenic strain unable to produce Type 1 fimbriae, colonized the
streptomycin treated mouse large intestine, when fed to mice simultaneously. This
result suggests that Type 1 fimbriae may not be expressed in vivo. In the present
study, we show that E. coli F18 does express Type 1 fimbriae in vivo. Furthermore,
our data suggest that Type 1 fimbriae may be a limiting colonization factor to a
specific niche (site) in the large intestine.

INTRODUCTION

Escherichia coli F18, which was isolated from the feces of a healthy human, is
an excellent colonizer of the streptomycin treated mouse large intestine (1).
Adhesion of this strain to receptors present in cecal and colonic mucus as well as
cecal and colonic brush border membranes is inhibited by D-mannose, suggesting
that Type 1 fimbriae mediate adhesion of E. coli F18 to intestinal mucosal
components (7). Recently, we reported that E. coli F18 is able to produce Type 1
fimbriae, when grown under laboratory conditions (5).

Mannose-sensitive binding of E. coli to intestinal epithelial cells has long been
recognized (2), however, the role of Type 1 fimbriae in intestinal colonization has
been controversial (3). Furthermore, there has not been no clear evidence of

Molecular Pathogenesis of Gastrointestinal Infections
Edited by T. Wädstrom *et al.*, Plenum Press, New York, 1991

fimbrial production by E. coli strains in the intestine. In this study, we show that Type 1 fimbriae are indeed produced in vivo.

RESULTS AND DISCUSSION

Six streptomycin-treated mice were colonized with either E. coli F18 or E. coli F18 FimA⁻ [an isogenic strain which does not produce Type 1 fimbriae (5)]. Colonization experiments were performed as previously described (6). Crude cecal mucus was collected from the six mice by scraping gently the cecum. The extremely viscous mucus was homogenized, diluted with saline buffer and solid debris was removed. Bacteria were then collected by centrifugation and fimbriae were purified (4). Fimbrial proteins were separated on SDS-PAGE, transferred to nitrocellulose filters and analyzed by Western blotting using specific antisera against E. coli Type 1 fimbriae (4). In Fig. 1 it is clearly seen that Type 1 fimbriae are produced in the E. coli F18 colonized mice but not in mice colonized with E. coli F18 FimA⁻ (Fig. 1, lanes 3 and 4, respectively). As a positive control, E. coli F18 was grown in broth, and then added to cecal mucus from mice colonized with E. coli F18 FimA⁻ and Type 1 fimbriae were purified (Fig. 1, lane 2). Interestingly, it appears that this strain overproduces Type 1 fimbriae in vivo relative to the amount produced in vitro. That is, judging from the immunoblot signal (Fig. 1), 10^6 CFU/ml of E. coli F18 obtained from the cecal mucus produced more fimbriae than E. coli F18 grown in broth and diluted into the mucus at 10^6 CFU/ml.

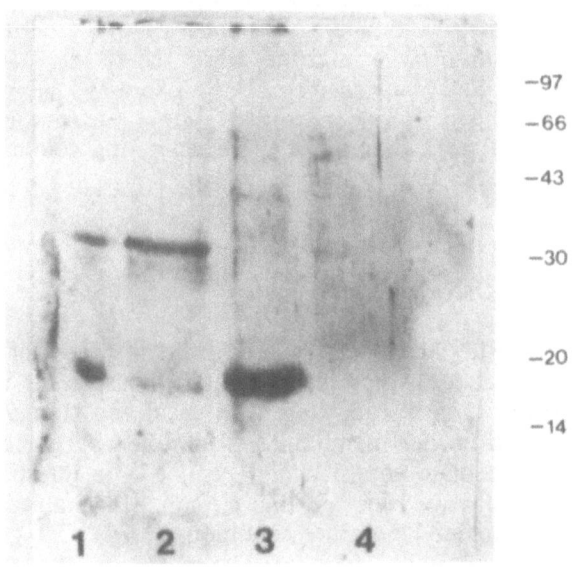

Fig. 1: Western blot of Type 1 fimbriae form E. coli F18 cultured in vitro and in vivo. Lanes: 1) Purified Type 1 fimbriae; 2) Fimbriae from E. coli F18 added into mucus from mice colonized with E. coli F18 FimA⁻; 3) Fimbriae form E. coli F18 isolated from colonized mice; 4) Fimbriae from E. coli F18 FimA⁻ isolated from colonized mice.

The relative colonizing abilities of the mouse large intestine of E. coli F18 and E. coli F18 FimA⁻ were also analyzed. Colonization of the large intestine was followed for 15 days. First, E. coli F18 and E. coli F18 FimA⁻ were simultaneously fed to streptomycin-treated mice at 10^5 and 10^8 CFU, respectively. The relative difference between these strains remained at approximately 3 logs. Next, the reverse experiment was done, i.e. mice were simultaneously fed 10^{10} CFU E. coli F18 and 10^4 CFU E. coli F18 FimA⁻. In this case, the latter strain grew very rapidly and at day 5 there was only a 1 log difference in the level of colonization between the two strains. These results suggest that the fimbrialess strain was able to colonize different niches (sites) in the intestine while the fimbriated organism was restricted to a certain niche in the intestine.

Furthermore, we showed that E. coli F18 FimA⁻ was able to colonize (at approx 10^7 CFU) the mice intestine, even when E. coli F18 was already well established (for 15 days). In contrast, a fimbriated E. coli F18 dropped to undetectable levels within 3 days. Our data indicate that in the large intestine there are two distinct niches which are colonized. The fimbriated E. coli F18 strain seems to occupy a primary niche, while the non-fimbriated E. coli F18 FimA⁻ is able to colonize both the primary and secondary niche. Since the primary difference between the two strains tested is their ability to produce Type 1 fimbriae, it seems reasonable to speculate that the binding of mannose-containing mucosal components restricts Type 1 fimbriated organisms to a specific niche in the intestine.

CONCLUSION

-    Type 1 fimbriae are produced in vivo.
-    There are two distinct niches in the large intestine which are colonized.
-    The Type 1 fimbriaeted E. coli F18 seems to be limited to a primary niche, while the non-fimbriated appears to colonize both niches.

REFERENCES

1.   P.S., Cohen; J.C. Arruda, T.J. Williams and D.C. Laux. 1985. Adhesion of human fecal Escherichia coli strain to mucus. Infect. Immun. 48:139-145.
2.   J.P., Duguid and R.R. Giles. 1957. Fimbrial and adhesion properties in dysentery bacilli. J. Pathol. 74:397-411.
3.   K.A., Krogfelt. 1990. Bacterial adhesion: Genetic organization, biogenesis and role in pathogenesis of fimbrial adhesins of Escherichia coli. Rev. Infect. Diseas. in press.
4.   K.A., Krogfelt and P. Klemm. 1988. Investigation of minor components of Escherichia coli type 1 fimbriae: protein, chemical and immunological aspects. Microb. Pathog. 4:231-238.
5.   B.A., McCormick, D.P. Franklin, D.C. Laux and P.C. Cohen. 1989. Type 1 pili are not necessary for colonization of the streptomycin-treated mouse large intestine by type 1 piliated Escherichia coli F18 and E. coli K12. Infect. Immun. 57:3022-3029.
6.   M.L., Myhal, P.S. Cohen and D.C. Laux. 1983. Altered colonizing ability for mouse large intestine of a surface mutant of a human fecal isolate of Escherichia coli. J. Gen. Microbiol. 129:1549-1558.
7.   E.A., Wadolkowski, D.C. Laux and P.S. Cohen. 1988. Colonization of the streptomycin-treated mouse large intestine by a human fecal Escherichia coli strain: role of adhesion to mucosal reseptors. Infect. Immun. 56:1035-1043.

FLAGELLAR COMPONENTS OF *Helicobacter pylori*

C.J. Luke, [*] T.S.J. Elliott, C.W. Penn

School of Biological Sciences, University of Birmingham, Birmingham U.K.
[*]Department of Clinical Microbiology, Queen Elizabeth Medical Centre, Birmingham U.K.

Helicobacter pylori has been shown to cause active chronic gastritis and has been implicated in the aetiology of peptic ulceration in both the stomach and duodenum. Although the subject of intense clinical investigation, *H.pylori* remains poorly characterised at the biochemical and molecular levels, and little is understood regarding possible mechanisms of pathogenesis.

The surface components of many pathogenic bacteria have been found to play significant roles in pathogenesis ; examples are adhesin molecules, outer membrane components and flagella.

This study reports the characterisation of the major flagellar and associated polypeptides of *H.pylori*.

The flagella of *H.pylori* have been shown to be essential for the organism to colonise the gastric mucosa of gnotobiotic piglets. The flagella also possess an unusual membrane-like sheath, which may protect the core from stomach acidity. It has also been suggested that the flagellar sheath is an extension of the outer membrane of the organism, although this is not evident from many ultrastructural studies. In *Vibrio cholerae* (P.Manning, this volume) the presence of lipopolysaccharide (LPS) has been reported in a similar flagellar sheath structure. If outer membrane or other proteins are present in the flagellar sheath they are potentially well placed to function as adhesins. The composition of the flagella of *H.pylori* has as yet not been fully determined (Refs. 1,2, & 3), and there is no reported study on the characterisation of the flagellar sheath.

In this study, flagella, complete with their sheaths, were mechanically sheared by homogenisation from intact cells. The sheared flagella and sheath material was then incubated in the presence of 4 % Triton - X 100 for 30 minutes at 37° C. The mixture was then ultracentrifuged at approximately 300,000 X g for 1 hour. The pellet and supernatant and also the pellet from unterated sheared flagella were analysed by SDS-PAGE, and it was found that the pellet from the detergent treated flagella consisted of a single 54 kD polypeptide, and that detergent treatment appeared to solubilise polypeptides of molecular weights 26, 30, 58, 62, 66 and 80 kD. The pellets from detergent treated and from untreated flagellar preparations were also examined by transmission electron microscopy after negative staining, and this showed that the treatment with detergent had completely removed sheath material. Immunogold labelling of naked flagella and of whole cells treated with Triton X - 100 (to remove flagellar sheaths), with a polyclonal mouse antiserum monospecific for the 54 kD polypeptide, confirmed that this polypeptide was the major flagellar core polypeptide.

These results show that the flagellar cores of *H.pylori* consist primarily of a single 54 kD polypeptide, and that material solubilised by detergent, containing polypeptides of molecular weights 26, 30, 58, 62, 66 and 80 kD may represent sheath material, and perhaps residual outer membrane material disrupted by homogenisation. In preliminary experiments to show whether or not the flagellar sheath is indeed core polypeptide. (See Figure 1.)

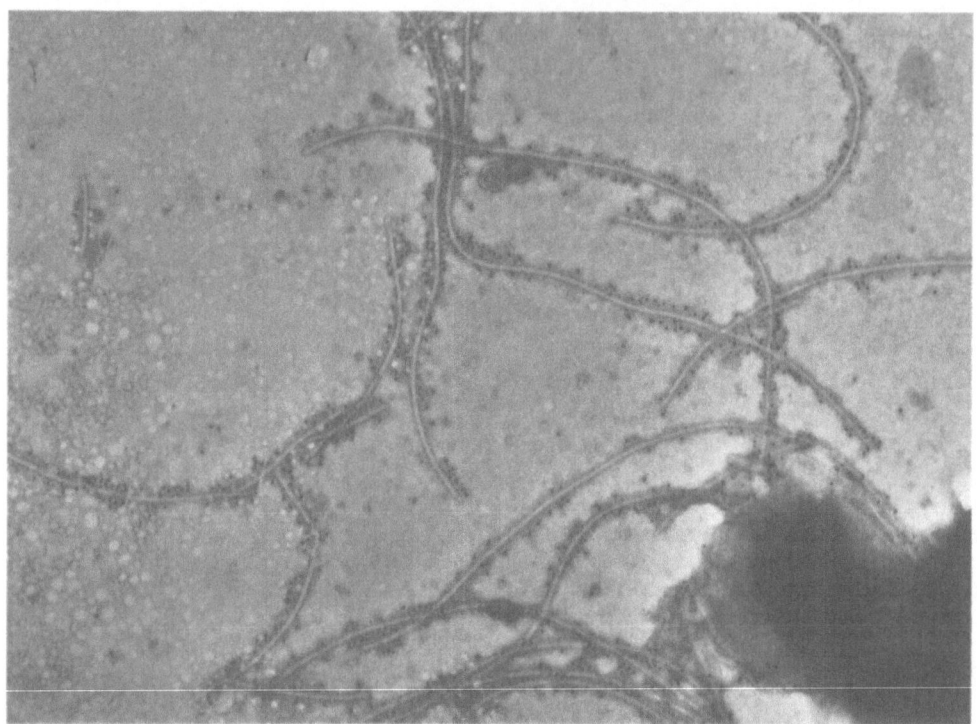

Figure 1

an extension of the outer membrane, proteinase K digests of flagellar and sheath preparations were immunoblotted with a rabbit polyclonal antiserum raised against whole *H.pylori* cells. A ladder pattern smooth type LPS was found to be present in whole cell proteinase K digests, in the molecular weight range of about 35 to 55 kD, and this was also present in Triton - solubilised sheath material, but not in flagellar core digests. Contaminating outer membrane vesicles may be present in the flagellar preparations, and so such results should be treated with caution. Further work is in progress.

It is clear that extensive characterisation of the outer membrane and other surface components of *H.pylori* is required so that impotant surface antigens can be identified and their location on the surface of the organism more precisely determined.

REFERENCES

1.   Newell D.G. (1987) J.Gen.Micro. **133** : 163-170

2.   Geis G., Leying H., Suerbaum S., Mai U., Opferkuch W. (1989) J.Clin.Micro **27** : 436-441

3.   Dunn B.E., Perez-Perez G.I., Blaser M.J. (1989) Infect.Immun. **57** : 1825-1833

THE INFLUENCE OF INTESTINAL MUCUS ON PLASMID ENCODED

ADHESION AND SURFACE HYDROPHOBICITY OF

YERSINIA ENTEROCOLITICA

Anders Paerregaard, Ole Mark Jensen and
Frank Espersen

Statens seruminstitut, Department of Clinical
Microbiology, Rigshospitalet, Copenhagen, Denmark

The mucus gel covers the epithelium throughout the intestine. Enteric invasive microorganisms such as Yersinia enterocolitica have to travel through this gel before adhesion to and subsequent penetration of the enterocyte can take place. Therefore, it is possible for constituents in the mucus layer to interact with the microorganism. This may lead to a change in the subsequent ability of the bacterium to adhere to the brush border membrane (1,2,5,7). We have previously shown that the outer membrane protein YadA, which is encoded by the Yersinia virulence plasmid, pYV, contribute to the adhesive potential of Y. enterocolitica (9). The presence of YadA also confers surface hydrophobicity on the bacterium (6). Mantle et al. (5) recently reported that binding of this organism to brush border membranes was reduced in the presence of mucus. In this study we have examined whether changes in bacterial surface characteristics such as hydrophobicity occurs concomitantly with the mucus-induced decrease in adhesion.

MATERIALS and METHODS

Bacterial strains. The pYV-carrying Y. enterocolitica O:3 strain YeO301P+ and its isogenic plasmid-cured derivative YeO301P- have previously been described (8). The bacteria were grown for 16 h at 37°C in Eagle minimal medium plus 10% fetal calf serum as described. Thymidine(methyl ^{3}H) was added for radiolabeling. The bacteria were harvested and washed in phosphate buffered saline, pH 7.4 (PBS), and adjusted spectrophotometrically to a concentration of about 8×10^8 cfu/ml.

Preincubation with mucus. Crude mucus was isolated from 4 rabbits as described by Laux et al (3). Equal volumes of bacteria and PBS (negative control), bovine serum albumin (BSA) or crude mucus (both 5 mg protein per ml) were incubated for 30 min at 37°C. Thereafter, adhesion to brush border vesicles and surface hydrophobicity were measured as described below.

Adhesion to rabbit ileal brush border vesicles (BBVs). Adhesion of preincubated bacteria to BBVs that were immobilized in the wells of polystyrene microtiter plates was determined as previously described in detail (8).

Surface hydrophobicity. Bacterial surface hydrophobicity was measured as previously described in detail (8), using the two-phase dextran-polyethylene glycol separation system of Magnusson & Johansson (4). Two ml of both phases and 0.1 ml of radiolabeled bacteria were mixed in a glass vial. An index value was calculated as the result of cpm in 0.5 ml of the bottom phase divided by cpm in 0.5 ml of the top phase (8).

Statistics. All experiments were performed in triplicate and repeated 6 times. A difference in adhesion or surface hydrophobicity was considered significant if present in all 6 experiments performed, making the probability, p, of the difference being coincidental $\leq (1/2)^6$ = 0.016.

RESULTS and DISCUSSION

Preincubation with BSA did not change adhesion of strains YeO301P+ or YeO301P- to BBVs as compared to adhesion after preincubation with PBS (the negative control) (Table 1). In contrast, adhesion of strain YeO301P+ was significantly reduced after preincubation with crude mucus in contrast to binding of strain YeO301P- which did not seem to change (Table 1).

Table 1. Adhesion[a] of two Y. enterocolitica strains to BBVs after preincubation.

Preincubation with:	BSA	Crude mucus
YeO301P+	106± 5	12± 5
YeO301P-	109±24	122±23

[a] The results are given in percentage of adhesion after preincubation with PBS (mean±SD).

Surface hydrophobicity of strain YeO301P+ changed simultaneously with the mucus-induced decrease in adhesion as shown in Table 2. After preincubation with PBS or BSA strain YeO301P+ was significantly more hydrophobic than strain YeO301P, but after preincubation with crude mucus both strains were hydrophilic and did not differ significantly.

Table 2. Surface hydrophobicity[a] of two Y. enterocolitica strains after preincubation.

Preincubation with:	PBS	BSA	Crude mucus
YeO301P+	1.8±1.4	2.0±1.9	0.2±0.1
YeO301P-	0.1±0.1	0.1±0.1	0.1±0.1

[a] The results are given as hydrophobicity index values (mean±SD).

In other experiments we found that strain YeO301P+ adhered to crude rabbit ileal mucus, in contrast to strain YeO301P- (10). The binding to mucus was mediated by the YadA protein, as was the binding to BBVs (10). Therefore, we believe that constituents present in mucus are able to bind to the YadA, resulting in a decreased adhesion mediated by this protein to BBVs. It is possible that the concomittant reduction in bacterial surface hydrophobicity may, at least in part, account for the effect of mucus on the subsequent adhesive abilities of Y. enterocolitica.

REFERENCES
1. Chadee, K., W.A. Petri, Jr., D.J. Innes, and J.I. Ravdin. 1987. J.Clin.Invest. 80:1245-1254.
2. Dinari, G., T.L. Hale, O. Washington, and S.B. Formal. 1986. Infect.Immun. 51:975-978.
3. Laux, D.C., E.F. McSweegan, and P.S. Cohen. 1984. J. Microbiol. Methods 2:27-39.
4. Magnusson, K.E., G. Johansson. 1977. FEMS Microbiol. Lett. 2:225-228.
5. Mantle, M., L. Basaraba, S.C. Peacock, and D.G. Gall. 1989. Infect. Immun. 57: 3292-3299 .
6. Martinez, R.J. 1989. Infect.Immun. 171:3732-3739.
7. McSweegan, E., D.H. Burr, and R.I. Walker. 1987. Infect.Immun. 55:1431-1435.
8. Pærregaard, A., F. Espersen, and N. Baker. APMIS (in press).
9. Pærregaard, A., F. Espersen, and M. Skurnik. 1990. APMIS (in press).
10. Pærregaard, A., F. Espersen, O. M. Jensen, and M. Skurnik. Infect. Immun. (in press).

THE BINDING OF BACTERIA CARRYING CFAs AND PUTATIVE CFAs TO RABBIT INTESTINAL BRUSH BORDER MEMBRANES

Christine Wennerås[1], Jan Holmgren[1], Moyra M. McConnell[2] and Ann-Mari Svennerholm[1]

Department of Medical Microbiology and Immunology Göteborg, Sweden[1] and Division of Enteric Pathogens Central Public Health Laboratory, London, UK[2]

INTRODUCTION

Enterotoxigenic *Escherichia coli*, ETEC, colonize the human intestine by means of fimbrial structures called colonization factor antigens, CFAs. On human ETEC strains three different types of CFAs have been described, i.e. the colonization factor antigens I, (CFA/I), CFA/II and CFA/IV. In addition a number of "new" putative colonization factor antigens have been identified, e.g. PCF0166, CS7, CS17 and CFA/III [M.M McConnell, this volume] although their prevalence and role in colonization have not been fully assessed.

We have previously shown that purified CFA/I and the three subcomponents of CFA/II (CS1, CS2 and CS3) bind to both shared and unshared membrane components of both human and rabbit intestinal cells[1]. In a subsequent study we have also found that CFA/II and CS4 of the CFA/IV complex bind to asialo GM1[2]. In the present study we have developed methods for studying the binding of bacteria expressing CFAs to rabbit intestinal membrane proteins, and compared the binding patterns of the bacteria with those of purified fimbriae. We have also studied the binding of the putative colonization factor antigens PCF0166, CS7 and CS17 to intestinal cell membrane components and compared their binding patterns with those of already "established" CFAs as well as evaluated whether their binding to rabbit brush border membranes correlated with their ability to colonize the rabbit intestine.

METHODS

The binding of CFAs to intestinal (glyco)proteins was studied by isolating membrane proteins derived from rabbit intestinal brush borders, by separating these proteins electrophoretically and transblotting them to nitrocellulose

Molecular Pathogenesis of Gastrointestinal Infections
Edited by T. Wädstrom *et al.*, Plenum Press, New York, 1991

and therafter incubating them with purified CFAs or live, CFA-carrying bacteria[1]. Bound CFA , either purified or expressed on bacteria, was visualized by an ELISA method employing MAbs or polyclonal antisera directed against the respective CFAs. Alternatively, CFA-carrying bacteria were biotinylated[3] and thereafter their binding to intestinal proteins was detected by the avidin-peroxidase system.

A rabbit non-ligated intestine model (RITARD) was used as previously described[4] to evaluate the colonizing ability of bacterial strains expressing different putative colonization factor antigens . The mean excretion time of bacteria in the stool was determined and compared with the time of excretion of *E coli* from the normal fecal flora, the latter excretion time is 2 days as has been previously shown[4].

RESULTS AND DISCUSSION

In the present study we evaluated whether CFAs expressed on whole bacteria recognize the same binding structures as the corresponding isolated fimbriae. E1392 bacteria which express CFA/II (CS1+CS3) bound to the same 120-140kD structures of rabbit brush border membranes as did purified fimbriae from the same strain. Bound bacteria were either detected by an ELISA method using a MAb directed against CS3 in the immunodetection step or first biotinylated and therafter visualized by the avidin-peroxidase system (See fig.1).

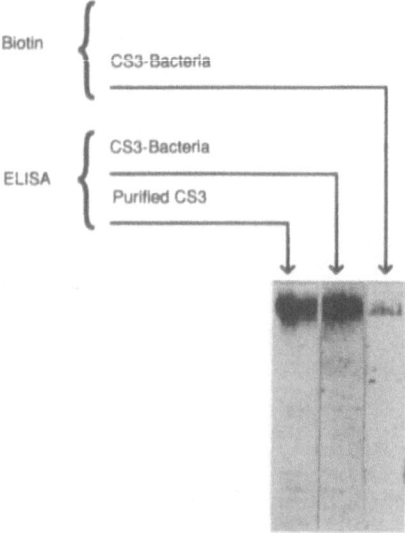

Fig.1

The results suggest that the binding specificity of the fimbriae is the same irrespective of whether they are sheared off the bacteria or expressed on the bacterial surface.

By using these methods we also examined whether strains carrying putative colonization factor antigens , e.g. PCF0166, CS7 and CS17 bound to the electrophoretically separated rabbit brush border membranes since some of these strains have been shown to colonize the rabbit intestine in vivo [Svennerholm et al, to be published]. Interestingly enough, it was found that the colonizing strains also bound well to components of the electrophoretically separated rabbit intestinal brush border membranes and vice-versa, that bacteria with poor colonizing capacity did not bind to the cell membrane structures (Table).

Table 1

The ability of ETEC carrying putative colonization factor antigens
to cause diarrhea and to colonize in the RITARD model

Infecting strain		Bacterial excretion	
Designation	Putative -colonization factors	No excreting > 2 days	Duration (days)
E29101A	CS7, LT+	4/4	8.5±2.2
E20738A	CS17, LT+	5/6	5.0±1.8
E7476C	PCF0166, tox-	3/10	2.8±1.1

Svennerholm, McConnell, and Wennerås, to be published.

It appears that strains expressing CS7 and CS17 bound to the very same 120-140 kD structures that have been identified as the binding structures for CS3 (Fig.2) whereas the non-colonizing PCF0166 positive bacteria did not bind to rabbit enterocyte membranes (not shown).

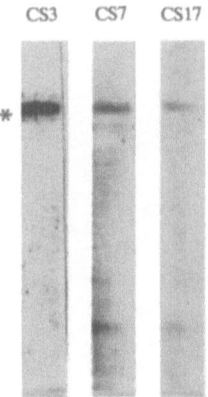

Fig. 2.   Electrophoretically separated rabbit brush border membranes were incubated with bacteria (10^9 bacteria per ml) expressing CS3, CS7 and CS17 respectively. All three strains bound to the 120-140kD structures (*).

These results suggest that antigenically and structurally distinct CFAs and putative CFAs do share the same binding specificities and that this binding is correlated to their ability to colonize the small intestine.

ACKNOWLEDGEMENTS

This work was supported by the Swedish Medical Research Council, grant 16X-09084.

REFERENCES

1. C. Wennerås, J. Holmgren, and A-M Svennerholm. The binding of colonization factor antigens of enterotoxigenic *Escherichia coli* to intestinal cell membrane proteins. *FEMS Microbiol. Lett.* 66: 107 (1990).

2. H.S. Orö, A-B Kolstö, C. Wennerås, and A-M Svennerholm. Identification of asialo-GM1 as a binding structure for *Escherichia coli* colonization factor antigens. *FEMS Microbiol. Lett.* (In press).

3. D.L. Hasty and W.A. Simpson. Effects of fibronectin and other salivary macromolecules on the adherence of *Escherichia coli* to buccal epithelial cells. *Infect. Immun.* 55:2103 (1987).

4. C. Åhrén, and A-M Svennerholm. Experimental enterotoxin-induced *Escherichia coli* diarrhoea and protection induced by previous infection with bacteria of the same adhesin or enterotoxin type. *Infect. Immun.* 50: 255 (1985).

FACTORS CONTRIBUTING TO THE PERSISTENCE OF E. COLI IN THE HUMAN

LARGE INTESTINAL MICROFLORA

Agnes E. Wold, Dominique Caugant, Gunilla Lidin-Jansson, Peter de Man, Catharina Svanborg

Department of Clinical Immunology
Guldhedsg. 10, S-413 46 Göteborg, Sweden

INTRODUCTION

Escherichia coli is an almost ubiquitous inhabitant of the large bowel flora of man, as well as many animals. Some of the *E. coli* strains persist for weeks, months or years (resident strains), while others are only found once, or on a few occasions closely spaced in time (transient strains) (1, 2).

It is not known what determines the ability of *E. coli* to colonize or persist in the colon. One such factor could be the ability to adhere to human colonic epithelium. A majority of E. coli isolates in the large intestinal microflora can express type 1 fimbriae and approximately 10% express P fimbriae (3). Both these specificities promote binding to human colonic epithelial cells (4). In the present study resident and transient colonic *E. coli* strains of schoolgirls presenting with asymptomatic bacteriuria were compared with respect to their serotypes, hemagglutination pattern, and ability to adhere to colonic epithelial cells.

METHODS

Thirteen schoolgirls presenting with asymptomatic bacteriuria were followed with rectal cultures at visits to the outpatient clinic, usually every third month (5). Individual E. coli strains were identified by electromorphic typing of cytoplasmic enzymes; all isolates with identical mobilities of 12 different enzymes were considered as belonging to a single strain (6). Strains occurring in one girl at more than one occasion were defined as resident strains, strains occurring only once as transient strains.

All isolates were serotyped and tested for their hemagglutination pattern and adherence to colonic epithelial cells. Serotyping was performed using antisera against 69 O types. For testing of hemagglutination pattern and adherence to colonic epithelial cells of the HT-29 cell line (4), all isolates were cultivated both on tryptic soy agar, and with three passages in static Luria broth, the latter to maximize the expression of type 1 fimbriae. The adherence assay was run both in the absence and presence of 1.5% methyl mannoside. The mean number of bacteria adhering per cell in the absence of mannose was named total adherence, and the adherence in the presence of mannose mannose-resistant (MR) adherence. The mannose-sensitive (MS) part of the adherence was defined as the total adherence minus the mannose-resistant part.

Some of the E. coli strains colonizing the colon also caused recurrences of bacteriuria in the girls, either as a new episode of asymptomatic bacteriuria or as cystitis or pyelonephritis. Theoretically, these strains could recolonize the bowel from the urinary bladder, and urinary strains would, thus, be overrepresented among resident strains. To exclude such a bias, strains colonizing the urinary tract were excluded and the statistical analyses were performed on this reduced number of strains, as well as on the total material (Tables 1, 2).

Table 1. Pattern of adherence to colonic epithelial cells in resident versus transient strains.

| | Adherence, mean no. of bacteria per cell (S.D.) | | | | | |
| | Broth-grown | | | Plate-grown | | |
	Total	Mannose-resistant	Mannose-sensitive	Total	Mannose-resistant	Mannose-sensitive
Total material:						
Resident	30 (18)	8.8 (16)	21 (18)	6.3 (11)	4.6 (8.1)	1.7 (6.8)
Transient	22 (21)	2.4 (4.9)	19 (20)	1.9 (5.3)	1.3 (4.9)	0.69 (1.3)
p value for difference	0.12	< 0.02	N.S.	< 0.05	< 0.05	N.S.
UVI strains excluded:						
Resident	29 (18)	9.0 (15)	20 (18)	8.2 (13)	6.0 (9.0)	2.2(8.0)
Transient	23 (21)	2.4 (5.0)	20 (20)	2.0 (5.3)	1.3 (5.0)	0.7(1.4)
P value for difference	N.S.	< 0.01	N.S.	< 0.01	< 0.01	N.S.

Table 2. Hemagglutination characteristics of resident versus transient strains.

| | | Hemagglutination pattern (% of strains) | | | |
| | | Broth-cultivated | | Plate-cultivated | |
	n	MR	MS	MR	MS
Total material:					
Resident strains	29	21	38	21	3.4
Transient strains	47	6.4	40	6.4	4.3
p for difference		0.062	N.S.	0.062	N.S.
UVI strains excluded:					
Resident strains	21	24	38	24	4.8
Transient strains	46	6.5	39	6.5	4.3
p for difference		< 0.05	N.S.	< 0.05	N.S.

MR = Mannose-resistant agglutination of human erythrocytes
MS = Mannose-sensitive agglutination of guinea pig erythrocytes

RESULTS AND DISCUSSION

Altogether, 29 resident strains were identified. The median colonization time of these strains was three months, although strains persisting for several years were found. When different isolates of the same strain were compared regarding serotype and hemagglutination pattern, these characteristics were remarkably constant between the isoates. Fourty-nine strains occurred only at one sampling occasion.

Strains resident in the colonic microflora exhibited an increased adherence to colonic epithelial cells from the HT-29 cell line. This was due to an increase in adherence mediated by a mannose-resistant mechanism, while there was no difference between the two groups of strains concerning the mannose-sensitive adherence (Table 1). The difference in adherence pattern between resident and transient rectal strains was seen as well when strains recolonising the urinary tract were omitted from the analysis.

Among the resident strains, a higher proportion likewise expressed a mannose-resistant agglutination of human erythrocytes than among the transient strains (Table 2). There was no difference in the proportion of strains exhibiting mannose-sensitive agglutination of guinea pig erythrocytes.

Fifteen and 16% of the strains in both groups, respectively, expressed a rough phenotype. Among the smooth strains, 60% in the resident group belonged to any of the following 10 serotypes: O1, O2, O4, O6, O7, O8, O16, O18, O25, or O75. The corresponding figure in the transient group was only 15% ($p < 0.001$). The difference between colonic resident and transient strains was seen regardless of if strains causing urinary tract recolonizations were included or not.

Thus, E. coli strains persisting in the large bowel microflora for an extended period of time exhibited characteristics in their serotype and adherence pattern that distinguished them from those only found on one occasion. The "clonality concept" which has been developed for E. coli causing infection in man (See Orskov and Orskov, elsewhere in this volume) may, therefore, also apply to strains that are resident colonisers of the human large intestine. Interestingly, the traits associated with clones successfully persisting in the large bowel microflora seemed to overlap with those previously known to characterize "pyelonephritogenic" clones of E. coli, such as the possession of particular O antigens and the capacity for mannose-resistant hemagglutination. These pyelonephritogenic characters cause inflammation in the urinary tract, which probably aids in clearing the infection. If carried for a long time in the urine, as in asymptomatic bacteriuria, E. coli tend to loose their O antigen as well as their capacity to adhere to the bladder epithelium. In the colon, however, the serotype as well as the adherence pattern were mostly remarkably constant during the period of colonisation, indicating a lack of pressure from the host response against these particular characteristics.

REFERENCES

1. H.J. Sears, I. Brownlee, and J.K. Uchiyama, Persistence of individual strains of *Escherichia coli* in the intestinal tract of man. *J. Bacteriol.* 59:299 (1949).

2. H.J. Sears, and I. Brownlee, Further observations on the persistence of individual strains of *Escherichia coli* in the intestinal tract of man. *J. Bacteriol.* 63:47 (1951).

3. L. Hagberg, U. Jodal, T. Korhonen, G Lidin-Janson, U. Lindberg, and C. Svanborg Edén, Adhesion, hemagglutination, and virulence of *Escherichia coli* causing urinary tract infections. *Infect. Immun.* 31:564 (1981).

4. A.E. Wold, M. Thorssén, S. Hull, and C. Svanborg Edén, Attachment of *Escherichia coli* via mannose of Galα1-4galb-containing receptors to human colonic epithelial cells. *Infect. Immun.* 56:2531 (1988).

5. G. Lidin-Janson, and U. Lindberg, Asymptomatic bacteriuria in schoolgirls. VI. The correlation between urinary and faecal *Escherichia coli*. Relation to the duration of the bacteriuria and the sampling technique. *Acta Paediatr. Scand.* 66:349.

6. D.A. Caugant, Enzyme polymorphism in *Escherichia coli*: Genetic structure of intestinal populations, relationships with urinary tract infection strains and with *Shigella*. Thesis, Göteborg, Sweden (1983).

THE ROLE OF PIGLET INTESTINAL MUCUS IN THE PATHOGENICITY OF

ESCHERICHIA COLI K88

P.L. Conway[*1], L. Blomberg[1], A. Welin[1], and P.S. Cohen[2]

[1] Department of General and Marine Microbiology, University of Göteborg, Carl Skottsbergs Gata 22, S-413 19 Göteborg, Sweden. [2] Department of Microbiology, University of Rhode Island, Kingston, Rhode Island 02881, USA
[*] Corresponding author

SUMMARY

The influence of mucus overlying the epithelial cells in the piglet ileum, on the colonization potential of enterotoxigenic E. coli K88 fimbriated cells was investigated. Using in vitro assays, colonization potential was assessed in terms of growth in mucus, presence of K88 receptors in mucus as well as penetration capacity of K88 fimbriated cells through mucus and subsequently adhesion to underlying epithelial cells. As piglet susceptibility to K88 strains is age related, ileal mucosa were sampled from piglets 0-45 days old. E. coli K88 cells grew well in mucus from all ages tested, however, the amount of K88 fimbrial receptor varied significantly with age, with almost undetectable levels in the newborn. The K88 specific mucus receptor could be removed by ultra centrifuging and therefore must be located on a large molecular weight component or fragment. Although the amount of receptor in the mucus had no effect on penetration of K88 bearing cells through the mucus, E. coli K88 cells could not adhere to epithelial cells after passage through receptor-rich mucus. Receptor-poor mucus had no effect on such adhesion. In considering the role of mucus in the pathogenicity mechanism, the questions can be posed: Will the mucus be a site of colonization? Can the presence of mucus receptor protect the piglet from disease?

INTRODUCTION

Strains of Escherichia coli bearing K88 fimbriae can colonize the small intestine and mediate diarrhoea in piglets during the first few weeks of life (Smith & Huggins, 1978; Smith & Linggood, 1971). It has been shown that the bacterial cells adhere by the K88 fimbriae to specific receptors on the brush borders of the epithelial cells (Bertschinger et al., 1972) and that these K88 fimbriae play an active role in the pathogenicity mechanism (Jones & Rutter, 1972). Piglets lacking the brush border receptor are resistant to infection by E. coli K88 strains. In phenotypically sensitive piglets, that is those with K88 receptors on the brush borders, susceptibility to infection has been shown to be age related. Unless protected by the colostrum of vaccinated dams, neonatal pigs are extremely sensitive (Rutter et al., 1976). Weaning piglets are also very sensitive.

In vivo, the epithelial cells are covered with a thick layer of mucus through which the E. coli cells must penetrate in order to reach the brush border receptor. We have used the fact that susceptibity to infection by E. coli K88 is age related, in order to study the role of the mucus layer overlying the epithelial cells, in the colonization capacity of E. coli K88 fimbriated cells.

RESULTS and DISCUSSION

Colonization capacity was assessed in vitro using a wild type porcine enterotoxigenic E. coli K88 strain Bd1107/7508 (supplied by O. Söderlind, National Veterinary Institute, Uppsala, Sweden). Ileal mucus and epithelial cells were collected from healthy piglets according Laux et al., (1984) and Deneke et al., (1984), respectively. The parameters monitored include bacterial growth in mucus, adhesion to mucus and epithelial cells (Laux et al., 1984) as well as the penetration of E. coli cells through mucus and their subsequent adhesion to epithelial cells (Nevola et al., 1987). For adhesion assays, the mucus or epithelial cells were immobilized overnight in polystyrene microtitre wells prior to adhesion of radioactively labelled bacterial cells. The penetration assay involved overlaying 0.5 ml mucus (3 mg per ml) onto immobilized epithelial cells and the labelled bacteria were added on top of the mucus. Initially mucus was collected from piglets 5, 26 and 47 days old, with weaning occurring at 35 d (Blomberg & Conway, 1989). Adhesion was significantly greater to 26 d mucus with 5 and 47 d old mucus containing 60% and 35%, respectively, of the amount of receptor in 26 d mucus. Using monoclonal antibodies, it was confirmed that this adhesion was K88 fimbriae mediated.

Subsequently, mucus and epithelial cells from newborn, but never fed, and 35 d piglets immediately prior to weaning were studied (Conway et al., 1990). The E. coli strain grew very well with a generation time 28 minutes in both 0 and 35 d mucus diluted in Davis broth minimal salts to yield 1 mg protein per ml. Substantial adhesion to epithelial cells from both age groups was noted, however, newborn mucus contained significantly less (about 16 times) K88 receptor than did mucus from 35 d piglets (Fig 1). This receptor could be removed by ultracentrifuging at 26,000 g for 9 h. The adhesion to mucus was confirmed to be K88 specific using monoclonal antibody inhibition studies and by absorbing out the receptor using purified K88 fimbriae prior to the adhesion assay. E. coli K88 fimbriated cells penetrated equally well through receptor poor (newborn) and receptor rich (35 d) mucus, however were unable to adhere to the epithelial cells only after passage through receptor rich mucus.

Figure 1. Adhesion of E. coli K88 to: (A) bovine serum albumin (control); (B) mucus from newborn piglet; (C) mucus from 35 d piglets; (D) mucus from 35 d piglet after ultracentrifuging 9 h; (E) mucus from 35 d piglet after absorbing with purified K88 fimbriae. Adhesion index (%) = x/y.100 where x = no. of adhering bacteria and y = no. of bacteria adhering to 35 d mucus. Results are expressed as mean of three piglets.

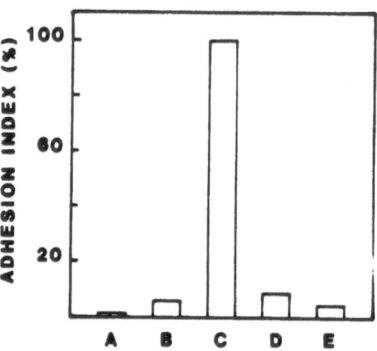

The presence of K88 receptor in ileal mucus is consistent with the fact that calf and mouse ileal mucus contain receptor for E. coli K99 fimbriae (Mouricout & Julien, 1987; Laux et al., 1984) and pig small intestinal mucus contains receptors specific for E. coli 987P fimbriae. Although epithelial cells from 0 and 35 d contained significant amounts of K88 receptor which is consistent with the fact that the pigs were phenotypically sensitive, the amounts of mucus receptor was age related. This is in agreement with Runnels et al 1980 who report that there are available epithelial cell receptors during the first six weeks of life. From the results, it may be proposed that the presence of receptor in the mucus may protect the 26 and 35 d piglet and that the small amount present in mucus from the newborn will facilitate adhesion to and subsequent penetration through the mucus to the underlying epithelial cells. Alternatively, could the presence of receptor in the mucus favour colonization? It is interesting to note that Söderlind et al (1988) reported that the frequency of detecting K88 strains infected piglets was greater in 1-6 week older piglets than those younger than one week and weaned pigs. This is consistent with the levels of mucus receptor reported here. It is plausible to suggest that the mucus layer may be a site for colonization of E. coli K88 cells if the generation time for these cells exceeds the rate of sloughing off and re-generation time observed for both newborn and 35 d old mucus. Diarrhoea and disease may not be a direct result of mucus colonization but rather related to the amount of mucus receptor if assuming adhesion to epithelial cells is a pre-requisite for disease. The effect of mucus in the action of the toxin produced by these K88 bearing E. coli cells will also need to be studied before definitive statements can be made on the role of mucus in the pathogenicity mechanism of enterotoxigenic E. coli K88. Characterization of the mucus receptor would also be useful in determining its role in colonization and disease.

REFERENCES

Bertschinger, H.U., H.W. Moon, and S.C. Whipp 1972. Association of Escherichia coli with the small intestinal epithelium. I. Comparison of enteropathogenic and nonenteropathogenic porcine strains in pigs. Infect. Immun. 5:595.

Blomberg, L. and P.L. Conway 1989. An in vitro study of ileal colonization resistance to Escherichia coli strain Bd 1107/7508 (K88) in relation to indigenous squamous gastric colonization in piglets of varying ages. Microb. Ecol. Health and Disease 2:285.

Conway, P.L., A. Welin, and P.S. Cohen 1990. The presence of K88-specific receptor in porcine ileal mucus is age-dependent. Infect. Immun. In press.

Deneke, C.F., K. McGowan., A.D. Larson, and S.L. Gorbach 1984. Attachment of human and pig (K88) enterotoxigenic Escherichia coli strains to either human or porcine small intestinal cells. Infect. Immun. 45:522.

Jones, G.W. and J.M. Rutter 1972. Role of K88 antigen in the pathogenesis of neonatal diarrhoea caused by Escherichia coli in piglets. Infect. Immun. 6:918.

Laux, D.C., E.F. Mc Sweegan, and P.S. Cohen 1984. Adhesion of enterotoxigenic Escherichia coli to immobilized intestinal mucosal preparations: a model for adhesion to mucosal surface components. J. Microbiol. Methods 2:27.

Mouricout, M.A. and R.A. Julien 1987. Pilus-mediated binding of bovine enterotoxigenic Escherichia coli to calf small intestinal mucins. Infect. Immun. 55:1216.

Nevola, J.J., D.C. Laux, and P.S. Cohen 1987. In vivo colonization of the mouse large intestine and in vitro penetration of intestinal mucus by an avirulent smooth strain of Salmonella typhimurium and its lipopolysaccharide-deficient mutant. Infect. Immun. 55:2884.

Runnels, P.L., H.W. Moon, and R.A. Schneider 1980. Development of resistance with host age to adhesion of K99 + Escherichia coli to isolated epithelial cells. Infect. Immun. 28:298.

Rutter, J.M., G.W. Jones, G.T.H. Brown, M.R. Burrows, and P.D. Luther 1976. Antibacterial activity in colostrum and milk associated with protection against enteric disease caused by K88-positive Escherichia coli. Infect. Immun. 13:667.

Smith, N.W. and M.B. Huggins 1978. The influence of plasmid-determined and other characteristics of enteropathogenic Escherichia coli on their ability to proliferate in the alimentary tracts of piglets. J. Med. Microbiol. 11:471.

Smith, N.W. and M.A. Linggood 1971. Observations on the pathogenic properties of the K88, Hyl, and Ent plasmids of Escherichia coli with particular reference to porcine diarrhoea. J. Med. Microbiol. 4:467.

Söderlind, O., B. Thafvelin, and R. Möllby 1988. Virulence factors in Escherichia coli strains isolated from Swedish piglets with diarrhoea. J. Clin. Microbiol. 26:879.

INDEX

Abrin, 148
Actin, 99
Actin-binding, 4
    polymerization, 99
Actinomyces, 12
Adenylate cyclase, 107
Adhesin, 55, 94
    gene, 103
ADP ribosylation, 107, 111, 116
Amoeba lectins, 296
    cytopathogenicity, 295
Antidiarrhoeal agents, 107
Arachidonic acid, 110
aro A mutation, 29, 280
aro A vaccine, 286
Attaching effacing E coli, AEEC, 49, 93
Attenuated Salmonella strains, 197, 279

Bacteroides, 38
Betagalactosidase, 38, 65
Bifidobacterium, 38
Bile salt, 1, 44
Binding epitope, 12
Blood group degrading enzymes, 39
Bordetella pertussis, 12, 280
Brush border enzymes, 3, 79, 93, 327
    cytoskeleton, 5

Calcium antagonists, 172
Calmodulin, 4, 108, 171
Calpactin, 6
Candida, 12

Canyon hypothesis, 9
Carbohydrate receptors, 9, 16
Catabolite repression, 35
Ceramide, 11
CFA/I, CFAII, 61, 72, 79, 287, 327
Chemotaxis, 32
Chimeric proteins, 120
Cholera-like toxin, 191
Cholera toxin in vivo, 122
    in vitro, 15, 17, 110, 11
    related antigen, CTRA, 191
Clathrin, 223
Clone concept, 49
Clostridium difficile, 35, 161, 169
    toxin A, 161, 169
    toxin B, 161, 169
Colicin E2, 289
Collagen, 213, 251
Colonic mucins, 32
Congo red, 225, 231
Cryptic receptor, 11
Cryptococcus, 12
CS 1, CS 2, CS 3, CS 4, CS 5, CS 6 genes,
    61, 79, 287
Cup-like pedestals, 94
Curli, 303
Cya- cyclic AMP, 181
Cyclic AMP, 107, 115, 175
Cyclic GMP, 107,125
Cystis fibrosis gene, 109
Cytoskeleton, 7
Cytotoxins, 55, 296

Detection of immobilized amplified
nucleic acid (DIANA), 269
Diphteria toxin, 148, 163

EAF probe, 98
Electrophoretic types, 50
Endoglycosidases, 38
Enlongation factor 1, 156
Enolase, 162
Entamoeba histolytica, 295
mutants, 297
Enterhaemorrhargic E coli, 51, 96, 155
Env Y gene, 62
Enterochelin, 180
Enteroinvasive E coli, 155
Enteropathogenic E coli, 49, 93, 103
adherence factor, 49, 95, 104
adherence gene, 61, 79
Enterotoxigenic E coli, 49, 79, 125, 287,
307, 327
heatlabile enterotoxin, 79, 107, 287
heatstable enterotoxin, 79, 94, 125,
287, 307
vaccine, 287
Epidermal growth factor, 6
Exoglycosidases, 38
Extracellular matrix, 249

FAS tests, 96
Fibronectin, 296
Fim A gene, 29
Fim B gene, 29
Fimbriae, 27, 61
Fimbrial antigens, 61
operons, 68
Fimbrin, 4
Fodrin, 4
Fucosidases, 38
Fur-like protein, 141

Gamma-interferon, 180
Gangliosides, 16, 43, 73, 116
receptors, 116
Gastric mucins, 24, 249
Gastritis, 249, 257
Gelsolin, 99
Gene probes, 158
Germfree animals, 33
Globotriaosylceramide, 13, 43
Glycolipids, 7, 10, 13, 33
Glycopeptides, 10, 24, 33
Glycoprotein glycans, 10

Glycosidases, 2, 37, 46
Giardia lamblia, 237
muris, 239
lectin, 237
GTP binding protein, GS, 107
Guanylate cyclase, 125

Haemolysin, 219
Haemolytic uraemic syndrome, HUS,
51, 96, 156
Heatlabile enterotoxin, LT, 57, 115
Hela cells, 103, 155, 223, 258
Helicobacter pylori, 249, 321
pylori flagellae, 249, 321
Hemagglutinin, 249
Hemin binding, 225
Hemolylytic uremic syndrome, 51, 96,
156
Henle 407 intestinal cells, 188
Hep2 cells, 93, 103, 187, 225, 304
Heparansulphate binding, 251
Histone-like protein, 56
Human embryonic lung fibroblasts,
170
Hydrophobicity, 27, 80, 224, 250, 304,
324

Immunotoxins, 152
Influenza virus, 13, 15
Intercellular junctions, 251
Intracellular growth, 180, 197, 198, 219
Invasion phenotype, 198, 219
Invasion plasmid, 198, 219
Iron starvation, 141
Isoreceptors, 16, 43

K88 antigens, 335
fimbriae, 335
receptor, 335

Lactocycleramide, 12, 42
Laminin, 304
Lectins, 243, 249, 296
Leucocyte adhesion molecules, 296
Limulus amoebocyte lysate (LAL) test,
232
Lipopolysaccharide responder 35
Lipopolysaccharides, 35, 143, 224, 231
Lipoprotein, 141
Localized adhesion, LA, 93
LPS Mutant, 32
Lysophospholipids, 172

M-cells, 176
Magnetic separation, PCR, 270
Mannose binding E coli, 16, 87
Mannose resistant
      haemagglutination, 80, 331
MDCK cells, 189
Microflora-associated characteristic,
      MAC, 35
Microvillar enzymes, 4
   glycoproteins, 5
Mucin degrading glycosidases, 44
Mucins, 10, 23, 30, 33, 38, 249, 317, 32?
   hydrophobic lining, 250
Mucus oligosaccharides, 23
   glycoproteins, 23
Multilocus enzyme electrophoresis
      type, 50

Neoglycoprotein, 11
Neuraminic acid, 34
Neurotransmitters, 2
Nonfimbrial adhesins, 71
Nonflagellated E coli strains, 71

Oligosaccharide cluster, 28
Osmolarity mutants, 55, 182
Outer membrane, 117
Outer membrane proteins, 51

pap-fimbriae, 72, 81, 329, 331
PCF 0166, 79
Peptidases, 2, 37
Peptide hormones, 2
Peptostreptococcus micros, 35
Periplasm, 117
Pertussis toxin, 111, 121
Peyers plaques, 176
Phage encoded toxin, 156
Phalotoxin, 96
Phase variation, 68, 91
pho P mutants, 198, 280
Phosphatase regulation, 280
Phosphatases, 2
Phospholipase A, 6, 112, 170
   C, 100, 108
   D, 108
Phospholipases, 3, 37
Picorna viruses, 16
Pili biogenesis, 55
Polymerase chain reaction, PCR, 159,
      267, 273, 282
   Clostridium difficile toxins, 273
   shiga-like toxins, 267

urease, Helicobacter pylori, 274
Porin regulation, 280
Protein kinase, 6, 100, 171
Proteoglycans, 10
Protoxin, 165
Pseudomonas aeruginosa, 12
Purine biosynthesis, 280

Quinacrine, 171

Regulation proteins, 83
   phoP/phoQ, 91, 182
   sequence, rns, 63, 66, 83
RITARD Model, 79, 288, 328
Rotavirus, 11
Ruminococcus gene, 40
   torques, 40

Salmonella cholaesius, 189
   enterica, 175
   enterotoxin, 186, 283
   TnphoA mutants, 187, 198
   typhimurium, 30, 185, 197, 280
Salt aggregation test, 231
Sereny tests, 231
Shigatoxin, 13, 147, 155, 272, 299
   subunits, 147, 155, 299
Shigella bodii, 232
   dysenteriae, 231
   flexneri, 217, 223
   haemagglutinins, 231
   hydrophobicity, 224, 231
Sialic acid binding, 13
Sialidases, 42
Silent regulatory gene, 66, 88
Site directed mutagenesis, 282
ST CTB hybrid proteins, 288
Synthetic ST peptides, 131, 288

Temperature-dependent
transcription, 67, 210
Temperature inducible proteins, 68,
      121, 224
Terminal web, 5
Tetanus toxin, 282
Thermoregulation, 55
Tight junction, 5
Toxin assembly, 115
   mutants, 279
   pentamers, 119
   regulated pilus TCP, 142
Tropomyosin, 5
Troponin, 162

Type 1 fimbriae, 31, 67, 87, 226, 317
   pili, 31, 87
Trypanosoma cruzi, 12, 97

Urease gene DNA probes, 2
Urease, Helicobacter pylori, 249, 259
Ussing chamber, 109

Vero cytotoxins, 96, 155, 163
   producing E coli (VTEC), 51, 96
Vibrio cholerae, 115, 126, 139, 313
   adhesins, 139
   haemagglutinins, 139, 313
   pili, 115, 139
   toxin, 115, 139
   toxin regulated pilus, TCP,120, 140,
      313

Villin, 4, 99
Vir B, Vir F genes, 218
   F genes, 62
   F protein, 62
   G gene, 217
   R gene, 218
Virulence plasmid, 179
Vitronectin, 303

Yersinia collagen binding proteins,
      213
   enterocolitica, 126, 213, 280, 323
   pseudotuberculosis, 96, 207, 213,
      280
Yop genes, 208, 213
   proteins, 208, 213